The Silwood Circle

A History of Ecology and the Making of Scientific
Careers in Late Twentieth-Century Britain

The Silwood Circle

A History of Ecology and the Making of Scientific Careers in Late Twentieth-Century Britain

Hannah Gay
Imperial College London, UK

Imperial College Press

ICP

Published by

Imperial College Press
57 Shelton Street
Covent Garden
London WC2H 9HE

Distributed by

World Scientific Publishing Co. Pte. Ltd.
5 Toh Tuck Link, Singapore 596224
USA office: 27 Warren Street, Suite 401-402, Hackensack, NJ 07601
UK office: 57 Shelton Street, Covent Garden, London WC2H 9HE

British Library Cataloguing-in-Publication Data
A catalogue record for this book is available from the British Library.

THE SILWOOD CIRCLE
A History of Ecology and the Making of Scientific Careers in Late
Twentieth-Century Britain

ISBN 978-1-84816-989-0
ISBN 978-1-78326-292-2 (pbk)

Printed in Singapore by Mainland Press Pte Ltd.

Acknowledgements

Many people have helped me with this book. I am especially grateful to those I have labelled members of the Silwood Circle for agreeing to be written about, for their trust, for being cooperative despite not knowing what I would write, and for giving much of their time. Thanks, therefore, to Bob May, Gordon Conway, Michael Hassell, Roy Anderson, Mick Crawley, John Lawton, John Beddington, John Krebs and David Rogers. I enjoyed meeting and talking with them, am grateful for the many e-mail exchanges that followed, and have learned much as a result. I am especially indebted to Bob May for patiently answering many queries, for the interesting discussions we had, and for being unfailingly courteous and supportive. I did not meet Richard (Dick) Southwood who died in 2005, but spent many hours in the Bodleian Library reading his papers. I would like to thank Colin Harris, superintendent of special collections, and Michael Hughes, senior archivist, for their friendly assistance. Thanks are due also to Joanna Corden, archivist at the Royal Society and, especially, to Anne Barrett, archivist at Imperial College London who, as always, was helpful and generous with her time. I have received much help also from others and would like to acknowledge and thank the following: Judith Anderson, Nigel Bell, Valerie Brown, Eric Charnov, Colin Clark, Susan Conway, the late Brian Flowers, Charles Godfray, Paul Harvey, Buzz Holling, Arne Mooers, Ronald Oxburgh, Randall Peterman, John Reynolds, Nancy Slack, Jeff Waage and Mark Williamson. My colleagues at the Centre for the History of Science, Technology and Medicine at Imperial College London heard me give a couple of talks on this project and gave much useful feedback. Especial thanks to David Edgerton, Andrew Warwick, Andrew Mendelsohn, Abigail Woods and Ralph Desmarais, as well as to students in the doctoral seminars of 2010 and 2011.

I would also like to acknowledge the late Michael Fellman, good friend and former colleague at Simon Fraser University, who read much of the manuscript. It was helpful to have the comments of a well published historian from a very different field. Thanks, too, to Norman Swartz for his comments on Chapter 10. Especial thanks to my husband Ian and son John for their help with copy editing. Finally, I would like to acknowledge the support of K. K Phua at Imperial College Press, and to thank editor, Kim Tan, for her fine work on this book.

Hannah Gay
hgay@sfu.ca or hgay@imperial.ac.uk

Contents

chapter one

Introduction

In the first half of the twentieth century ecology was situated on the scientific periphery. It was widely seen as a domain for amateurs. By the end of the century it had become more centrally placed. This shift in fortune is the backdrop to an account of the careers of a close-knit group of ecologists. Since the group began to coalesce at the Silwood Park campus of Imperial College London I have named it the Silwood Circle. Tellingly perhaps, its father figure, T. R. E. Southwood, occasionally referred to it as the 'Silwood Mob'.[1] Together with other young ecologists of the 1960s and 1970s, members of the circle were determined to promote their discipline. They believed that the way forward was to combine experiment and field observation with a mathematically informed theoretical approach. As environmental and epidemiological issues came to the fore, the new mathematized ecology came to be recognized as a discipline central to modern existence. Members of the circle were sought for their expertise on matters related to the conservation of ecosystems and biodiversity, resource management, food policy, genetically modified crops, sustainable agriculture, international development, defence against biological weapons, and infectious disease control. By the end of the century, aided by luck and a changing political climate, they had seized the political as well as the scientific high ground.

The circle was not a research school, nor was it a team; rather it was a loose generational and fraternal group of scientists, with similar but non-competing research interests. There was, however, sufficient overlap of interest for individuals to cooperate on a number of projects and to remain scientifically and socially engaged with each other for over forty years. Among those to be discussed in the following chapters are ten people who can be seen as belonging to an inner circle: Richard Southwood, Robert May, Gordon Conway, Michael Hassell, Roy Anderson, Michael Crawley, John Lawton, John Beddington, John Krebs and David Rogers.[2] Some others, more loosely associated with this core group, will be mentioned later. Given the high proportion of men among academic scientists during the 1960s and 70s, it is not surprising that this was an all-male fraternity. The group lived through the feminist movement of the later twentieth century and witnessed women entering and succeeding within the academic ranks, but no woman was part of the inner circle.

I have long been interested in generational groups and in how they make their way in science, and am curious about successful networking — in its social dimensions, its effect on how and what science gets done, and in its socio-cognitive dimensions, in how it affects what gets accepted as true. Also of interest is how scientific expertise influences the making of decisions in areas of public importance. I hope these various themes, along with an account of some developments in modern ecology, are brought together in a way that will interest also others. If there is a single question this book raises it is how scientists achieve success. The story of a single group can make only a small contribution to answering it, but historical studies of the kind undertaken here will be indispensable in doing so.

The Silwood Circle caught my eye because it was identifiable over a long period and because the individuals associated with it have all had successful careers, some outstandingly so. A conventional, albeit loose, measure of that success is that of the ten people mentioned nine were elected to the Royal Society, one as president. Seven received knighthoods and two with knighthoods were later given life peerages. There is a faint hint of feudalism in the state deeming that at least some scientists belong within the contemporary aristocratic order. Such inclusion, as indeed fellowship in the Royal Society, raises some interesting questions. Among them, what kind of scientific activity is rewarded, and how do scientists achieve professional and state recognition? In the case of the Silwood Circle, was being part of the group a factor in individual success? Or were its members simply in the right place at the right time? If so, what exactly does it mean to have been so well placed? Small generational groups, where close relationships predate any claim to fame, have been important in many creative areas.[3] The Silwood Circle is a further example.

Bruno Latour is representative of the backtracking. Like others, he is concerned with the ways in which some people use uncertainty as a weapon to manipulate opinion, often against hard-won evidence, in important debates such as those having to do with climate change.[11] In 2004, he wrote:

> I myself have spent some time in the past trying to show "*the lack of certainty*" inherent in the construction of facts. I, too, made it a primary issue. But I did not exactly aim at fooling the public by obscuring the certainty of a closed argument ... I'd like to believe that, on the contrary, I intended to emancipate the public from prematurely naturalized objectified facts.

Latour is a creative thinker who has enriched the conceptual repertoire of historians and sociologists of science. But, as the above quotation suggests, his work has led also to some obscuration. In this he has not been alone. One problem has been a muddling of the meanings of uncertainty and indeterminacy. Our knowledge may be uncertain, but truth, *per se*, is not necessarily indeterminate.[12]

Although in some ways insightful and revealing, the deconstruction of scientific claims by social scientists has led to confusion and, understandably, has drawn criticism. Especially disturbing has been the broadening of the Marxist predilection for materialist explanation to include the idea that scientific observation and reasoned argument are simply *post hoc* ways of justifying, not always consciously, positions favoured for social or political reasons.[13] However, as Latour pointed out in his 2004 paper, the landscape of criticism has changed since the 1980s. While not abandoning a critical stance, he now claims that we need to pay attention to 'hard-won evidence that could save our lives'. Having deconstructed the older positivist image of science it would appear that the new problem is to identify a positive role for science in modern society. In practice this means finding institutional and democratic ways of allowing the public its say on technical matters of concern, while encouraging also trust in ideas that meet basic standards of rationality.[14] This book cannot engage with the complicated boundary issues this problem entails, but it recognizes their relevance to the period when members of the Silwood Circle were part of an advisory elite.[15]

Latour and his colleagues at the Centre for the Sociology of Innovation at L'École Nationale Supérieure des Mines, in Paris, had earlier argued that to understand the nature of science we should look at it in a very broad context, while focussing especially on its relationship to the state. Latour encouraged scholars to examine what he termed 'centres of calculation' — places such as major laboratories and museums where observational data, often gathered from elsewhere, is codified and organized into what counts as knowledge.[16] Such sites, he argued, are worthy of ethnographical and historical study since they help to fashion and lend privilege to the pronouncements of the scientists working in them. In this, Latour helped to motivate scholars who were exploring the centres and peripheries of the scientific universe — the geographies of science.[17] Silwood Park is a site where, since World War II, certain kinds of biological data have been collected from around the world. It privileges people working there by enabling their voices to be heard. As we will see, the ecological data of interest to members of the Silwood Circle came from many sources, local and international. It included data collected in the field, both passive observations and those resulting from contrived experiment, as well as observations made during laboratory experiments, and those made by doctors and veterinarians when treating and studying parasitic and infectious diseases.[18]

Much current sociological and philosophical debate over the cognitive status of science has roots in the early twentieth century. The overthrow of Newton's theory of gravity was a shock. If even the best of scientific theories can be shown to be false (while often remaining useful), how should we describe the scientific enterprise? This is a question addressed loosely in chapter ten. Some of our present difficulties stem also from ideas promoted during the 1940s and 50s. For example, Robert Merton, a sociologist of science to whom we will return below, argued that scientists were interchangeable workers who, provided they followed the proper method, would all arrive at the same conclusion.[19] Uncertainty, for Merton, represented methodological failure, a view that no longer carries conviction.[20] During the more positivist period of the 1950s and 60s, however, schoolchildren and undergraduate science students were taught to respect *the* scientific method, something for which Merton and his contemporaries bear some responsibility. It is perhaps not surprising that, today, people often fail to respect scientific claims that lack full consensus. Dissident scientists can have a disproportionate influence by pointing to what they see as others' failure in method. We need a more realistic view of science, one where many approaches are recognized as viable, conceptual repertoires of their disciplines and have thus provided new ideas for their contemporaries to work with. As this book demonstrates, heuristic leadership, showing a way forward, is important in academic life. Regardless of whether the concepts and approaches being championed have any lasting value, something is learned.

To my mind the history of science pendulum has swung a little too far from the older 'internal' way of doing things, even allowing for new work on laboratory practice. For that reason this book includes more discussion of scientific ideas than is usual in today's socio-cultural histories of science, and in some recent ethnographic studies. I want to convey, even if only superficially, the contributions of the people I write about. One reason for the swing to the 'external' was that early twentieth-century history of science was seen as overly hagiographical. Many of the so-called 'internal' histories of science were quite literally internal — written by scientists about their heroes. Some portrayed science as an almost superhuman activity, and scientists as noble seekers after truth. Later historians reacted by bringing scientists down to earth, portraying them as ordinary human beings, and science as the human activity it surely is.[27] Paradoxically, in making scientists only too human much of their humanity was lost. While not wishing a return to hagiography I am sensitive to a complaint I have heard from scientists: that, biography aside, they are unable to recognize themselves or what they do in much recent history of science.[28] I hope this book meets the recognizability criterion. It focuses on a just small group of scientists, but its scope is fairly broad. At one level it is a type of prosopography, or multiple-career-line story.[29] At another level it is an intellectual history of a branch of modern ecology and, at yet another, a socio-cultural history of a science that gained enormous political influence in Britain during the second half of the twentieth century. First and foremost, however, it is a work in the humanities, an account of human lives, and of how a group of people found their way in the world.

Fraternal groupings of scientists, 'invisible colleges', are not uncommon in science and inclusion in them is often of advantage in the furthering of careers.[30] Such groups were especially important in the days before the internet made instant long-distance communication possible. While conferences remain popular sites for the exchange of information, for much of the twentieth century, including the period in which the Silwood Circle formed, frequent face-to-face meetings with those sharing common interests were seen as necessary. The type of association favoured by members of the Silwood Circle distantly resembles the Arthurian round table as well as the contemporary, Arthurian-inspired, Rotary.[31] Important features of the round table include each person bringing something different (knowledge, skill, trade) to the table, group loyalty, and cooperation toward shared goals. The research school is an associational model well studied by historians of science, but the round table model, while not uncommon in science, is less well documented.[32] The Silwood Circle lasted over forty years and, as will be shown, a round-table ethos aided both the exchange of scientific ideas and the making of careers. Members fostered each others careers, something made easier because the circle existed during a period in which ecology was well funded. The funding was largely the consequence of political pressure related to a growing number of environmental concerns, and to new epidemiological problems associated with the spread of disease caused by the increasing movement of people, plants, and animals during the second half of the twentieth century.

The Silwood Circle resembles a much earlier fraternity, the nineteenth-century X-Club. The X-club included Joseph Hooker, Thomas Henry Huxley and William Spottiswoode, all future presidents of the Royal Society. Robert May, a central figure in the Silwood Circle, was similarly elected president. As mentioned, May and John Beddington served as chief scientific advisors to the British government, and others in the Circle rose to prominence. The earlier X-Club consisted of nine like-minded men, some of them prominent, who wanted to further the cause of science and thought they knew best how to do so.[33] Unlike the X-Club, the Silwood Circle has never identified itself as a club; it remains an informal association. Further, its members have been more scientifically focussed than were most members of the X-Club. Like the earlier club, however, members met frequently at various events, both professional and social. Similarly, they organized regular walking and hiking expeditions. And at their various meetings they, too, strategized over career advancement, the distribution of awards, internal science politics, and science policy.

The Silwood Circle also displays some features possessed by a group that Gary Werskey labelled the 'Visible College'.[34] This small group of left-wing scientists, J. D. Bernal, J. B. S. Haldane, Hyman Levy, Lancelot Hogben and Joseph Needham came to prominence in the 1930s and 40s. Among their beliefs was that science was objective and, if used well, was a force for progress. Indeed, they held the utopian view that science would promote democracy were it not continually being hijacked by capitalist interests. They feared that, unless things changed, capitalist-driven science would lead to democracy's demise. According to Werskey the group saw professional advancement as much 'a political duty as a personal aspiration' and that being elected to the Royal Society would give them a platform from which to voice their views.[35] After World War II and the birth of the atomic bomb, their socially progressive take on science became more difficult to defend. By the 1950s awareness of the damaging effects of chemicals in the environment, and the spectre of chemical and biological warfare, only added to public concern. Bernal was the inspirational force behind the establishment of the British Society for the Social Responsibility of Science in 1969.[36] The Society's approach, leftish but not too much so, held much appeal for a new generation of scientists, among them members of the Silwood Circle.[37] While born too late for the left-wing politics of the 1930s, members of the circle were born too soon to become seriously influenced by postmodern ideas. They inherited Bernal's Enlightenment view of science as objective, and as something that could be used for both good and evil. They also inherited his concern that scientists be socially responsible and that they work for the good of humankind.

The Silwood Circle has been a more cohesive group than the one discussed by Werskey. Further, members of the Silwood Circle have not been openly political and, in writing about them, I am not politically motivated. But like Werskey, I look at ways in which a group of people thought alike, how they sought common goals, and at how they integrated their scientific activity with other, more social, pursuits. Another group of scientists, one that similarly combined the social with intense scientific activity, gathered each summer at a private laboratory located on an estate in Tuxedo Park, New York, owned by the banker-scientist Alfred Loomis. The group included several of America's best young physical scientists.[38] Social life is in part organized around what people believe they need to know. For those with ambition it also provides a way of collectively influencing their time and place, something that both the Tuxedo and Silwood groups were to do — the former with their important contributions to military technology during World War II and to the ways in which American science was organized and funded after the war.

As to scientific interests, most members of the Silwood Circle began with a strong interest in ecology and with the seeking of mathematical models that had some bearing on species populations in laboratory or natural settings. They would probably have agreed with John Maynard Smith who, though not a member of the inner circle, was a close associate. In 1974 he wrote:

> ecology is still a branch of science in which it is usually better to rely on the judgement of an experienced practitioner than on the predictions of a theorist [and] ecology will not come of age until it has a sound theoretical basis, and we have a long way to go before that happy state of affairs is reached.[39]

Members of the circle were engaged in trying to reach that happy state. It is not clear how Maynard Smith would have judged the progress made in the years since 1974, though there is little doubt that some has been made.

The next two chapters help to set the scene for the story of the Silwood Circle. Chapter two is a brief overview of some early history of ecology with an emphasis on aspects that anticipate the kind of work engaged in by members of the Circle. It includes a section on the theoretical ideas of Alfred Lotka and Vito Volterra (articulated further in appendix one). Their equations were foundational to the ecological modelling of the later twentieth century. Chapter three looks briefly at the history of the Silwood Park campus of Imperial College and outlines some of the work carried out there in the years before Richard (Dick) Southwood became professor of zoology and director of the field station.[40] Institutional culture is important in the forming of scientific identity. Southwood, himself a product of Silwood, was key to the formation of the Silwood Circle. His role in bringing people together is the principal focus of chapter four which shows how he promoted ecological work, encouraged younger scientists and helped in the furthering of their careers. The chapter will also give some detail of the wider cultural and political climate of the 1960s and 70s, especially in relation to a growing environmental movement. The fifth chapter includes discussion of ecological work carried out elsewhere from the 1940s to the early 1970s. While important forerunners to the work of the Silwood Circle can be found in many places, especially significant was work carried out at Oxford University in the Bureau of Animal Population under Charles Elton and at the Edward Grey Institute of Field Ornithology under David Lack. Connections between

Oxford ecologists and those at Imperial College began early in the twentieth century and continue to this day. The fifth chapter also introduces some North American ecologists: the Yale University zoologist G. Evelyn Hutchinson, his students Lawrence Slobodkin and Robert MacArthur, and the zoologist Edward O. Wilson. Their work, too, was foundational to that of the Silwood Circle. In chapter six I discuss work carried out by Robert May and Richard Southwood during the early 1970s. The chapter shows how May conceptualized current problems in ecology, put them into elegant mathematical form, and aided others in the Silwood Circle in their work. The chapter briefly touches on the work of some other ecologists, notably that of the Canadian, C. S. Holling, whose research influenced some members of the circle. It considers early reception of the new ideas, how members of the Silwood Circle began to defend them, and how this led to further consolidation of the group. Finally, the chapter illustrates, and to a degree explicates, Southwood's growing reputation as an effective science administrator and committee man. It covers his election to the Royal Society and describes how, once elected, he worked to elect his close associates. Chapters seven and eight cover the progress of individual careers from the mid-1970s to the mid-1990s. However, while focussing on individual pathways, the chapters do not overlook the place of group dynamics. The ninth chapter covers the Silwood Circle when its members were at the peak of their careers. It shows how they became public scientists much in demand for their expertise in the kinds of matters mentioned above. In chapter ten some of my own philosophical views as they relate to this history are outlined. Chapter eleven includes some overall analysis and draws some conclusions.

In writing this book I used information gained during interviews with a number of people, including nine of the main protagonists. However, over the course of research I learned most about the one person I did not meet, T. R. E. (Dick) Southwood. This is because he left a large body of papers, including much personal correspondence.[41] Without his archive, writing this book would have been close to impossible. Southwood kept just about every scrap of paper that arrived on his desk, as well as copies of letters that he wrote to others. He belonged to a generation not yet fully reliant on electronic mail, and was therefore among the last keepers of personal history in tangible form. Further, after his death people were very willing to talk about him. I learned much from listening to them, and not just about Southwood.

But the old saw is true — witnesses do not always agree. Memory is not always reliable. It is uneven, sometimes vague, sometimes deliberately so in the telling, and sometimes it is enhanced. Nonetheless I enjoyed the many discussions and found them helpful and informative. I should also add that I admire members of the Circle and cannot claim to be totally distanced from them in writing this book — not that a distanced stance is fully realisable even when writing about those long in the past.

Ecology is a highly diffuse science. Yet, like other sciences, it has its own situational logic. This is implied by the fact that very similar ideas often appear at roughly the same time, and in more than one place. Merton described this feature of science well.[42] Ideas are often not recognized as being similar until later. This poses a problem for the historian, not simply in assigning credit for priority but, more importantly, for understanding the flow of ideas. Further, there are socio-cultural dimensions both to the assignation of credit and to the flow. Some people, for example, are better than others at attracting the attention of granting agencies, at furthering their ideas, and at impressing their colleagues at conferences. Perhaps the best known ecologists are the synthesizers, people able to make sense of this very complex discipline. Ecology requires some among its many practitioners to explain exactly what is going on. Chapters five and six will show how Hutchinson, MacArthur and May successively played that role. All were instrumental in furthering theoretical approaches to ecology. In my view their character and the nature of their work mattered in this. All three were charismatic leaders whose work was more important in the heuristic sense of suggesting to others what would be worth pursuing next, than in delivering any truly descriptive or long-lasting scientific theory.[43] Heuristic leadership, namely having ideas that open up new fields, being able to suggest new avenues for others to explore, and showing how existing ideas can be brought together, helps to build reputation in science — at least among one's contemporaries. Historical recognition and reception depend on a different set of contingencies. While May is central to my story, Hutchinson, MacArthur and May were all skilled in persuasion and each had a major influence on the ecological thinking of the later twentieth century. However, as others have said before, ecology has yet to find its Newton or its Einstein.[44]

Merton, who took an empirical approach to the sociology of science, demonstrated that in cases of simultaneous discovery, as well as in collaborative work, recognition usually goes to the person already best known in the field. It is one of many gifts that come their way and, as can be seen with the Silwood Circle, it helps in leveraging reputation. Well known people become even better known, and their papers receive more citations than do those of others who make equal contributions to a particular discovery. Merton labelled this the 'principle of cumulative advantage' or the 'Matthew Effect' (from the gospel according to St. Matthew in which it is stated, 'unto everyone that hath shall be given'). Indeed the paper in which the Matthew Effect was first proposed is a case in point. It was co-authored, but the idea has long been associated with Merton.[45]

While Hutchinson, MacArthur and May produced syntheses and provided heuristic guidance, there are other forms of leadership. One such is illustrated in this story by Richard Southwood who, toward the end of his career, was appointed Vice-Chancellor of the University of Oxford. He was a good scientist, in my view better than some of his younger associates acknowledge.[46] Unlike the three people just mentioned, however, he did not come up with ideas suggestive of new and exciting pathways. What was special about him, aside from his considerable entomological and ecological expertise, was that he was an enabler *par exellence*. He wanted ecology to have a more central place in British science and, being politically astute, played a leading role in the achievement of that goal. He supported people with new ideas and promoted the interests of a younger generation of ecologically minded scientists. For those reasons he plays a major role in this story.

Endnotes

1. I came across the group while carrying out research on the history of Imperial College London (Gay, 2007). Some outsiders have referred to the circle as the 'Silwood mafia'. I do not wish to imply that there were no academic or professional ecologists in the first half of the twentieth century, but there were not many. Silwood Park is located near Ascot, Berkshire.
2. Biographical details of the central figures will be given in later chapters. Rogers was closely associated with the circle only in its early years.
3. This point has been noted often. See, for example, (Collins, 1998), 36. Other examples of such groups: Max Born and his theoretical and experimental physicist colleagues at the University of Göttingen in the 1920s, Ernest Rutherford's students at the Cavendish Laboratory in Cambridge during the

1920s and 30s, the Bloomsbury Group which included the luminaries Virginia Woolf, John Maynard Keynes and E. M. Forster; the Group of Seven, Canadian landscape artists extraordinary; and the theatrical and media talent that emerged from the Cambridge University Footlights Theatre in the late 1950s and early 1960s (among others, Jonathan Miller, Peter Cook, David Frost, Eleanor Bron, Germaine Greer and John Cleese).

4. The contraction 'biodiversity' is relatively new. It is a term that seems to have taken off and, like many fashionable terms, it is semantically unstable. For an early usage, (Wilson and Peter, 1988).

5. (Dobzhansky, 1973).

6. (McIntosh, 1980), 197.

7. Why Hawking thought that $E = mc^2$ was the one equation that he had to include in his *A Brief History of Time* (1988) is unclear.

8. Network theory is a major field in sociology, not easily accessed by outsiders. My own reading is limited to a rather cursory look at the work of Harrison C. White, and that of his student Mark Granovetter. The circle, as I use it, would be labelled a type of net by White, and is only one of the many types of social formation he considers. Another social theorist I have found useful on interpersonal transactions is Charles Tilly. (White, 2008; Granovetter, 1983; Tilly, 2002)

9. (Shapin, 2008), 1–2. Shapin supports his claim well, especially as it relates to new high-tech forms of industrial science and entrepreneurial networks. He shows how trust and personal virtue are still important to science in the late modern age. For a philosophical treatment of trust, (Code, 1987).

10. The UK outbreak of Bovine Spongiform Encephalopathy (BSE), the associated Creutzfeldt Jakob disease (v-CJD), and the consequences for the framing of scientific advice, will be discussed briefly in chapters 7 and 9.

11. For an interesting account of the denial of global warming in the United States, (Oreskes and Conway, 2010). The authors point to some physical scientists who used uncertainty as a political weapon. Global warming is a topic on which May and Beddington advised the government, as will be discussed briefly in chapter 9.

12. I leave aside here the quantum world and the uncertainty principle.

13. Some of this thinking can be traced to the Institute for Social Research founded in Frankfurt in the 1920s (it moved to New York in 1934 and returned to Frankfurt in the 1950s). The ideas are explicit in (Mannheim, 1936). Mannheim placed some stress on generational groups and their attitudes, as I do in this book.

14. For example, (Bijker, Bal and Hendriks, 2009; Collins and Evans, 2007; Jasanoff, 2005). These authors are seeking ways in which expertise can be better understood, delivered, and accepted in democratic societies.

15. For a discussion of these issues see references in note 14; also (Gieryn, 1999). For the scientific/socio-cultural/political boundaries relevant to ecological science see, for example, (Taylor, 2005). Like May, but a few years later, Taylor was drawn to ecology through his engagement with environmental activism in Australia.

16. (Latour, 2004), quotation, 227. Latour's article does not advocate giving up on social constructionism; rather it looks at ways to avoid past errors. For 'centres of calculation', (Latour, 1987). For more on Latour see chapter 10.

17. For example, (Livingstone, 2003).

18. The nature of observation is historically very varied, and also contentious, as can be seen from a recent collection of essays: (Daston and Lunbeck, 2011). While seeking a larger picture, the editors make no claim for comprehensiveness — something that would be impossible even in a collection focussed on a much shorter period than the fifteen centuries their book covers. The methods of scientific observation associated with the Silwood circle bear little resemblance to anything described in the book. Perhaps closest in type were those of the Russian zoologist, N. P. Vagner (1829–1909). Known for his theory of pædogenesis, Vagner, like many in the Silwood group, was an entomologist. Entomologists have their own ways of making and validating observations, some of which have survived since Vagner's time. Further, like some in the Silwood group, Vagner 'did not want to be a worker bee. He wanted to throw his weight behind a monumental general law'. And, like those in the circle, he published early and often. (Gordin, 2011), quotation, 141.

19. (Merton, 1973). In chapter 13, 'The normative structure of science' (originally published in 1942), Merton wrote of the 'moral equivalence' of scientists. Seeing scientists as interchangeable workers appealed to many thinkers in the 1940s and 50s. It was a view held by members of the 'Visible College' and Tots and Quots (note 34 below) and has a long historical pedigree. Immanuel Kant, for example, admired Newton enormously but saw him less as a genius, than as someone who worked especially hard to uncover truths open to anyone following the same method. For Kant on Newton see (Kant, 1952), 168–170 (47: 308–309).

20. This point of view is in the broad Kantian tradition. But Ernst Cassirer (1874-1945), working earlier and in the same tradition, differed from Merton in arguing that increasing diversification in the sciences would result in an

increase in methodological approaches, even different epistemologies. An older contemporary of Fleck (see below), Cassirer had a genetical/historical approach to the development of knowledge. For him, as for Robert May later, the power of mathematics to increasingly and symbolically represent our world was a mark of progress. (Cassirer, 1956).

21. Mary-Jo Nye has argued that the roots of a new social conception of science lie in 'the scientific culture and political events of Europe in the 1930s, when scientific intellectuals struggled to defend the universal status of scientific knowledge and to justify public support for science in an era of economic catastrophe, the rise of Stalinism and Fascism, and increasing demands from governments for applications of science to industry and social welfare. ... [They] arrived at a strong new conception of science as a socially based enterprise that does not rely on empiricism and reason alone but on social communities, behavioural norms, and personal commitments that ultimately strengthen rather than weaken the growth of scientific knowledge'. (Nye, 2011), xv–xvi. Nye's biography of Michael Polanyi provides a further example of a successful generational group — Hungarian physical scientists of mainly Jewish parentage, including Polanyi, Eugene Wigner, Edward Teller, Nicholas Kürti and Leó Szilárd, who emigrated first to Germany to further their careers, and then elsewhere to escape from Germany during the 1930s. For my own views on some of the intellectual roots of social constructionism in science see chapter 10.

22. (Fleck, 1979), quotations, 38, 39, 42, 43, 82, 106, 107. Fleck drew on ideas of the French sociologist Emile Durkheim, though the groups of interest to Durkheim were not scientific. For a discussion of Fleck and Durkheim, (Douglas, 1986), 11-14 and *passim*. The 'fact' of Fleck's title was that the Wassermann Test was indicative of syphilis. Ludwik Fleck (1896-1961) was born in Lwów (Lemberg) in the Austro-Hungarian empire (today Lviv, Ukraine). An eminent Jewish immunologist and bacteriologist, he was incarcerated by the Germans during World War II but survived due to his knowledge of typhus and to his having pioneered a vaccine for the disease. I do not know how effective it was, but Fleck prepared his vaccine in the Auschwitz and Buchenwald concentration camps. For some more recent treatments of the social construction of scientific facts, (Latour and Woolgar, 1986; Thackray, 1995; Golinski, 1998). The construction of facts was not a new idea, even in the 1930s. Fleck may have been influenced not only by French sociology, but also by French philosophy of science of the early twentieth century. For this see chapter 10.

23. Fleck was not alone in challenging the positivist take on science, others working in the continental philosophical tradition (Emile Meyerson, Ernst Cassirer and

Gaston Bachelard, for example) did so too, and in a variety of ways. Together these thinkers anticipated ideas that emerged later, for example at the Science Studies Unit founded by David Edge at the University of Edinburgh. The faculty working in the Unit were mainly ex-scientists. The original purpose was to help young scientists to better understand their discipline and its role in society. But what soon emerged was a new academic discipline, the Sociology of Scientific Knowledge (SSK). David Bloor, one of the discipline's founders, does not deny the existence of an external world open to study by scientists, nor that observation has a role to play in knowledge formation. But for him, as for Fleck, and later Latour, it is not a privileged role. (Bloor, 1976). In my view the role of observation, albeit problematic, is privileged. The sociological approach was much in evidence also at the University of Sussex when I was a doctoral student there in the late 1970s.

24. (Kuhn, 1970).
25. Note that, as mentioned in note 23 above, the Edinburgh school used the term 'sociology of scientific knowledge' rather than 'sociology of science' to describe what they were doing. This was a way of distinguishing themselves from Kuhn who retained an Enlightenment view of knowledge as objective and cumulative, and from Merton who studied how scientists worked and were rewarded, rather than how they constructed knowledge.
26. (Latour and Woolgar, 1986; Rheinberger, 2010). Earlier Rheinberger wrote, 'my argument has been that the genesis and development of scientific facts, to use the words of Ludwik Fleck, is less a matter of convention or of negotiation and thus, in the end reducible to a relation between subjects, neither does it result from a relation between subjects and objects; rather, in a way difficult to explain and describe in a lucid fashion, it amounts to a relation between objects themselves: between the experimentally produced traces that are to be taken as the material form of concepts, and the elusive objects which they are presumed to be traces of' (Rheinberger, 2005). In his thought-provoking 2010 book, Rheinberger presents some interesting studies of experimentation in the life sciences, but whether they are cases of what, according to the above quotation, he found difficult to express lucidly, is very much an open question. For Rheinberger the objects in question are those of the experimental system: the model organisms (fruit flies, viruses etc.), instruments, apparatus, etc. From these, scientists (their minds in some sort of symbiotic relationship with their tools) configure new concepts — products of the material conditions of the experiment. Up to a point this makes sense but I am not persuaded that materiality is all, that 'nature' should be so sidelined, or that

sociality, about which I have much to say in this book, has nothing to do with 'fact' construction.

27. One area in the history of science that expanded as a consequence has been the history of subalterns. Much work has been carried out on the 'heroic' underclass of science: assistants, technicians, women, native (aboriginal) informers and so on. It is a field to which I have made a small contribution. (Gay, 1996, 2000, and 2008).

28. While interested in the socio-cultural, I recognize that individual histories play an important role in what scientists do. For some discussion of biography as it pertains to the history of science see 'Focus: Biography in the history of science', *Isis*, 97 (2006), 302–329. As Joan Richards writes, 'the biographical focus on individuals will always trouble the broad sweep of institutional and cultural history'(304). This is an aspect of a more general truth about syntheses and generalizations, namely that there will always be exceptions to the rule. See also (Söderqvist, 2007). In his biography of the Danish immunologist and Nobel Laureate, N. K. Jerne, Söderqvist is critical of historians of science who take what he sees as too sociological an approach to scientific discovery. Individual histories play an important role in what scientists do; therefore biographical material is included in this book.

29. Lawrence Stone defined prosopography as 'the investigation of the common background characteristics of a group of actors in history by means of a collective study of their lives'. However, the kinds of studies he had in mind were different from the one engaged in here. Earlier studies looked at the families, marriages, and economic ties of political actors and how this affected wider political activity. There has been some work in this genre in the history and sociology of science. A classic is Robert Merton's study of seventeenth-century English intellectuals. Merton found a burgeoning interest in natural philosophy among them as the century progressed. He also noted that many among the growing number of natural philosophers came from puritan protestant backgrounds. Steven Shapin looked at 495 ordinary fellows of the Royal Society of Edinburgh from 1783–1820, and Arnold Thackray at the membership and development of the Manchester Literary and Philosophical Society, founded in the 1790s. Their interests were in showing how 'natural knowledge served as a key ingredient in a deep transformation of British thought, society and culture'. I am more interested in social circles and in how they operate within the larger cultural context, an interest not unlike that found in some work on the X-Club, to be mentioned below. The older prosopographies were approached from a largely social science perspective, and tended to look, statistically, at largish numbers of individuals. (Stone, 1971), quotation, 46; (Shapin and Thackray, 1974), quotation, 21; (Merton, 1958).

30. The term 'invisible college' was coined by Robert Boyle in reference to the seventeenth-century group of natural philosophers with which he associated.

31. Rotary International is a body of service clubs. The first was founded in Chicago in 1905.

32. Three excellent examples of studies on research schools: Robert Kohler has written about two generations of scientists who worked with Thomas H. Morgan on the gene-mapping of *Drosophila* in his 'fly-room' at Columbia University, and later in his laboratory at the California Institute of Technology. His book also describes work carried out at Cold Spring Harbour, NY. Kohler was interested in material culture, on the role played by the 'fly' — standard specimens — as well as that of laboratory instruments and other material agents in defining what went on, and in determining knowledge outcomes. People's careers are described, but the humanistic aspects of the story are secondary to the programmatic ones; (Kohler, 1994). Jack Morrell has written about the research schools of three important chemists, Justus Liebig, Thomas Thomson and W. H. Perkin Jr.; (Morrell, 1972, 1993).

33. For the X Club and the making of professional careers, (Jensen, 1970; Barton, 1990, 1998, 2003; Desmond, 2001). T. H. Huxley is reported as having overheard a conversation at the Athenaeum Club: someone asked, 'I say, do you know anything about the X-Club?' and received the reply, 'Well, they govern scientific affairs, and really, on the whole, they don't do it badly'; (Browne, 2002), quotation, 249.

34. (Werskey, 1988). The scientists Werskey wrote about were all highly visible through their speeches, journalism, and books. They also joined a group called Tots and Quots, founded in London in 1931 by the South African zoologist Solly Zuckerman. This left-wing group took its name from the Roman dramatist Terence's aphorism, *quot homines, tot sententiae* (so many men, so many opinions). Its purpose was to apply science and the scientific approach to social problems. The group included at least one politician, Hugh Gaitskell, as well as some of his fellow economists. For more on Tots and Quots', (Desmarais, 2004–2011).

35. (Werskey, 1988), 330.

36. (Bernal, 1950; Brown, 2005). The society was launched at the Royal Society with Bernal, Hogben, Needham and Levy in attendance (Haldane died in 1964). Bernal, already seriously ill in 1969, died in 1971.

37. When I was a chemistry student at Imperial College in the early 1960s, Hyman Levy, by then an emeritus professor, was a much revered figure. Even before 1969 there was a student society which met to discuss the social responsibility of the scientist. I remember hearing Bernal give a talk on the topic to a rapt audience in a large, and packed, lecture theatre. Many of our teachers expressed

the view that science was objective and could be used for good or ill, and that was true also of nuclear science. A related argument was made to claim that scientists were divorced from how their science was used; and that use was something best left to politicians. It was this latter idea that Bernal challenged in his lecture, and was one that brought him and the younger generation together. As we will see, as a young man in Australia Robert May was much influenced by the social responsibility in science movement; it may have been a factor in his move from physics to ecology.

38. (Conant, 2002). Loomis used his fortune to cultivate physical scientists and engineers — young people like Ernest Lawrence, Luis Alvarez, and George Kistiakowsky with big ideas and little money to pursue them. With the advent of war Loomis turned his mind to military science. A gifted amateur physicist, with connections to political, industrial and scientific elites in the U.S., he was the organizing genius behind the MIT Rad Lab. In 1940 Loomis entertained Henry Tizard (recently resigned as Rector of Imperial College) and other members of the UK government's secret scientific mission to the U.S. Among them was Edward (Taffy) Bowen, the Welsh physicist and radar expert who was appointed guardian of the University of Birmingham's famous magnetometer. Bowen stayed with Loomis at Tuxedo Park. After the war, with Loomis's financial help Bowen realized his dream and became the master mind behind the construction of the Parkes radio telescope. (Loomis also persuaded the Carnegie Institution to contribute funds and the Australian government to match them.) Conant shows something of the camaraderie, dedication, and competition among the physicists that Loomis brought to Tuxedo Park, features recognizable also among those who gathered later at Silwood Park.

39. (Maynard Smith, 1974), xi. For more on Maynard Smith see chapter 6 and *passim*. It is worth noting that Maynard Smith was one of Haldane's students and one of his most ardent admirers.

40. The term 'field station' was later dropped. Today Silwood Park is named a campus of Imperial College.

41. The T. R. E. Southwood papers are held in the Bodleian Library, Oxford University.

42. (Merton, 1973), chapters 15–17. As Merton noted, often a scientist will recognize that they have been beaten into print by a competitor. In some cases this will cause them to abandon their line of work, in other cases the idea will be recast so as not to appear identical.

43. I use the term 'charisma' in the tradition of the sociologist Max Weber for whom it was one form of authority. Weber saw Jesus as the archetypical charismatic

leader, someone able to persuade people to give up old orthodoxies and to follow a new path. He also argued that modernity entailed the loss of charismatic authority; but given the events of the twentieth century on that point he was probably wrong. Charisma is important in all forms of leadership, science included; (Shapin, 2008), 4–5; (Weber, 1968, 1978).

44. Whether ecology should follow in the footsteps of physics is much debated, but mathematical models that help to account for large bodies of data are a mark of advance in science more generally (see note 20). Ironically it is now physics that needs to find its Darwin. What, exactly, is the origin of all the fundamental particles?

45. (Merton and Zuckerman, 1968; Merton, 1988). These papers were based in part on Zuckerman's PhD thesis for which she interviewed a number of Nobel Laureates. Although similar, the Matthew Effect is not the same idea as 'winner takes all'. I am not sure how much credence we should give to the Matthew Effect since many people with seemingly good starts in life fail in the kinds of competition found among scientists. Not everyone fully exploits their cultural capital.

46. In talking with people I repeatedly heard the opinion expressed that Southwood was a good scientific leader/organizer but not an outstanding scientist. His papers, however, show considerable observational and conceptual ability, as well as wide entomological knowledge.

chapter two

Some Ecological Ideas that Anticipated those of the Silwood Circle

2.1 From Linnaeus to Lotka and Volterra

Ecological ideas have been around for a very long time but, as a discipline, ecology is still fairly young.[1] The term *oekologie* was coined in 1866 by Ernst Haeckel in order to define a body of knowledge related to what, until the late nineteenth century, was commonly termed the 'economy of nature'.[2] This semantically unstable expression was used by Linnaeus as the title for his 1749 essay, *The Economy of Nature*.[3] The essay included much practical advice on matters relating to plant and animal husbandry but, principally, it was a work in ecology and natural theology. Its main point was that the economy of nature is fixed. Nature, for Linnaeus, was perfectly ordered within an almost machine-like universe designed by God for human use. Each species had a purpose and a functional form to match.[4] Further, inspired by the Stoics, especially by Seneca's claim that everything is permanently in flux, he wrote of the interaction and co-habitation of species, of cycles in nature, and claimed that there was a balance in nature achieved primarily, though not exclusively, through the cycle of birth and death. As a consequence, he claimed, matter flows not only through living things but also through the earth, rocks, water and air. Some of these ideas were anticipated not only by the Stoics, but even earlier by Plato in his *Timaeus*.[5] Indeed, many ecological principles, including those of food chains and cycles were widely, albeit only roughly, understood well before Linnaeus. Shakespeare, for example, had Hamlet amusingly state, 'a man may fish with the worm that hath eat of a king, and eat of the fish that hath fed of that worm.' Linnaeus brought such ideas together, and influenced others to think along lines that would later be labelled ecological. One whom he influenced was Gilbert White, sometimes identified as England's earliest ecologist. White acknowledged his debt to Linnaeus and to Daniel Solander, Linnaeus's student and, later, naturalist on the voyages of James Cook.[6] Linnaeus's anticipation of geo/bio/chemical cycles was to be given new expression in the twentieth century; for example, by G. E. Hutchinson whose ideas will be discussed in chapter five.[7]

Haeckel knew of Darwin's ideas when he coined the term *oekologie*. He recognized that Darwin had given new meaning to the interaction and co-habitation both of species and of individuals within species. Indeed, ideas of community and population as understood by Darwin were included in Haeckel's umbrella-like definition. Given the complexity of the natural world it is not surprising that his definition was a functional one. *Oekologie* was to include any approach leading to a greater understanding of the relationship between organisms and their environments. Therefore natural history, physiology, the larger environment, and evolutionary theory all came together in the new field of ecology. Haeckel's closely associating Darwin with the new discipline helped to give it legitimacy.[8] While this association was fully justified — Darwin, after all, had discussed the 'conditions of the struggle for existence' — it made many early ecologists think of their discipline as one primarily concerned with the study of adaptation in various environments. Today's ecologists have a wider perspective but continue to link their work to Darwin and to natural selection. Indeed, many claim Darwin as the father of their discipline. Beyond Germany, ecology was perhaps first recognized as an academic discipline at the University of Uppsala where Linnaeus had worked. A chair in ecological botany was founded there in 1897.

The idea of a biological community was not new with Darwin; it had been given expression by earlier naturalists, among them Alexander von Humboldt.[9] But the problem with terms such as community or ecosystem is that they are abstractions, and their meanings are only roughly spelled out. Nonetheless they are good heuristic guides. Many who joined the growing number of natural history clubs in the nineteenth century found the idea of biological community useful. In turn, early twentieth-century ecologists were indebted to members of such clubs for the data they collected. By then some ecological activity existed also at universities and within government agencies. At first, academic and professional work was largely oriented towards aquatic studies (limnology and marine biology) and was largely zoological. Terrestrial work in ecology tended more toward the botanical. In Britain it was the botanist and ecologist Arthur Tansley, together with members of his British Vegetation Committee, who founded the British Ecological Society in 1913.[10] The Ecological Society of America was founded in 1914 with a zoologist, Victor Shelford, as its first president.[11]

While much early twentieth-century ecological work was descriptive, it was not carried out in a theoretical desert. For example, questions were raised as to whether the old idea of the 'balance of nature' was plausible and, if so, whether balanced natural communities were being permanently upset by human interference. Further, did species find their ideal habitats by dispersal and natural selection, much as Darwin had supposed? Was chance an important factor in the making of communities? The idea of the biome, a seemingly natural association of plants and animals (as on a grass prairie or in a boreal forest) was debated. To what degree were such associations determined by environmental factors? Or could such associations be the result of the chance invasion and then succession of species? Do local environments sustain the maximum number of species (and of individuals) possible? Is the food and living space available in a given habitat being used to its maximum? In addressing these and other such questions new terms such as 'food chain', 'biomass', and 'energy flow' through biological communities entered the vocabulary. These terms were suggestive and prompted a range of quantitative work. They also reflected some of Linnaeus's earlier ideas and their utility reinforces the heuristic importance of good metaphors and analogies in the furthering of science.

Many of the ideas just mentioned were discussed by Charles Elton in his *Animal Ecology* published in 1927.[12] Elton, a pioneer in terrestrial animal ecology, wanted to move from what he thought was a too Darwinian emphasis on adaptation and selection by his fellow ecologists, and turn instead towards what he called 'the sociology and economics of animals'. He was interested in the ways in which populations fluctuated, and in the connections this might have to climate change (he even considered the effect of sunspot cycles), to physiology, and to evolutionary selection mechanisms. In recognizing that population sizes among animals oscillated, Elton was far from alone. The food chain, too, was central to Elton's view of animal community and he had much to say about herbivores and carnivores, and about predation as a means of regulating community life.[13] He used the term *trophic* (Gr. food) when discussing food chains, but the term took on a wider meaning in the 1940s when used by the Yale University zoologist, R. L. Lindeman, to encompass ideas relating not only to food chains or webs, but more generally to the flow of matter and energy through ecological communities.[14]

Elton's field work on northern animals, and on the voles and mice local to Oxford University's Wytham estate, led to much debate. For example, his view that there were optimum population densities for community stability was contested. But in looking at what was behind birth and death rates, Elton prompted a considerable body of work. Factors such as disease, migration, predation, parasites, changes in environment and weather patterns — and the associated effects of these on the food supply — came increasingly under study. Overall a wide range of quantitative data on biological communities was gathered in the first half of the twentieth century, but with little formalization. Naturalists determined the total numbers of species within given areas; and species populations, density and diversity were considered in relation to community stability.[15] Interestingly the mathematical tools to approach some of these relationships were already in existence but ecologists were largely ignorant of this fact. In the 1960s and 70s ideas proposed by earlier theorists were slowly incorporated into the discipline and new ones added. Members of the Silwood Circle were among the first to appreciate these new developments and so were able to ride the crest of the new theoretical ecology wave.

By the 1950s biologists were seeking new ways of thinking about dynamical processes in the natural world. But the divide between those seeing ecological communities in Darwinian terms as relatively stable assemblages of co-adapted species, and those, such as Elton, who saw communities as continually in flux, and with individual histories of species immigration and emigration, continued to fuel debate throughout the twentieth century.[16] One of America's leading ecologists, G. E. Hutchinson, was to set the agenda for much of the 1960s. Hutchinson held an interest both in the so-called trophic approach to ecological communities, and in the historical question of how communities are constructed and how they maintain stability. People who studied at Yale University under Hutchinson became leaders in these different areas of study.

Hutchinson's generation of ecologists inherited earlier speculations on species abundance within geographic areas, and on the spatial distribution of individuals within given populations. One source was a classic paper on species abundance, published in 1943, in which data collected by scientists working at the agricultural research station at Rothamsted was analysed by the statistician Ronald Fisher.[17] Fisher looked at the relative abundance of species within collections of moths and, on the basis of rather limited evidence, came up with some general distribution curves for the number of species one might expect from the total number of individuals present. This and other data led to the view that a lognormal pattern (one in which the variable itself does not have a Gaussian distribution but its logarithm does) worked for relative species abundance. The lognormal curve appears to have been popular among those mapping empirical results on abundance during the 1940s and 50s, a fact that led later theoreticians, notably Hutchinson's student Robert MacArthur, to seek some underlying cause for this observed pattern. His ideas, to be discussed in chapter five, were to have a major influence on Robert May and others in the Silwood Circle.

The earlier pioneers in mathematical ecology were people who approached the field from outside the discipline. Fisher was by no means alone in turning to biological problems. Other mathematicians as well as some people with backgrounds in demography, physics and chemistry did the same. The mathematical aspects of this work, work beginning roughly in the 1920s, is covered in a little more depth in appendix one. It shows how an equation introduced in 1838 by the Belgian mathematician and demographer, Pierre-François Verhulst, was later applied to an ecological problem by the Italian mathematician Vito Volterra. Verhulst had recast Thomas Malthus's idea on population growth in mathematical form to arrive at the well-known logistic equation. Its characteristic sigmoid curve roughly predicts population growth over time.[18]

For Volterra, Verhulst's equation was the starting point for thinking about animal populations.[19] In this he was not alone, but he made a conceptual breakthrough in thinking about two populations interacting.[20] Volterra had been alerted to a problem in the North Adriatic fishery by his son-in-law Umberto d'Ancona, a fisheries biologist.[21] d'Ancona had made a statistical study of the fishery during the period 1905–23. He noted that for much of that time there were typical fluctuations in fish populations. However, during World War I, and with a marked reduction in fishing, the numbers among some predator fish species increased rapidly at the expense of some of their prey. Since some of the prey fish were those sought also by the fishers the problem had commercial significance. In order to help his son-in-law understand these phenomena, Volterra first considered the interaction of just two fish species, one a predator, the other its prey. His approach was statistical, in that he considered the likelihood of an individual prey fish bumping into a predator and being eaten. He came up with two equations (see appendix one), envisaged as representing the two populations, prey and predator, over time. The equations correctly predicted that fishing would aid the prey species and that cessation of fishing would aid the predators. The fishers had long been part of the ecological community and their absence resulted in a significant change that paradoxically reduced their catches.

Volterra's two equations had been arrived at a few years earlier by Alfred Lotka working independently, and using a different approach. As mentioned in chapter one, roughly simultaneous developments of this kind are not unusual in science. And, as was the case here, priority disputes often ensue. Lotka was disturbed by the lack of recognition for his work. Today the equations are known as the Lotka-Volterra equations, but during Lotka's lifetime the scales were tipped in favour of the better known Volterra who had both academic standing and the mathematical ability to generalize his ideas — a good example of Merton's Matthew Effect (see chapter one). Lotka, however, had the broader scientific outlook and his approach appears to have anticipated some later system ideas in biology.[22] Indeed he appears to have anticipated many ideas in modern ecology, albeit in rather indirect ways. It is interesting that he used both a chemical kinetic and a thermodynamical approach in his work and that the tools of those disciplines continue to underlie much of what goes on in theoretical ecology today. For this reason I think it worth taking the space to discuss Lotka a little further. Some readers may wish to avoid this digression, and the philosophical one that follows it. They should move directly to the next chapter.

2.2 Alfred Lotka and the source of his ideas

Lotka's 1907 and 1920 chemical papers, where some of his ideas were developed, and his wide ranging 1924 book where they were applied to biology, must have appeared somewhat abstruse to his contemporaries. It is one reason why Volterra's version of the equations gained quicker recognition. However, Lotka's training in physical chemistry, and his serious reading of the works of Herbert Spencer, make for an interesting intellectual history. He thought more deeply than Volterra about biology as a whole, and was a system theorist *avant l'heure*. The case of Volterra and Lotka illustrates that sciences have their situational logics. Lotka was a true original, but what was useful among his ideas cropped up also elsewhere, and in far more accessible form.

Lotka's parents were Polish-American Jews who converted to Christianity. They returned to Europe and worked for a London missionary society among the Jews of Poland and Ukraine.[23] Lotka was born in Ukraine and educated at schools in France and Germany. In 1898 he entered the University of Birmingham to study chemistry. The university had a good chemistry department which, at the time, was being enlarged by the professor, P. F. Frankland. There was also a strong physics department at the university and Lotka learned much from J. H. Poynting to whom he later dedicated his book, *Elements of Physical Biology* (1924).[24] Lotka's leaning toward physical chemistry led to his spending a postgraduate year in Wilhelm Ostwald's laboratory in Leipzig.[25] Ostwald was born in Riga, and educated at the University of Dorpat (now University of Tartu, Estonia). He later claimed that had he been educated in Germany he would have become an organic chemist and, by implication, boring.[26] Interestingly, his chemical education appears to have been not unlike the one Lotka later received in Birmingham, and included much physics. Ostwald was famous, perhaps infamous, for his energeticist views. An opponent of the atomic theory, he claimed that energy was the fundamental basis of all matter, and that an understanding of thermodynamics would, in turn, lead to a clearer understanding of the chemistry of both living and non-living matter. This was another missionary idea, and one that must have appealed to Lotka. But Ostwald was also a pioneer in chemical kinetics and Lotka took from Leipzig the dual kinetic and thermodynamical approach that was to characterise much of his later work.[27]

Lotka, an American citizen, decided to look for work in the United States, even though he had never lived there. He was able to find a number of relatively well paying jobs over his career, but the ideal academic position eluded him. At first, and later during World War I, he worked for the General

Chemical Company. His work there was varied and included research in photography and on nitrogen fixation. But he also began applying some of Ostwald's ideas to biology. According to Sharon Kingsland, Lotka was discovered almost accidentally by the ecological community.[28] This seems an overstatement since his ecologist contemporaries were largely ignorant of his work. But he was fortunate in gaining the attention of the energetic Raymond Pearl of Johns Hopkins University. Pearl had moved to Johns Hopkins as professor of biometry and vital statistics after having served as chief statistician in the wartime Food Administration Program. At Johns Hopkins he joined the new School of Hygiene and Public Health from where he advocated the use of mathematics in both medicine and biology. He was a keen promoter of Verhulst's logistic equation in the study of human and animal populations.[29] Lotka was applying mathematical ideas to biology in ways that appealed to Pearl, and was given associate status at the School so that he could publish his papers from a university address. In 1922 Pearl offered him a fellowship and Lotka spent two years at Johns Hopkins working on the book that he was to publish in 1924. According to Kingsland, Lotka's book 'expressed the synthetic spirit of the nineteenth century'.[30] This is a fair assessment; Lotka drew on ideas from many Victorian sources, notably the works of Herbert Spencer, Charles Darwin, Hermann von Helmholtz and Ludwig Boltzmann, as well as those of Wilhelm Ostwald.

Lotka's overall evolutionary vision was more Spencerian than Darwinian in that it included everything — living and non-living matter.[31] Like Spencer, he believed in chemical evolution, the evolution of life forms, of human social behaviour, and of societies. He wanted to give all of this a foundation in an Ostwaldian-inspired thermodynamic and kinetic system. But, as Giorgio Israel has pointed out, Lotka's book with its 'uncontrolled mingling' of many disciplines held little appeal for more specialized scientists.[32] Indeed, by the early twentieth century scientists were increasingly expected to specialize. Lotka's outlook was too holistic, even for those ecologists who later labelled themselves holists. But his system ideas did have a small following and were later rediscovered.[33] Volterra's reductionist, statistical mechanical, approach was far easier to follow, especially by other mathematicians entering the biological field. But this approach, too, had its limitations and for many biologists Volterra's overreaching mechanism held little appeal.[34]

Lotka is best understood as someone thinking along two parallel tracks. On the one hand he wished to apply the principles of physical chemistry, especially as articulated by Ostwald, to the biological world. On the other, he was a devotee of Herbert Spencer, had read Spencer's *First Principles* (1862) and *Principles of Biology* (1864), and wished to capture some of Spencer's ideas in his own theory. He appears to have accepted Spencer's utopian idea that nature, left to itself, will result in some final equilibrium, with harmony and cooperation operating at the social level.[35] Lotka's book contains a number of quotations from Spencer, including the following which has direct ecological significance.

> Every species of plant and animal is perpetually undergoing a rhythmical variation in number — now from abundance of food and absence of enemies rising above its average, and then by a consequent scarcity of food and abundance of enemies, being depressed below its average ... amid these oscillations produced by their conflict, lies that average number of the species at which its expansive tendency is in equilibrium with the surrounding repressive tendencies.[36]

The passage illustrates well one of the ideas that Lotka was trying to capture for ecology. But while living and non-living matter changes over time, according to Spencer it does so in a purposeful direction. The slow process of evolution toward some final stable state is punctuated, he believed, by oscillations of the type alluded to in the above quotation, and by a wide range of both unstable and relatively stable intermediate states. Lotka accepted this view of things and that the evolving natural world consisted of heterogeneous aggregates of matter with varying life spans. These aggregates could also be seen as finely graded series of objects moving towards relatively stable evolutionary states such as 'plant' or 'animal', 'oak tree' or 'dog', 'tribe' or 'society'. But for Lotka such identities were never totally firm, and he viewed disputes over what constitutes a species, for example, as pointless. Similarly, the distinction between living and non-living aggregates was, he believed, obscure.

Ideas such as these began to form in Lotka's mind while he was working in Ostwald's laboratory in 1902. Some were expressed in his 1907 paper where he looked at two cases, one chemical and the other demographic. The chemical dynamics used in the paper was applied to what he claimed was a special case of inorganic evolution, namely the reorganization of matter over time.[37] In his later book, written for both biometricians and biologists, Lotka stated that he wanted to build on this and his other early papers to elucidate the 'broad application of physical principles and methods in the contemplation of biological *systems*' (emphasis in original).[38] The emphasis on systems was intended to set his work apart from biophysics which he saw as a discipline in which physical principles were applied only to individual biological entities.[39] As he put it, nature had to be seen as a whole before it could be understood in any detail. In this I believe he was inspired by the Swedish physicist, J. R. Rydberg, and his theory of state.[40]

Individual biological organisms were seen by Lotka as assemblages of chemicals capable of growth based on environmental conditions (food sources, sunlight etc.). By allowing that matter was in continual motion, that sometimes it slowed down, remaining in stable configurations for very long periods, and that at other times it underwent rapid rearrangement, Lotka believed he could account for the number and variety of biological entities, and the systems to which they belonged. He recognized that organic evolution is a very slow process characterized by time lags, implying that evolution is the accumulation of a series of irreversible processes. Thus, Lotka's definition of evolution had both a thermodyamical and a statistical mechanical aspect. He envisaged a system undergoing irreversible change, in the direction of increasing global entropy, and that each step on the way was one of increasing probability. Science relies on good analogies to progress. However, the way in which Lotka expressed his thermodynamical, kinetic, and statistical mechanical analogies appealed more to demographers than to biologists attuned to the Darwinian idea of difference at the individual level. Nonetheless, in the years that followed some ecologists did take the thermodynamical route and looked to see where and how energy was distributed within ecosystems.

Others took a route more akin to Lotka's approach to statics. Here he was trying to understand the nature of biological systems in steady states. Lotka thought about the problem topographically, in terms of peaks and valleys and energy flow. Even with a minimum of dependent variables the range of possible steady states is large. In order to find some general principles, Lotka looked again to chemistry for inspiration. Consider, for example, a chemical equation. The two sides of the equation represent stable states but what actually happens in moving from one side to the other? How is the peak negotiated to arrive at the valley on the other side? Little was known about this, though presumably fleeting chemical intermediaries existed. As Lotka put it,

> in a material system in which physical conditions vary from instant to instant and from point to point, certain individual constituents (molecules, organisms) may have a transitory existence *as such*, each lasting just so long as its conditions and those of its neighbourhood continue within certain limits.[41]

But the final equilibrium between the two sides is determined by thermodynamic laws.

This kind of thinking allowed for overreaching ecological speculation. The final sections of Lotka's book in which he applied the principles of dynamic equilibria are difficult to follow. They range far and wide with speculation following upon speculation. Lotka's topics include human populations both on their own, and interacting with populations of other species — in one case with herds of sheep! He discussed the seasonal stomach content of various birds which, in turn, he related to insect populations, and to the equilibria between birds and insects in different environments. He recognized that long food chains are energy wasteful and therefore unlikely on evolutionary grounds. And he had some applied ecological advice on that score. For example, he recommended feeding algae to cattle, and that human beings should eat less meat and more shellfish.[42] He also anticipated G. E. Hutchinson in recognizing that because food chains were short they could not, alone, account for the large number of animal species. He discussed atmospheric and geochemical ideas and focussed on the cycles of a few basic elements. In discussing the nitrogen cycle, for example, he explained how increasing the use of artificial nitrogen fertilizers would result in evolutionary change. He also discussed distributed and local sources of energy and the scrimmage for energy among plants and animals. He had much to say about the role of plants and animals in the grand design. For example, he stated that a grazing cow is able to subsist because 'grass is hungry for oxygen', but the cow is also a catalyst, oiling the machine and assisting energy on its path to lower availability (higher entropy). According to Lotka, as long as there is an abundant surplus of energy flowing to waste, there will be a marked advantage to any species able to use some of that energy in adaptive ways. Indeed nature works, he believed, to increase both the total energy flow and the biomass within the biosphere. However, while evolution could be explained along such lines, Spencer's survival of the fittest and Darwin's natural selection were important ideas, not deducible from thermodynamics. Spencer and Darwin had introduced principles that, at a different level, gave meaning to the entire evolutionary process. Lotka recognized this, and on that point, at least, was at one with many biologists. How to integrate the idea of natural selection with the theoretical ideas of Lotka and Volterra (and ideas coming from genetics) was a problem for future ecologists and one that was to engage members of the Silwood Circle.

For the Circle we can sum up the direct importance of Lotka and Volterra's work by stating that it showed how to predict population size as a function of time in situations where two species either competed for resources,[43] or in situations where one species was prey and the other predator. But, as the discussion of Lotka implies, there is more to the inheritance than just the mathematical models that he and Volterra produced (see appendix one). Variations on these models soon appeared and a couple of important early ones are worth mentioning.[44] In 1927 W. O. Kermack and A. G. McKendrick considered what would happen in the case of a pathogen attacking an animal population. Indeed, McKendrick, in applying the law of mass action to such problems more generally, helped to lay the foundations of modern epidemiology.[45] In 1935, A. J. Nicholson and V. A. Bailey looked at the situations where parasites or parasitoids attack insect herbivores. While Volterra's model assumes the death of the prey, in these other cases it is important for the continued existence of pathogen and parasite that not all the infected hosts die, and that not all prospective hosts become infected.[46] These early pioneers showed that by varying the Lotka and Volterra equations in light of such considerations progress could be made.

2.3 What should we make of this? A philosophical aside

Differential equations describe the evolution of continuous variables. Difference equations are approximations to differential equations and can give numerical solutions when analytical solutions cannot be found.[47] The differential equations used by Lotka and Volterra express patterns of various kinds which only partially mirror the real world. However, whenever there is some mirroring, mathematical models are worthy of further attention. This is something that early mathematical biologists understood, and is something understood even better today. While planetary motion can be seen as continuous, the world of biology with its births, deaths, population explosions and crashes seems far from being so. But differential equations can be used to describe collectivities of discrete events. As Lotka understood, they are used in the study of chemical kinetics because the populations of the atoms and molecules being considered are sufficiently large for the many small-scale discontinuities to be smoothed out. Similarly in demography, one of Lotka's other main interests, differential equations work reasonably well because the populations being examined are of a certain size, say at least a thousand or more. As we have seen, Lotka did not confine himself to chemical kinetics and demography. He believed that the principles underlying those disciplines were universal and could be applied to all areas of life, to population biology, to biological evolution, to social evolution, and to an evolving bio/geo/chemical planet. Drawing inspiration from thermodynamics and statistical mechanics, he saw life as the interplay of organisms moving over a grand topographical map. As individuals move they collide with a succession of other life forms and with other environmental features in their path, exchanging energy in the process.

Lotka's reductionism was steeped in the newly fashionable areas of statistical mechanics, kinetics, and thermodynamics. The relative newness of these disciplines in the early twentieth century makes it not altogether surprising that his eager application of them in so many different areas of biology was little appreciated. Volterra, too, was a reductionist. But his ideas were well entrenched in the mathematical culture of his day. In retrospect it is easy to see why Volterra was seen as sensible and measured, and Lotka as muddled and overly ambitious; though, in so far as their famous equations go, the results were much the same. Lotka's type of far-reaching ambition has not gone away. For a while in the mid to late twentieth century the systems approach to biology was very fashionable and projects emerging from that tradition still carry on. Let us look at a recent example of a more reductive approach to be found in a 2002 paper by Simon Levin.[48] Levin's paper is no less ambitious in its intent than was Lotka's book. Further, his approach bears some resemblance to Lotka's. The abstract of Levin's paper reads as follows:

> The study of complex adaptive systems from cells to societies is a study of the interplay of processes operating at diverse scales of space, time and organizational complexity. The key to such a study is an understanding of the interrelationships between microscopic processes and macroscopic patterns, and the evolutionary forces that shape systems. In particular for ecosystems and socioeconomic systems, much interest is focussed on broad scale features such as diversity and resiliency, while evolution operates most powerfully at the level of individual agents. Understanding the evolution and development of complex adaptive systems thus involves understanding how cooperation, coalitions and networks of interactions emerge from individual behaviours and feed back to influence those behaviours.

In my view the abstract is a fair summary of the situation facing today's ecological theorists, though the program Levin envisages is hugely ambitious. Lotka would surely have been interested in it — though the language of diversity and resiliency would need translation. Lotka anticipated Levin in seeing the evolving biosphere as well as its parts as heterogeneous, complex, adaptive systems of the highest interest. He understood that there were different scales of space, time and organizational complexity and that new entities are forever forming, and that others are disappearing. Chemistry had taught him that some substances are very stable, and can be arrived at by different chemical routes. Others are less stable. So, too, in biology, some life patterns are stable and can be arrived at in many ways. Others are more transient. He would have fully agreed with Levin's conclusion, namely that while not everything about evolution and ecology is knowable,

> the central problem is to develop an appropriate statistical mechanics that allows one to separate the knowable unknown from the truly unknowable. Such a mechanics will have to deal with heterogeneous ensembles of interacting agents and with the continual refreshment of that ensemble by novel and unpredictable types.

Today's scientists know far more than Lotka could ever have imagined and use more sophisticated mathematical tools. Levin's reductive vision, however, is one that both Volterra and Lotka would have appreciated. Further, the goal of the physicist of finding continuum descriptions of the various phenomena of interest is one that Volterra, Lotka, Levin and many other of today's theoretical ecologists share. But biologists also understand that living populations are unlike collections of atoms or molecules in that they have many more degrees of freedom. To what extent ecological models should reflect this greater complexity is a matter of debate, something to be discussed in later chapters.

The historian cannot take sides on many of the issues raised by these reflections; indeed, wearing the philosopher's hat is difficult. I am not much concerned with the cognitive status of the various concepts that Lotka or his modern counterparts have dreamed up, something I understand only in outline. I am, however, interested in the historical role that such concepts play. If the reductions sought by mathematical modellers help to give meaning to the biological world, then that is a good thing. If the models go further and allow for important predictions, then so much the better. If they can be of help in solving some of the major problems that we face today in areas such as infectious disease epidemiology, biodiversity and the environment, then what more can we ask for? Whether they are truly descriptive is of secondary importance. Today's biological theorists have little time to think about their intellectual roots, or about the cognitive status of their models. Historians, too, cannot worry much about the latter, something best left to philosophers. The role of the historian is not to uncover 'Truth' but to recover something of the past, to place people and their ideas in a broad context, to attempt an understanding of how they developed their ideas, and to tell stories that give some meaning to their activities. It is also important to consider why people believe what they do, and why some ideas gain credibility while others do not. In the above I have attempted to show the source of some of Lotka's ideas and to give meaning to some of his work. Following others, I have claimed that, by and large, his many speculations failed to gain credibility. In his case he was in the wrong place at the wrong time. In what follows I hope to give meaning to the work of members of the Silwood Circle. As will be shown, they were luckier in their place and time. But first we will look at some other early history, and at the discipline of entomology as practised at Imperial College. It was there that some of the foundations were laid for later work at Silwood.

Endnotes

1. Among the general histories I have used (Bramwell, 1989; Worster, 1994; McIntosh, 1985; Sheail, 2002). Good bibliographic sources, (Egerton, 1977 and 1983). For the history of theoretical ecology, (McIntosh, 1980; Kingsland, 1995).

2. Ernst H. P. A. Haeckel (1834–1919) was professor of comparative anatomy at the University of Jena. He was also an evolution theorist, a gifted naturalist, and a fine nature illustrator. He introduced the term *oekologie* (*oekos* Gr. house) in his *Generelle Morphologie der Organismen* (2 vols., Berlin, 1866). One could argue that his wonderful drawings of marine life played a role not unlike mathematical models; they were abstractions that allowed for an understanding of the organisms they portrayed.

3. A witty short essay (Hestmark, 2000) informs us that Linnaeus wrote his *Economy of Nature* as a dissertation for one of his students, and that he wrote similar theses for about 180 other students. In those days students had to defend their teachers' dissertations, not their own!

4. Linnaeus was not alone in thinking this way. (Glacken, 1967) plausibly claims that natural theology prompted a range of early work such as Linnaeus's and that it anticipated much that came later. However, Linnaeus was forced by the evidence of seemingly new species to modify his views. In later editions of his *Systema Naturae* he claimed that God had fixed the higher taxa but that genera and species were the 'offsprings of time'. Linnaeus shared some geological views with his compatriot Emanuel Swedenborg. Noting that the sea levels of the Baltic were declining, he assumed that ocean levels more generally had been dropping ever since the creation — though it is unclear where he thought the water had gone. The creation, he believed, occurred on an edenic island, which, at the time, was the only existing bit of dry land, and that it was located somewhere near the equator; (Oldroyd, 1980), 19–21; quotation, 21.

5. Others, too, abstracted Plato's ideas, including that of a chain of being, another important antecedent of modern biological ideas, (Lovejoy, 1936). Alexander Pope's *Essay on Man* (1733) was written a little before Linnaeus's essay, and is metaphysical in approach. It, too, illustrates the wide currency of Stoic ideas within the natural theology of the eighteenth century, and makes use of the chain of being.

6. (White, 1993). This work was first published in 1788/89. Linnaeus corresponded with Solander. May and McLean (2007) claim White as the 'father' of ecology in the introduction to their book. They especially note White's inclusion

of quantitative information on the steady population of eight breeding pairs of swifts nesting in the church at Selborne. In North America, Henry David Thoreau plays a similar role to White in the historical memory of ecologists.

7. For similar ideas, (Lovelock, 1979).

8. Darwin's ecological ideas are discussed by Vorzimmer who claims that Darwin did not fully appreciate the connection between his idea of natural selection and his own ideas in ecology. This is not surprising. The connections are subtle and took time to be recognized (Vorzimmer, 1965).

9. Humboldt characterized different regions by their distinctive life forms and helped give rise to the serious study of biogeography, (Browne, 1983). A later contribution to the idea of natural plant associations is Warming (1909). Warming's translator, Percy Groom, later taught botany at Imperial College. It is probably fair to say that Warming's work was more influential in popularizing the idea of ecology than was Haeckel's earlier. Indeed the term 'ecology' was hardly used between Haeckel's introduction in 1866 and Warming's use in 1895. Among others, Warming's book influenced the Scottish brothers and pioneer plant ecologists, William and Robert Smith. Robert Smith studied both with Bayley Balfour in Edinburgh and in Montpellier with another pioneer plant ecologist, Charles Flahault, who studied the vegetation of the Rhône delta. He published his important *La répartition géographique des végétaux dans un coin du Languedoc* in 1893.

10. For Tansley, (Hope Simpson, 2004; Fischedick, 2000) also Anker (2001), ch. 1. Sir Arthur George Tansley (1871–1955) was Britain's preeminent plant ecologist of the early twentieth century. The British Vegetation Committee members (notable among them also Robert Smith, see note 9) spent many hours mapping the vegetation of Britain, before reinventing themselves as the British Ecological Society. Tansley became its first president. In 1917 he founded the *Journal of Ecology* and was its first editor. He was appointed to the Sherardian chair of botany at Oxford in 1927. Tansley's best known work is *The British Islands and their vegetation* (1939), for which he was awarded the Gold Medal of the Linnean Society. See also (Tansley, 1923). The BES remained relatively small until the second half of the twentieth century with a membership of just under 500 on the eve of World War II. Tansley was also a founder of the Nature Conservancy in 1949.

11. Victor Ernest Shelford (1877–1968) was an aquatic entomologist and professor at the University of Chicago. He was one of the moving figures in the important Illinois Natural History Survey. See (Clements and Shelford, 1939) for the promotion of an organismic view of biological community.

12. (Elton, 1927). Elton was a university demonstrator when asked to write this book by Julian Huxley. Huxley, his former tutor, had by then had moved to a professorship at King's College London and was editing a series of textbooks. He wanted something on ecology suitable for undergraduate zoology students. In 1932 Elton became the first editor of the *Journal of Animal Ecology*. For Elton, (Paviour-Smith, 2004; Hardy, 1968). Elton's work will be discussed further in chapter 5.

13. Elton used the term 'niche' to describe an animal's relationship to its food source and to its predators. Niche is another ecological term with little semantic stability; though it was stabilized, to a degree, by G. E. Hutchinson (as is discussed in chapter 5). E. P. Odum listed several meanings of the term, (Odum, 1953). Elton looked at the population numbers of the snowshoe rabbit and Canadian lynx in the Hudson's Bay Company records. The numbers of both species oscilated with approximately ten-year cycles.

14. (Lindeman, 1942). Lindeman, a post-doctoral fellow who worked at Yale University with G. E. Hutchinson, was a limnologist who died prematurely in 1942. His paper was published posthumously. Lindeman thought of a lake as a 'biotic community' with a 'constant organic-inorganic cycle of nutritive substance'. See also chapter 5.

15. At this time most naturalists believed that high species density implied greater stability, though not all held this view. Tansley, for example, believed that maximum stability was arrived at after a period of settlement and followed one of maximum density; (McIntosh, 1985), 141.

16. For some early treatments, (Willis, 1922). See also the views of H. A. Gleason and F. E. Clements on the question of plant associations. Gleason appears to have favoured a more stochastic view of the history of plant communities while Clements believed in rules of assembly with species gradually accumulating into communities that could be seen as some sort of ecological superorganism, the various species forming its co-adapted parts. Clements was a Lamarckian and so placed a heavy emphasis on physiological studies relating to how species adapted to their environments; (Clements, 1916; Gleason, 1926). As will be shown, in more general terms this debate continued throughout the twentieth century, as can be seen also in (Hubbell, 2001). Clements was an influence on A. G. Tansley who, while rejecting the idea of a superorganism, nonetheless saw something in the idea. Tansley preferred the more neutral term 'ecosystem'. His may have been the first usage, but the term took on a new meaning when system theory entered biology in the 1960s.

17. (Fisher, Corbet and Williams, 1943). Their paper was an analysis of the number and variety of moths trapped at Rothamsted. Earlier, Corbet had studied

butterfly and moth populations in Malaya and that data, too, was subjected to statistical analysis. Fisher who held a number of different academic posts moved his University College (London University) department to Rothamsted during the war. He was a major statistician working in the areas of eugenics and genetics, and brought together Mendelian and Darwinian ideas in his work (Fisher, 1930). The mathematical foundation for the 1943 paper had been laid out even earlier in his classic paper, (Fisher, 1922). For more on Fisher see (Yates and Mather, 1963). Sewall Wright and J. B. S. Haldane were further contributors to the mathematical foundations of population dynamics in the early twentieth century. Sewall Wright is also of interest in light of Alfred Lotka's ideas to be discussed below. Like Lotka, he introduced a topographical way of looking at living things as they negotiate life's pathways. To the best of my knowledge, Lotka did so earlier, but Wright's work was more extensive and his influence on population biologists far greater. C. B. Williams, Fisher's co-author on the moth paper, was the PhD supervisor of T. R. E. Southwood, the father figure of the Silwood Circle.

18. Thomas Robert Malthus (1766–1834) was an Anglican priest and economist. He published his *Principles of Population* in 1798 in response to Enlightenment thinkers who were predicting a rosier future than he thought possible. P. F. Verhulst (1804–49) was a younger contemporary of the Belgian demographer and statistician A. L. J. Quetelet (1796–1894). Like Malthus, Verhulst and Quetelet were interested in population checks, and in why population growth rates decrease as the population increases. Verhulst expressed these ideas mathematically. Interestingly, his equation (see appendix 1) was soon picked up by chemists thinking about autocatalysis, a situation where the products of a chemical reaction themselves catalyse the reaction, leading to exponential growth in the reaction product. It is likely that Alfred Lotka, to be discussed below, was introduced to the equation through his chemical studies. The situation in demography is not unlike that in autocatalysis; children/product of the first generation contribute to the overall increase in population/product which, given successive generations, leads to exponential growth unless checked. As with chemical checks, the question of checks in human and other animal populations is still very much alive and is a principal interest of ecologists.

19. For Volterra, (Whittaker, 1941). Volterra, professor of mathematics at the University of Rome, was not only a first-rate mathematician, but also a remarkable man. For Whittaker's comments on Volterra's work in biology see especially pp. 707–14. Whittaker notes that Volterra was interested in some biological applications of mathematics already by 1901, but that his first major paper on

the subject was published by the *Accademia dei Lincei* in 1926, later translated and published in *Nature*; (Volterra, 1926). See also (Volterra with Umberto d'Ancona, 1935). It is Whittaker's summary of the ideas in their book that I use in appendix 1. Volterra published over twenty articles on biological mathematics. For more recent discussion of the work of Volterra and Lotka, (Kingsland, 1995; Israel, 1988). Israel discusses well Volterra's philosophical view that the methods of classical physics are appropriate also in biology.

20. The Verhulst equation had been used, for example, by the entomologist/ecologist W. R. Thompson and by two pioneers in malaria research, Ronald Ross and Emilio Martini. For Ross, (Nye and Gibson, 1997). Already in 1911 Ross was thinking along the lines later developed by Lotka and Volterra. Thompson was the first director of the Farnham Royal Laboratory, opened by the Imperial Institute of Entomology in 1927.

21. (Whittaker, 1941), 710–711 for details on d'Ancona, the fishery, and Volterra's model as fully articulated in his 1935 book. See also G. Israel, 'The scientific heritage of Vito Volterra and Alfred Lotka in mathematical biology': giorgio. israel.googlepages.com/Art82.pdf.

22. Lotka's contributions are recognized in (Bertalanffy, 1968).

23. For Lotka's father, Jacob Lotka, and his missionary work, (Gidney, 1908). Alfred Lotka was born in Lwow (Lemburg, now Lviv, Ukraine), a town which then had a large Jewish population and, as mentioned in chapter 1, was the birthplace also of Ludwick Fleck. Lotka ended his professional life as a statistician with the Metropolitan Life Insurance Company in New York. For an obituary by one of his colleagues, (Dubin, 1950).

24. Percy Faraday Frankland (1858–1946) was the son of the chemist Edward Frankland (a member of the X-Club, see chapter 1). P. F. Frankland had interests in agricultural chemistry and bacteriology. John Henry Poynting (1852–1914), like his teacher James Clerk Maxwell, had interests in electromagnetic theory. Lotka's 1924 book was later renamed (Lotka, 1956).

25. A. Wilhelm Ostwald (1853–1932), professor of chemistry at Leipzig, was a founder of classical physical chemistry, and winner of the Nobel Prize for chemistry in 1909 for his work on catalysis and chemical kinetics. It is interesting that someone highly sceptical of the existence of atoms should have been a pioneer in kinetics.

26. (Brock, 1992), 378–379. Classes at the University of Dorpat were given in both Russian and German. Ostwald came from a German family living in Latvia. Organic chemistry flourished in German universities during the later nineteenth century leading to the rise of the German chemical industry.

27. It is possible that Lotka will have learned something about autocatalysis (and the Verhulst/logistic equation) already as an undergraduate, but it was one of Ostwald's interests. In his book Lotka noted that he read Ostwald's lectures in natural philosophy, published in 1910.

28. (Kingsland, 1995), 25–26. Kingsland gives an account of Lotka and Raymond Pearl in her book.

29. As Kingsland (1995) points out, the mathematization of biology was fashionable at the time. For example, D'Arcy Wentworth Thompson published his *On Growth and Form* during the war (Wentworth, 1917). At the time, however, his ideas met with limited approval and he was criticized by both strict Darwinians and geneticists.

30. (Kingsland, 1995), 32.

31. For Spencer's scope, (Gay, 1999).

32. Israel, see note 21.

33. Among his contemporaries, Lotka's system ideas appealed to the Russian biochemist Vladimir Kostitzin. While Kostitzin was a friend of Volterra, he was not entirely keen on the reduction of biological systems to mechanics. Nonetheless he used the Lotka-Volterra equations to look at the carbon cycle as it related to plants and animals. In doing so he assumed that the population growth of animals was regulated by the photosynthetic production of plants and that photosynthesis depended on the production of carbon dioxide by animals. Similar but more expansive views were held by Vladimir I. Vernadsky. He recognized that carbon dioxide levels depended also on many geochemical factors. Interested in the biosphere, he believed that all the major atmospheric gases were organic in origin. See chapter 5 for Vernadsky's influence on G. E. Hutchinson.

34. As Israel (note 21) makes clear, Volterra raided the arsenal of mechanics in seeking equations to apply in biology. Using Newtonian mechanics recast in Hamiltonian form, Volterra sought biological parallels for particle interactions and the laws of motion. Just as Newtonian mechanics allows the positions of particles to be tracked through time so, perhaps, entities making up biological populations could be similarly tracked. Today's modellers, while indebted to Volterra, no longer seek the kind of exactness implicit in Newtonian mechanics.

35. This is surprising. By 1924 Lotka must have been aware of the view that radioactivity in the interior of the sun was a continuing heat source and that the sun could still have billions of years left in its lifespan. One might think he would have seen more clearly the possibilities for future evolution of life on earth, and that evolution well beyond that of the human species was likely. Further, at

thermodynamic equilibrium there will be no human beings, let alone human societies. Darwin, less certain of the lifespan of the sun was more understandably, albeit similarly, shortsighted. Like Spencer and later Lotka, he clung to an idea of progress even while recognizing that progress was not necessary. As he wrote to Joseph Hooker, 'I quite agree how humiliating the slow progress of man is ... . To think of the progress of millions of years, with every continent swarming with good and enlightened men, all ending in this [the death of the sun]'; quoted in (Browne, 2002), 313.

36. (Spencer, 1862), ch. 22, section 173, as quoted in (Lotka, 1956), 62.

37. (Lotka, 1907). Further papers along these lines appeared in the 1920s.

38. (Lotka, 1956), viii.

39. I think he may have had in mind (Wentworth, 1917) in which mathematical models are applied to the growth and form of individual organisms.

40. Lotka mentions Rydberg in chapter three of his book. J. R. Rydberg (1854–1919), professor at the University of Lund, famed for his foundational work in atomic and molecular spectroscopy, was also a promoter of the use of equations of state. He believed that their use was the simplest way forward in physical chemistry. He introduced the term *allgemeine zustandlehre* which roughly translates as a general theory of state.

41. (Lotka, 1956), 154.

42. I have no idea whether algae are more efficient at capturing the solar flux than grass, but clearly Lotka believed so.

43. This aspect of Lotka and Volterra's work was better articulated later by Gause with his idea of competitive exclusion. Gause provided some laboratory and field evidence in support of his idea; (Gause, 1934).

44. For more detail on these and other early models, (Kingsland, 1995), 116–150.

45. McKendrick served under Ronald Ross in the 1900 malaria campaign in Sierra Leone. For more on McKendrick, (Harvey, 1944).

46. (Kermack and McKendrick, 1927; Nicholson and Bailey, 1935).

47. Nicholson and Bailey used difference equations since they were looking at insect herbivores and their parasites which displayed discontinuity in their life patterns. Both insect and parasite were assumed to have only one generation per year and to die after laying eggs.

48. (Levin, 2002).

chapter three

Entomology and Ecology at Imperial College, 1907–1965

3.1 Entomology at Imperial College prior to the acquisition of Silwood Park

Scientists are in many ways bound by the histories of the institutions from which they graduate, and by those where they begin their careers. Silwood's early history was especially important to T.R.E. Southwood's intellectual and social development as, to a lesser degree, it was to some others in the Silwood Circle.

The botany and zoology departments of Imperial College had very few students during World War II and the Rector of the college, Richard Southwell, thought seriously of closing them down. He was dissuaded from doing so, largely by James Munro the professor of applied entomology. Munro was also the moving force behind the 1947 purchase of the Silwood estate, intended as a field station for all the biological sciences, though used at first mainly by him and other applied entomologists. We will return to Silwood below, but before doing so will take a brief look at the strong tradition in entomology (with its major emphasis on applied work) that existed at the college prior to the Silwood purchase. This tradition owed much to the efforts of an earlier professor, Adam Sedgwick.[1] Before moving to Imperial College Sedgwick held the chair of invertebrate morphology at the University of Cambridge. During an extended medical leave he travelled to India where he had something of an epiphany. He decided to devote the rest of his life to the cause of empire, resigned his Cambridge chair and, in 1909, accepted the chair in zoology at the recently founded Imperial College of Science and

Technology. His purpose was to train young people for overseas work in applied entomology.[2] At that time those studying zoology at the other London colleges were mostly medical students. Imperial had no medical and very few zoology students.[3] But Sedgwick was optimistic that more students could be attracted to the department by emphasizing the need for entomologists trained for work in the Empire. He wrote to the Rector of the college 'we shall be expected to have a supply of properly trained men to take up posts in connection with [colonial and domestic] agriculture'.[4] He appointed an excellent staff, including Harold Maxwell Lefroy as professor of applied entomology.[5] Lefroy was a good teacher and attracted many students. As a result, applied entomology soon dominated the zoology offerings, and enrolment in the department increased significantly.

When Lefroy arrived there was no field station and research space was almost non-existent. But his appointment as honorary entomologist to the Royal Horticultural Society allowed his students to carry out work at the Society's gardens at Wisley. Lefroy's main base of operations, however, was his home in Heston, West London. Much insect breeding and insecticide experimentation was carried out there, with an emphasis on orchard and market garden pests. Students also studied the insects causing defoliation of the oaks in Richmond Park, something that would later interest Richard Southwood. Munro, who came to study with Lefroy on a fellowship for advanced research students, worked successfully also on the control of the death-watch beetle that had infected the timbers of Westminster Hall.[6] This was good for public relations and helped with departmental funding. Lefroy's students were among the leading entomologists of their generation, and many found work in the empire.[7] A few students came from India, including P. V. Isaac who, like Lefroy earlier, was appointed Imperial Entomologist.[8] Professional lineages are important. One of Lefroy's students was C. B. Williams. Williams was to work in Trinidad before becoming head of the entomology department at the Rothamsted Agricultural Research Station. He was the author of a standard work on forest insects. Southwood, whose work at Silwood will be discussed in the following chapter, was his doctoral student.

During World War I Lefroy's work turned toward the control of the house fly (*Musca domestica*) and to pests that were infesting domestic food stores, especially grains. This work included chemical research to replace arsenical insecticides widely recognized as too toxic for use on stored foods. Despite moving toward lower toxicity Lefroy was poisoned by one of his fumigants and died in 1925.[9] His successor in the chair was Frank Balfour-Browne who, although only a part-time professor, brought entomology instruction back from its more extreme focus on applied work. A specialist in water beetles, he is said to have carried out what was 'arguably the first intensive, community-based, survey of an insect group'.[10] Applied entomology continued under the energetic James Munro who was appointed assistant professor.[11]

The late twenties and early thirties saw a great increase in the number of students interested in applied entomology. The cramped space available at Imperial College's South Kensington campus could barely accommodate them.[12] But this was to change. Shortly before his death Lefroy was asked to help in combatting an infestation of cocoa moths (*Ephestia elutalla*) in the stocks of cocoa beans held at the London docks. Munro took over this work and, looking for a sponsor, asked Stephen Tallents, Secretary of the Empire Marketing Board, to view the infestation.[13] It was through Tallents and the marketing board that Munro was able to raise sufficient funds for the college to acquire an eleven-acre site at Hurworth, near Slough. It became the college's first field station.[14] Financial help came also from the Imperial Bureau of Entomology and the Board of Agriculture. Balfour-Browne was not supportive of the applied work that Munro engaged in, believing it unsuited to the academy. The resulting dispute led to Balfour Browne's resignation in 1930. Munro succeeded him in the chair of entomology, and four years later succeeded Ernest MacBride as head of the newly named department of zoology and applied entomology.[15]

Before World War II a BSc course in applied entomology was given at the Hurworth field station and scholarships were offered by the Colonial Office, the Ministry of Agriculture, and the Empire Cotton Growing Corporation. The course was well publicized and the college calendar stated that promising careers for entomologists were available in the colonies. Indeed, many of the course graduates found such work, a few of them at the Imperial College of Tropical Agriculture in Trinidad which had been set up in 1922 with the help of John Bretland Farmer, the professor of botany.[16] But all was not smooth sailing at Hurworth. For a start, the Empire Marketing Board, which provided much of the funding, suffered during the Depression and was closed down in 1933. The Rector, Henry Tizard, was able to persuade the Department of Scientific and Industrial Research (DSIR) to step in and help keep Hurworth running, allowing work to continue there until World War II.[17]

Munro headed the department of zoology and applied entomology for close to twenty years and his legacy is still apparent today. On taking up the chair, he appointed some new people, further adding to the department's entomological expertise. O. W. Richards, who came from Oxford in 1927 was, however, already in place. Richards was responsible for much of the undergraduate teaching in systematics and physiology, helped later by a German Jewish refugee, G. S. Fraenkel, an expert in insect physiology appointed in 1933.[18] Another of the new appointees was W. S. Thompson. Munro wanted Richards and Thompson to help with the taxonomy of insect pests in stored foods. In this they worked closely with staff at the Natural History Museum, using the national collections.[19] As Southwood put it, Richards' 'wide taxonomic knowledge ensured that the "early days" of stored-products entomology were not confused with a mass of misidentifications'.[20] Munro appointed G. V. B. Herford to be in charge of insect stocks, and M. J. Norris to rear large quantities of caterpillars. Norris and Richards married in 1931 and worked together on a number of projects.[21] Chemists, too, were added to the department, including A. B. P. Page who moved from the chemistry department, and O. F. Lubatti who had worked for the Colonial Medical Services in Hong Kong. While their main focus was on new chemical insecticides, they also developed pyrethrum sprays first used in some food warehouses at the London docks.[22]

By the eve of World War II the department was recognized as a leading centre in applied entomology, outside of the medical field.[23] It took in research students and gave short graduate courses for people already working, many from overseas. Insect ecology, which became an important research area in the department during the 1950s, was a defined departmental field already before the war, but few students were then interested in it. It was outweighed by work in systematics, physiology, and most notably by work on the control of insect pests. By the 1930s, Munro was thinking about wartime food supplies and, in a 1936 lecture given at the Royal Society of Arts, he noted that 'the storage of foodstuffs and other commodities will be a question of vital importance'.[24] He wanted better government oversight (the food industry was reluctant to admit to its pest problems) and more vigilant pest control. At the start of the war Munro gave advice to Sir William

Beveridge, seconded to Whitehall from Oxford to head the Food Defence Plans Department. Those in government slowly came round to Munro's views on food stocks and, in 1941, decided to take over the Hurworth field station and place it under the DSIR. It became not only the government research centre for pests in stored foods, but also the headquarters for a national grain survey programme. The takeover of Hurworth led to Munro losing control of the stored-food work, and forced him in a new direction. He was unhappy that his staff had been seconded to the government and that, aside from some minimal teaching duties, were no longer answerable to him.[25] Richards and Nadia Waloff, employed at Hurworth on the biology of grain weevils, managed to carry out some ecological work.[26] In his grain weevil work Richards anticipated some later ecological ideas. For example, in considering the effect of environmental variation on the reproduction and adult weights of the weevils he wrote:

Perhaps the chief conclusion to be drawn from the present work is that population studies should consider not only the numbers of individuals which are produced, but also their weight and other properties. Weight may be determined not only by genetic factors and by the limiting effects of the environment, but also by the efficiency of conversion of food, as is perhaps the case in *Calandra*. It may be suggested that a detailed population study should deal first with the numbers of individuals produced under varying conditions; secondly, with the weights of individuals produced; and finally with the efficiency of the process.[27]

The kind of ecological ideas expressed in the above quotation were central to his research after the war. In another study on grain moths Richards and Waloff emphasized the development of 'life-budgets' based on field work, and laboratory work in insect physiology.[28] Before the war Richards and Norris had carried out some ecological work together. A trip to British Guiana (now Guyana) in 1933 led not only to a major collection of social and solitary wasps, but also to work in behavioural ecology and evolution.[29] This work was to be of interest to the evolution theorist, W. D. Hamilton, who Richards brought to Silwood in 1964. Of a major publication by Richards and Norris on South American wasps Hamilton wrote, it was sought and valued by generations of students aspiring to study the unusual polygynous organization of the group. ... [Their] suggestion of ant-predation in the tropics as the key evolutionary maintainer [of such organization], stands much more strongly now than when [they] made it.[30]

Later Richards invited the Australian government to set up a laboratory at Silwood for work on wood wasps that were threatening pine plantations in that country.

While Richards and Waloff worked at Hurworth, Munro, offended at being sidelined by the DSIR, turned his back on much of the work carried out there. He began new work on mosquito control and, in doing so, laid the foundation for what was to become another major research area at Imperial College. Mosquitoes as disease vectors were to be of interest to a range of future scientists, including ecologists within the Silwood Circle. Munro continued, however, looking for the perfect insecticide and, according to Richards, 'raised the art of fumigation to a science'.[31] He decided to work alongside the professor of organic chemistry, Ian Heilbron, and persuaded Heilbron of the necessity of finding a substitute for pyrethrum, the supplies of which had almost dried up since the start of the war.[32] At the time Gesarol, a known insecticide, was one of several Geigy products being tested at Rothamsted. Munro persuaded Heilbron to take an interest in it, and to study its chemistry. Heilbron discovered that only one of the chemical isomers of Gesarol's main component was an active insecticide. He also arranged for the government to obtain a licence from Geigy for Gesarol's manufacture in the United Kingdom. Munro organized a team to test the new product, by then known as DDT, on malaria mosquitoes in British Guiana.[33] The success of DDT as an insecticide was such that Munro was able to start afresh after the war with major grants from industry, from the Colonial Office, and from the ARC.

Munro set his sights on using DDT and other insecticides to help curb insect disease vectors in Africa. It was the start of the zoology department's various activities on that continent.[34] Before Munro could begin his post-war work, however, he needed new research space. The college's new Rector, Richard Southwell, was an engineer who downplayed the importance of the biological sciences and, as mentioned above, thought seriously of closing down the botany and zoology departments. Munro insisted that closure of zoology was not acceptable. Southwell appears to have been surprised that, in addition to Munro, the zoology department had three readers and five lecturers all intending to return to work in South Kensington. Since A. C. Chibnall, the professor of biochemistry, had left for a chair at Cambridge, Southwell decided instead to close down biochemistry. It was as shortsighted a decision as shutting down the biological sciences would have been; but, as a result, the biochemistry space, as well as some of the space previously used by botanists, was handed over to Munro.[35]

3.2 The purchase of Silwood Park

Munro also insisted that Southwell take up the problem of getting the government to return the Hurworth field station. The DSIR stated that since it was now doing the work initiated by the Empire Marketing Board the college no longer needed a field station. Munro, persistent as ever, gathered support and persuaded the Treasury that since the DSIR was keeping Hurworth, the college should receive monetary compensation with which to purchase a new site.[36] He was successful and Silwood Park was purchased in 1947.

Army personnel remained in occupation at Silwood and Munro had to have them removed before work could begin.[37] He asked Henry Tizard, the former college Rector, by then chief scientific advisor to the ministry of defence, for help. Tizard, who shared Munro's displeasure with the DSIR takeover of Hurworth did, indeed, help. Within two weeks of his hearing of the occupation problem, the army began moving out. A few weeks later Sir John Fryer, Secretary of the ARC, and Sir Edward Salisbury, Director of the Royal Botanic Gardens at Kew, came to look at the Silwood site and, impressed with its potential for ecological work, urged Munro to expand in that direction. They also encouraged the college to add more land should any adjacent property come up for sale. Munro already knew that the neighbouring Ashurst Lodge was for sale. Fryer and Salisbury persuaded the University Grants Committee to support its purchase by the college.[38] There were many problems in starting research at Silwood, including squatters who moved in just as the army moved out. Much restoration was needed and, given post-war austerity, it was difficult to purchase furniture and scientific equipment. Munro, it appears, was up to the challenge.

3.3 Early work at Silwood

Munro realized that if people were to enjoy working in the relatively isolated location of Silwood, they would need some social outlets. He and his wife held parties for the students and invited women students from the nearby Royal Holloway College to attend. Research students were invited to dine with the Munro family. As student numbers increased dinner dances were organized. The creation of a collegial and good social atmosphere was important in the shorter, but also the longer, term. It helped people to settle and to carry out good work. The chemists were among the first to get going. Page and Lubatti moved into Ashurst Lodge where they carried out insecticide work in newly built laboratories. They also conducted field trials outdoors. The huge expansion of the insecticide industry after the war meant that many grants came to the college. Silwood took on work in spraying technology for the Colonial Office and, in 1953, the Colonial (later Overseas) Spraying Machinery Centre opened. It received major funding from the World Health Organization. Students came from around the world to learn about the new insecticides and how to use a wide range of spraying technologies.

During the 1950s DDT's deleterious ecological effects became better known, inspiring some of the younger scientists to find alternatives to chlorohydrocarbon insecticides. Some looked to new chemicals, but a younger generation of entomologists, a generation that included Southwood, turned increasingly toward biological control. As a consequence interest in parasitology and insect ecology increased. Munro had long lobbied for a chair in parasitology and, in 1954, one year after his retirement, B. G. Peters, a plant nematode specialist came from Rothamsted as the first professor of parasitology.[39] Peters, together with J. Desmond Smyth who would later succeed him in the chair, influenced many students to take up parasitology.[40] As head of department, Richards welcomed more botanists to Silwood. Like the entomologists, many of the botanists working at Silwood were interested in tropical agricultural problems.[41]

As an administrator Munro was a hard act to follow. Richards was seen as the better scientist, but his organizational skills were viewed unfavourably. He had a shaky start as head of department and his ambitious plans for expansion at Silwood were not supported by the University Grants Committee in 1955.[42] Nonetheless, he laid the foundation for Southwood's future work and, by association, for that of the Silwood Circle. It is also worth noting that Richards anticipated some of Southwood's methodological views. He was no mathematical theorist but, as Southwood put it, recognizing nature's enormous complexity he believed that 'the holistic approach to animal ecology was not profitable [and that] a reductionist approach focussing one's study on the population dynamics of a single species was the way forward'.[43] This was not simply propaganda on Southwood's part. Richards was a reductionist at a time when more holistic approaches to ecology were in fashion. Also, in his bias toward the study of populations, Richards was at odds with his friend and Oxford contemporary, Charles Elton. According to Southwood, Elton told him that 'Richards was very hard on the Wytham work; he told me just after the war that I was going to waste the best 20 years of my life'.[44] Earlier in his career Richards, like Elton, was more of a community ecologist and had studied insect communities on heathlands. Both approaches proved useful.

Like Lefroy and Munro before him, Richards continued training people for work overseas, but under his leadership applied entomology was balanced by work in the central areas of the discipline, and by a growing interest in evolutionary and ecological matters. Richards was recognized as a world expert in insect ecology, but was little known in the wider ecological community. Southwood speculated that this was due to his unwillingness to 'interpret and extrapolate or generalize'. Relatedly Richards, and Waloff with whom he worked closely, was reluctant to engage in controversy or debate. By the late 1960s they were viewed as rather old-fashioned. Nonetheless, Southwood believed that future ecologists, interested in including environmental variation in their models, would look back profitably at their work. Richards must have felt pressured by the new developments in ecology since, on retiring in 1967, and seemingly relieved, he told Southwood that he could leave ecology 'with its theories and mathematics' behind, and return to being a straightforward entomologist.[45] Perhaps he was also bewildered by the ideas of W. D. Hamilton, the evolution theorist he had appointed to his staff in 1964. Today many see Hamilton as one of the leading Darwinians of the twentieth century.[46] Richards' own contributions were also considerable. Without his pioneering work in insect ecology it seems unlikely that Southwood would have been in a position to further promote ecology at Silwood. It is probably also fair to state that Southwood learned much from

Munro on how to organize a department, and on the importance of social life to the success of a research community.

Endnotes

1. Adam Sedgwick FRS (1854–1913) was educated at King's College London and Trinity College, Cambridge. He was the great nephew of Darwin's geology professor of the same name. After gaining an MD he began research under Michael Foster and Francis Maitland Balfour at the School of Physiology in Cambridge.

2. Imperial College was founded in 1907 by the amalgamation of three already existing colleges. The intent was for the new college to give young people the technical expertise needed to support industry both at home and throughout the British Empire, (Gay, 2007). For science and empire see especially chapter 7. For Sedgwick and empire, (Gardiner, 1913).

3. T. H. Huxley, founder of the Royal College of Science's department of biology, believed that he needed to teach future schoolteachers in order that science be better taught in schools. As a result little research was carried out in the department and many of the students did become teachers. The situation changed after the founding of Imperial College. Applied botany was promoted by the professor of botany, John Farmer. He carried out research and prepared students for positions in imperial agriculture, even before the arrival of zoologist Adam Sedgwick.

4. Imperial College archives (ICA), KZ9/1.2; Sedgwick to Bovey, April (no day), 1909.

5. Harold Maxwell Lefroy (1877–1925) was educated at King's College, Cambridge where he was a research student under David Sharp, curator of the university's zoological museum. On leaving Cambridge, Lefroy worked in Barbados as Entomologist to the Imperial Department of Agriculture for the West Indies. In 1901, after four years there, he was appointed Imperial Entomologist in India. While in India he published some major works on Indian insect pests and taught occasionally at Imperial College when on leave. His permanent appointment began in 1912. Among the other Sedgwick appointees were the parasitologist C. C. Dobell FRS (1886–1949), and his former demonstrator at Cambridge, E. W. MacBride FRS (1866–1940). MacBride became head of the zoology department after Sedgwick's death in 1913.

6. This work was the start of the firm Rentokil Ltd. Lefroy set up a small manufacturing plant in Hatton Garden where one of his assistants, Elizabeth Eades, supervised the production of 'woodworm fluid'. The defoliation of oak trees was

a later research interest of the Oxford entomologist George Varley, doctoral supervisor of both Michael Hassell and David Rogers. As will be discussed in chapter 7, Hassell carried out work on one of the defoliators for his DPhil.

7. One of Lefroy's wartime students was Evelyn Cheeseman (1881–1969) who left before completing her degree. She led an adventurous life as world traveller, author and entomological collector. For much of this early history I am indebted to (Davies, 1995). R.G. Davies was an important mentor to Roy Anderson and other students of his generation.

8. That many of Lefroy's students took up positions in the empire was to have consequences for the future of applied entomology at Silwood. Among his students were the brothers Harry and Ernest Hargreaves, appointed Government Entomologists in Uganda and Sierra Leone, respectively, W. H. Edwards who was Government Entomologist in Jamaica and T. A. M. Nash who became eminent for his work on the tsetse fly. Some of Nash's undergraduate notebooks are held in the ICA. A very negative view of P. V. Isaac is presented by his daughter; (Roy, 1999).

9. I think this may have been a suicide. On two earlier occasions Lefroy had been found close to death after inhaling his fumigants.

10. Quotation, (Foster, 2004). Balfour-Browne was the author of a three-volume monograph, *British Water Beetles* (Ray Society, 1940, 1950, 1958).

11. The appointment of Balfour-Browne was intended to bring entomological education more in line with that being offered at Cambridge. The undergraduate curriculum had been somewhat attenuated by Lefroy's interests in the applied aspects of the discipline. James Watson Munro (1888–1968) was educated at the University of Edinburgh, and at the forestry school at Tharandt near Dresden where he took a doctorate under a leading forest entomologist, Karl Leopold Escherich. Munro's work with Lefroy was interrupted by war service. After the war, he worked briefly in medical entomology (the scabies mite) at the Lister Institute and then at Cambridge. In 1920 he returned to forest entomology at Oxford's Imperial Forestry Institute and was instrumental in helping to set up the Forest Products Research Laboratory at Princes Risborough. The position of assistant professor was roughly equivalent to that of reader elsewhere. When Imperial College joined the University of London in 1929 most of its assistant professors became readers.

12. During the 1920s and 30s entomology occupied the top floor of the old Huxley Building (now the Henry Cole Wing of the Victoria and Albert Museum). Lefroy and Munro kept insect colonies in huts and greenhouses on the roof. Other zoologists occupied the floor below. In those days the college

charged students per course. In 1926 the fee for Munro's course of 60 lectures
was £2.00. As with Lefroy, many of Munro's students found work in the
empire.

13. The Empire Marketing Board was established by the Baldwin government to
assist trade between countries within the British Empire.

14. Several people wanted the college to purchase Darwin's old home, Down House
in Kent, which was then on the market. But Munro preferred Hurworth and
persuaded the Rector to go along with him. See ICA, Zoology Department correspondence file, KZ/9/5 letters from 1928.

15. There was much tension between Balfour-Browne and Munro. The dispute over
the direction of entomology within the zoology department is discussed in (Gay,
2007), 151. The College Rector, Thomas Holland, supported Munro's work,
seeing it as useful to both the domestic and imperial agricultural industries, and
as being in accord with the college charter. He was also a realist and saw the
potential of funds coming to the college from Munro's type of work. Indeed,
Munro had research contracts also with the Australian Dried Fruits Board, the
Government of Southern Rhodesia, the Imperial Tobacco Co., and Imperial
Chemical Industries, and was therefore able to support a wide range of applied
work at the college.

16. Some of the library and written resources of the Trinidad college came to
Silwood many years later. The college itself became part of the University of the
West Indies. Gordon Conway was to take a course in tropical agriculture there.

17. Further help came from the University Grants Committee and from the
Carnegie Foundation.

18. Owain Westmacott Richards FRS (1901–84) was a contemporary of Charles
Elton at Oxford and, like Elton, was tutored by Julian Huxley. On graduating
he became a research student under E. B. Poulton. (Oxford University did not
yet offer the D. Phil., Richards was awarded the DSc. in 1934.) He became a
major systematist who worked also on insect ecology, the ecology of heath lands
and on problems of sexual selection — the latter being one of Poulton's interests. Insect physiology was brought to the fore in the 1930s with the work of
V. B. Wigglesworth at the London School of Hygiene and Tropical Medicine.
His highly influential *Insect Physiology* was published in 1939 (expanded as
Principles of Insect Physiology (1942)) and he moved to Cambridge during the
war. G. S. Fraenkel (1907–1984) left Imperial College for a chair at the
University of Illinois (Urbana) in 1948. O. W. Richards wrote a memoir of the
zoology and applied entomology department (1927–1965) which I have used
in reconstructing its early history. Typescript in ICA, KZ.

19. See (Richards and Robson, 1936) for a collaborative work with a member of the Natural History Museum staff. The book shows that, in the 1930s, Richards while not a Lamarckian was also not strictly Darwinian. He did not accept adaptionist arguments for certain complex structures. While the book was a well reasoned product of its time, it was also untimely. The rise of neo-Darwinism resulted in Richards being viewed with suspicion, something that probably cost him the Hope chair at Oxford.

20. (Southwood, 1987), 543; (Waloff. 1986).

21. Maud J. Norris had a BSc in zoology from King's College London and came to Imperial College to take the Diploma of Imperial College in entomology. She gained a doctorate while working for Munro. Later she was appointed to a lectureship and worked alongside her husband while also raising their two daughters.

22. By then the earlier arsenical insecticides had been replaced. Munro, however, was still using carbon disulphide which, in addition to other problems, is highly flammable.

23. Medical entomology was the domain of the London School of Hygiene and Tropical Medicine. The other major applied entomology centre in Britain was at Cambridge where the emphasis was on domestic crop pests. Before and during World War II Munro directed work on human pests such as bed bugs and fleas. He wanted the college to move into the veterinary and medical fields and wrote to the Rector stating that it would be a good idea to follow the lead of Charles Elton at Oxford who was studying land animals as disease vectors. See, for example, Munro's report to the Rector, March 13, 1939; ICA, KZ9. One of his staff members, Humphrey Hewer, carried out work on rats as disease vectors and was Britain's Chief Rodent Officer during World War II.

24. (Davies, 1995), quotation, 27.

25. G. V. B. Herford was appointed director of the DSIR laboratory.

26. Nadia Waloff was educated at Birkbeck College and was appointed to work at Hurworth by Munro in 1941. After the war she was given a lectureship in the department where she worked for over fifty years. In addition to the Hurworth work on grain pests, Richards and Waloff began some ecological work on insects on hawthorn. After the war Waloff worked also on locust infestations in Cyprus and, later, she and Richards were to carry out a major population study of four species of grasshopper found at Silwood; (Richards and Waloff, 1954). They also did important ecological work on the broom beetle; (Richards and Waloff, 1961 and 1977; Waloff, 1968).

27. (Southwood, 1987), quotation, 555. Original in (Richards, 1948).

28. (Richards and Waloff, 1946a and 1946b). Flour moths were later studied by Rogers and Hassell.
29. This was not Richards' first visit to the region. He was part of the 1929 Oxford University expedition to British Guiana under the leadership of R. W. G. Hingston.
30. Quoted in (Southwood, 1987). I have inserted gender neutral pronouns where Hamilton used the masculine, since the paper he refers to was a joint publication: (Richards and Richards, 1951).
31. Richards typescript, ICA, KZ, p.3.
32. Pyrethrum is extracted from *Chrysanthemum cinerariaefolium*. The main sources were Kenya, Japan and Macedonia. The United States had first rights to the Kenyan crop and Japan and Macedonia were enemy territories.
33. DDT is the acronym for dichlorodiphenyl trichloroethane. However this name, not strictly correct, is also a shorthand.
34. After the war O. W. Richards worked in Africa both on mosquito and locust populations — the latter following earlier work by Waloff on Cyprus. His wife, Maud Norris, accompanied him on several expeditions including to Africa, and on the Royal Society's Mado Grosso Expedition to Brazil in 1971. She died shortly after returning from Brazil.
35. At Cambridge Chibnall trained a stellar cast of biochemists including two Nobel laureates, F. Sanger and R. R. Porter.
36. ICA, Governing Body minutes, vol. 39 (1945). Also, (Gay, 2007), 344.
37. Silwood Park House was requisitioned as a military hospital during World War II. The college inherited also a number of rather ugly outbuildings constructed during the war as troop dormitories. These 'temporary' buildings were to prove useful for various purposes and remained in existence until the end of the twentieth century.
38. The purchase of house and estate cost £24,000. The Silwood Park manor house was built in 1878. The architect was Alfred Waterhouse who also designed the Natural History Museum in South Kensington.
39. In 1960 a new parasitology laboratory for work in plant nematology was built at Ashurst Lodge with money from Shell. Parasitology had been taught earlier and there had been a readership in the field.
40. One whom he influenced was Roy Anderson, (Esch, 2007). Anderson was not so keen on Smyth's approach. When he succeeded Smyth in the chair of parasitology, he brought his own vision to the department.
41. The applied focus in botany was the legacy of John Bretland Farmer FRS (1863–1944), Imperial College's first botany professor. After the war Munro invited the botanical ecologist Arthur John (Jack) Rutter (1917–2011) to work

at Silwood. Rutter had been a research student under G. E. Blackman, the earliest botanical ecologist at Imperial College. Together they conducted a study of bluebell ecology, concluding that the bluebell had become a woodland plant largely because sheep had grazed it close to extinction in its earlier habitats. Blackman and Rutter developed new forms of statistical analysis in their studies which included also work on weed control. Rutter began to work at Silwood, as well as at Forestry Commission land at Bramshill Forest in Hampshire. There his interests were in wet heaths, in water conservation in soils, and its loss from forest canopies.

42. (Southwood, 2007).

43. (Southwood, 2007), quotation, 555.

44. (Southwood, 2007), quotation, 556. Elton must have learned something from Richards whom he acknowledged in the preface to his *Animal Ecology* (1927); 'I am indebted to Mr. O. W. Richards for a great deal of help and criticism. Many of the ideas in this book have been discussed with him and gained correspondingly in value and in particular his extensive knowledge of insects has been invaluable in suggesting examples to illustrate various points'. For Elton's work in Wytham Woods see chapter 5. See also (Strong, Simberloff, Abele, and Thistle, 1984) for a later articulation of some views held by Richards.

45. (Southwood, 2007), 557–558. On retirement, together with R. G. Davies, Richards carried out the 1977 revision of what was then the entomologists' Bible, namely Imms' *General Textbook of Entomology*.

46. See, for example, (Dawkins, 2000). Hamilton will be discussed further in chapter 4.

chapter four

T. R. E. Southwood and the Early Years of the Silwood Circle

You played the game as to the manor born...[1]

4.1 Southwood's youth and his arrival at Silwood

By the 1960s Silwood had become one of the foremost centres for entomology in the world and was known also for insect ecology. T. R. E. Southwood who did much to carry on the tradition was an undergraduate in the department.[2] His PhD was also awarded by Imperial College (University of London), though his doctoral research was carried out away from the department at the Rothamsted experimental station. Southwood returned to Silwood in 1955, as a research assistant under W. P. Jepson. He was also appointed warden of the student hostel.[3] One year later he was given a lectureship. Shortly before taking up his new position Southwood married Alison Harden whom he met at Rothamsted where she worked as secretary to the entomology department. Together they were to purchase a house with a garden backing on to the Silwood estate. It was to become an important site for the social gathering of biological scientists.

Southwood grew up in Gravesend, Kent. His childhood appears to have been close to idyllic despite the attendant difficulties of wartime. An only child and a precocious young naturalist, his dairy-owning parents lavished attention on him, promoted his interests, and saw to it that he was helped by friends and relatives. The vicinity of Gravesend was then still largely rural and Southwood was able to wander freely. He was encouraged to collect specimens, to join a number of field clubs, give talks, and to publish his observations. At the age of sixteen he became a Fellow of the Royal Entomological Society of London.[4] His personal memoir, which includes much information on his early life, has been used to very good effect by Robert May and Michael Hassell in their Royal Society biographical memoir. I agree with them that pedagogues, including historians, could learn much about the 'well-springs of creativity and collegiality' by looking more closely at the early lives of successful scientists.[5]

The collection of Southwood's papers at the Bodleian Library is large and reveals much of his character, even to someone who never knew him. He kept a lifetime of papers and correspondence, starting at a very young age. Included are school reports, early natural history talks, childhood correspondence, as well as correspondence and papers related to his later career. There are also lecture notes that he took as a student, and notes made later for his own lectures. Scattered throughout the collection are short memoranda that Southwood wrote as guides for future readers, suggesting that legacy was important to him. Even a disinterested reader will infer that Southwood helped many people in their careers, and that those within the Silwood Circle owe him a special debt. This is something I hope to illustrate below and in later chapters. It would also appear that his principal coworkers on several projects, especially his long-term studies, were women. Southwood's professional relations with women were good but, as to patronage, young men within his orbit fared better than their female contemporaries. On a social level Southwood enjoyed the company of women and, when it came to career advancement, was more supportive of those he found attractive.[6]

Southwood was very collegial. Perhaps he had learned something as witness to the collective effort and spirit of wartime. During his tenure as head of department (1967–1979) he successfully brought people together. Notable early on was his encouraging the consumption of tea, coffee and biscuits in the communal tea room, and his actively discouraging the same in people's offices. He also continued the practice of having 'dinners in hall' three times a year, allowing staff and students at Silwood to come together in a formal yet collegial social setting.[7] Guy Fawkes night was an enjoyable occasion, though instead of burning an effigy of Fawkes the students would burn effigies of their instructors — a different person was chosen for this fate each year. Another social event was the Silwood summer ball. Gordon Conway remembers a fancy dress ball that he attended dressed as Al Capone, and that he used the occasion to toast his 'godfather', Dick Southwood.[8] Intended in fun, the toast carried an element of truth in that Southwood did orchestrate people's lives, at least to a degree. However, he appears to have been a well-intentioned 'godfather' who helped many to advance their careers; and regardless of area of interest, or whether they were engaged in applied, field, laboratory, theoretical, or other work, students and staff were made to feel that they were part of a noble enterprise.

As mentioned, Southwood had earlier moved to Rothamsted for his doctoral research. The move was, in part, because he did not wish to be drawn into Munro's ambit.[9] His doctoral supervisor was C. B. Williams.[10] Another person at Rothamsted who helped to guide his early career was Kenneth Mellanby. Later, after Southwood was given a chair, Mellanby nominated him for membership at the Athenaeum Club where he was elected in 1970.[11] This was to be an important venue for Southwood, a meeting place, and somewhere to invite his favourites to lunch or tea, to discuss their career plans and other topics of the moment.

After returning to Silwood as a young lecturer, Southwood began a study of a cereal crop pest, the frit fly, *Oscinella frit*.[12] His senior co-worker on the project, W. F. Jepson, a reader at the college, was later to find employment with the agricultural chemicals division of American Cyanamid. For Southwood the work led elsewhere. The frit fly population appeared to be regulated by the growth stages of the oat plant, invasion of the fly by parasitic mites, and by the fly's own migration patterns. These observations suggested further work on plant-insect interactions, population dynamics, parasitology, and in finding new methods of biological, rather than chemical, control. In collaboration with others, Southwood published papers on all these topics. He also kept in close intellectual touch with his applied entomology colleagues and worked on problem insects of interest to them, such as the mosquito *Aedes aegypti*, a dengue fever vector, and the types of borers that infested sugar cane plantations. These studies were in the Imperial College tradition of seeking overseas problems. A pragmatist, Southwood knew that financial support and research grants for colonial, and later Commonwealth, projects were relatively easy to come by. He also initiated some long-term studies not unlike the ones he had seen in operation at Rothamsted. One on

grassland management, and a later one on the viburnum white fly, were
undertaken with Patricia Reader.[13] In these studies one aim was to develop
procedures for the more precise measurement of insect populations.

Already as a child Southwood had developed a love for the *Hemiptera* (true
bugs) and he became a major systematist in that field.[14] It was, perhaps, in
thinking about the great structural variety of these insects, their different
habitats in water and on land, and their modes of dispersal, that much of
Southwood's early evolutionary ecological thinking developed. Also important to his later work was his early experience at the field studies centre at
Flatford Mill in East Bergholt, Suffolk. Southwood first went there in 1948
when a pupil at Gravesend Grammar School.[15] As an undergraduate, he
taught classes there during the summer. Among the ongoing projects at
Flatford was one comparing insect damage in different types of tree.
Southwood learned how to trap insects in trees, to sort and to classify them.
Some tree species showed more damage than others, but why? Related to this
was Southwood's later interest in the four dominant species of insect defoliators on British oak trees, an interest of Lefroy's many years earlier. More
generally, Southwood looked for patterns of abundance and distribution
among herbivorous insects, work entailing also some understanding of plant
biochemistry and plant distribution.[16] With Reader and Valerie Brown, a former student who had returned to Silwood as a lecturer, he began a long-term
study suggested by his observations on the frit fly. Instead of looking at how
the maturation stages of an individual plant affected an associated insect
population, they looked more generally at the succession of a plant community and its effect on the associated community of insect herbivores.[17] It
would seem that Southwood, unsure of where he could make a breakthrough,
was looking in several directions. Great scientists are usually more focussed
and single minded than he was. Southwood had ability, and was to shine as
an entomologist, but he was a man in a hurry; determined and ambitious, he
was seeking a way to the top.

Southwood's ambition was helped by a good strategic mind. As will be
shown, he rightly saw the interface of ecology with new environmental concerns as a route to advancement. He also saw possibilities in new theoretical
approaches. Like Richards he was no mathematical biologist, but he was more
open than Richards to those who were. Southwood's 1977 presidential address
to the British Ecological Society, 'Habitat the templet for ecological strategies?',

illustrates this openness. An overview of recent ecological developments, it linked descriptive approaches to the study of life histories with new theoretical ideas.[18] It also illustrates increasing cross fertilization between ecologists and the mathematical biologists invited to Silwood during his term as Director.[19]

When Southwood succeeded Richards as head of the department of zoology and applied entomology in October 1967 there was already one person at Silwood with some mathematical skill, namely William Hamilton. Southwood recognized Hamilton's brilliance but did not yet understand the importance of his work. They had a number of disagreements.[20] Southwood thought Hamilton a poor supervisor of research students and a poor lecturer. Because of this, or possibly because Southwood sensed that he might be on to something, Hamilton was allowed to spend much of his time carrying out research, both as travelling naturalist and as resident theorist. In 1974 Southwood wrote a letter to the Rector, Brian Flowers, supporting Hamilton's promotion to a readership. Hamilton, he wrote, 'has thought very deeply about the evolution of altruistic behaviour and animal societies', but the letter was only faintly positive in tone and did not fully capture Hamilton's major contribution to evolution theory. Hamilton was not promoted and, in 1977, left the college for North America after thirteen years as a lecturer.[21] As will be discussed in chapter six, Southwood was far more aggressive in promoting the interests of his favourites, especially those whom he saw as good departmental citizens. Later he made amends and invited Hamilton to Oxford where he was to hold a Royal Society research professorship.

Hamilton wrote about his time at Silwood and remarked that no one shared his interests in evolution theory and that he did not engage in any collaborative work. 'Few questioned me or passed me their manuscripts to check.' He wrote that he worked alone in his laboratory, in the field, and in the computer room of the experimental nuclear reactor located on the Silwood site. He also spent a fair bit of time working in Brazil with Warwick Kerr. However, he must have caught the Imperial College interest in parasitology, since later in his career he worked on the impact of parasitism on evolutionary history.[22] Shortly after Southwood's letter to Flowers the tide turned in Hamilton's favour. With the publication of E. O. Wilson's *Sociobiology: The New Synthesis* in 1975 and Richard Dawkins's *The Selfish Gene* one year later, Hamilton began receiving much attention and was to receive many awards.[23]

Overall, however, Southwood was alert to what was happening around him and thought clearly about how to take advantage of new situations. For example, earlier he had thought about the looming 1968 closure of Sir Vincent Wigglesworth's ARC Unit of Insect Physiology at Cambridge. The paper trail suggests that others were thinking about it too. Well before Southwood was appointed head of the zoology department, correspondence between Sir Gordon Cox FRS, Secretary of the ARC, and Imperial College shows that Cox was trying to find a new home for two senior scientists from the Unit, J. S. Kennedy and A. D. Lees.[24] Both had been offered chairs in Australia but neither wished to leave Britain. Cox began negotiations with the Acting Rector of the college, Sir Owen Saunders. Part of their correspondence dealt also with who should succeed Richards as head of department. Cox suggested that Saunders appoint Southwood and that C. P. Whittingham succeed W. O. James as head of the botany department. In the event, Saunders agreed with everything Cox suggested, though he may well have arrived at the headship decisions independently.[25]

As to Kennedy and Lees, it took about a year to negotiate all the details of their transfer. Kennedy was appointed professor of animal behaviour in 1967. Lees was also appointed, but given a chair only after being elected a Fellow of the Royal Society in 1968. Lees was a more traditional insect physiologist but Kennedy engaged also with Southwood's ecological interests and drew many behavioural ecology visitors to Silwood where he ran a lively seminar series. Many years earlier Kennedy had been an undergraduate at Imperial College but, finding the applied entomology lectures boring, had dropped out. He later completed his degree at University College. During World War II he worked in the Colonial Office's Middle East Anti-Locust Unit organizing crop dusting aircraft. When the war was over he moved to Cambridge with Wigglesworth and continued his research on locusts. Kennedy had been of great help to Southwood at the start of his career, had edited some of Southwood's papers and had taught him much about insect migration.[26] Southwood was delighted to have Kennedy come to Silwood, despite being warned that he was a former member of the Communist Party — something that had prevented his advancement at Cambridge. Not only did Southwood wish to repay his intellectual debt, he was keen to have a Fellow of the Royal Society join the department. Kennedy was given Munro's old space and converted it into a fine laboratory. He was finally able to build his own research group and supervised work on locusts, aphids and moths — much of it related to insect flight and migration patterns.[27] Later, when Southwood moved to Oxford he invited Kennedy to join him there.

When Southwood gave his inaugural lecture as professor in 1968, W. R. Jepson with whom he had collaborated on the frit fly work, was a little critical. There was much to admire about Southwood and the lecture, he wrote, but when he 'launches into the realm of pest management, he seems at times to be succumbing to the elbow jogging influence of the Carsonites, the integrationists and the conservationists, amateur as well as professional'.[28] On these issues, as in so much else, Southwood had highly sensitive political antennae. He recognized a growing environmental concern among the public and understood that universities needed to respond. As will be discussed below, he thought of ways to do so while building on the department's existing strengths. He also kept an eye out for new opportunities in applied entomology. One example of this was that in 1968 the department received a grant of £45,000 from the Ministry of Overseas Development. It was used to set up a centre at Silwood for the identification of insect blood meals (by the analysis of stomach contents). The idea was to gain a better understanding of diseases such as malaria, dengue fever, sleeping sickness and yellow fever. The laboratory, which handled about 25,000 mosquitoes and 5000 tsetse flies each year, was under the direction of P. F. L. Boreham. There was much international collaboration on the mosquito work, as well as collaboration with scientists at the Animal Virus Research Station at Pirbright. A year later Michael Way, the director of the Overseas Spraying Machinery Centre, was given a chair.[29] Way was interested also in the biological control of aphids and studied both the population structure of aphid colonies and aphid behaviour. Southwood was building bridges to people in applied fields and included them in his expansion plans. But expansion was not easy. By the early 1970s the economy was in recession, made worse by the OPEC crisis of 1973 and the huge inflation in oil prices that followed. Compared to some other areas of science, however, the situation for ecological and environmental research was relatively good. This was because of political pressure on governments to do something for the environment.

4.2 Environmentalism: some cultural and political events of the 1960s and 1970s

As is still the case today, during the 1960s and 70s environmental issues were discussed almost daily in the media. It was impossible for a thinking person to be unaware of them. One notable event was the publication of Rachel Carson's *Silent Spring* (1962 in the US, 1963 in the UK). The book became a major talking point, especially as the ecological damage caused by the heavy use of agricultural pesticides was increasingly apparent.[30] Another book to receive much press coverage was *The Population Bomb* (1968) by the University of California biologist, Paul Ehrlich. It portrayed a looming population crisis and argued that environmental conservation and limits to the earth's resources both be taken seriously. A further expression of the population problem, also published in 1968, was Garrett Hardin's essay, 'The tragedy of the commons'. Hardin had earlier been a student at the University of Chicago and was influenced by the ecologist Warder Clyde Allee who, already in the 1930s, was worried about future overpopulation.[31] Also in 1968 the Club of Rome was founded by the Italian industrialist Aurelio Peccei and the British civil servant Alexander King. Their concerns were similar to Ehrlich's and their purpose was to bring scientists and business people together to work on environmental and conservation issues. The Club of Rome's first major publication, *Limits to Growth* (1972), was commissioned from scientists, including social scientists, at the Massachusetts Institute of Technology. A computer model led them to predict the collapse of the industrial West by 2100 unless major steps were taken to conserve resources. Earlier King had been a chemistry student, then lecturer, at Imperial College. He was able to draw many Imperial College people, including Southwood, into Club of Rome activities.

One notable British resource about to be exploited in that period was North Sea oil and gas. Like the Torrey Canyon oil-spill disaster of 1967, the new offshore industry fed into the environmental debate.[32] There was much concern over what drilling would mean for the ecology of the Scottish coastline, and for offshore marine life. Also politically important was the growing nuclear disarmament movement. It set a pattern for other fast-growing organizations such as Friends of the Earth (founded in the United States in 1969) and Greenpeace (founded in Canada in 1971).[33] In 1966 the Council of Europe declared that 1970 would be European Conservation Year and, to start the year off, a major conference was held in Strasbourg. On the 22nd of April 1970 the first Earth Day was celebrated in the United States.[34] In Britain people were working also on a range of domestic rural and marine conservation projects such as protecting fenlands, conserving seal populations, and protecting Exmoor. The Wildlife and Countryside Act (1981) was an attempt to address concerns of that kind.

Even before Carson's book appeared there was public concern over a postwar increase in the use of chemicals in agriculture; and that land-use decisions were being made without taking ecological, environmental and health considerations into account. The issue of chemical usage in the countryside prompted the minister of agriculture, John Hare, to set up a research study group in 1959 under Harold G. Sanders, his chief scientific advisor.[35] The British Ecological Society made a presentation to the study group. It was drafted by O. W. Richards, his younger brother Paul W. Richards, an ecological botanist at Bangor, and his former Imperial College colleague Geoffrey Blackman (by then Sibthorpian professor of rural economy at Oxford). The study group's report was an important factor in the creation of the Natural Environment Research Council (NERC) under the Science and Technology Act of 1965. The Council had responsiblility for supporting research in the earth sciences, ecology and environmental science. A new source of funding opened up for those in universities with ecological interests and their numbers began to grow.

In the run-up to the United Nations Human Environment Conference held in Stockholm in 1972, a group of environmentalists co-authored a document titled *Blueprint for Survival*. Importing some ideas from *Limits to Growth*, the document took up the entire January 1972 issue of *The Ecologist*, a journal founded by Edward (Teddy) Goldsmith in 1970.[36] Given the journal's content, a more appropriate title might have been *The Environmentalist*. Indeed, the title is illustrative of a conflation in the meaning of the terms 'ecology' and 'environment' that continues to this day. Goldsmith was the principal author of *Blueprint*, a curious manifesto that was both progressive and regressive in its radicalism. Regressive in that it suggested the need to totally dismantle modern industrial society and move back to a simpler way of life; progressive in that it was a serious wake-up call, pointing to problems that governments needed to address, and arguing for the need to include ecological and environmental ideas in school curricula. Published later that year as a small book, it sold close to a million copies; its American edition had a supportive foreword by Paul Ehrlich. Goldsmith was also a founder of the Green Party of England and Wales.[37] The Stockholm conference for which *Blueprint* was intended as an unofficial manifesto, was energetically chaired by Gro Harlem Brundtland, the future Prime Minister of Norway and future Director of the World Health Organization. It was successful in helping to place the environment on the international political agenda. Many things flowed from the conference including the founding of the UN Environmental Programme (UNEP) with headquarters in Nairobi, and the founding of the International Institute for Environment and Development (IIED) headed by the economist Barbara Ward.[38] Its headquarters soon moved to London. Governments represented at the conference promised funding for environmental and ecological science and, by and large, they delivered despite the recession.

Those were heady times and Southwood saw a number of opportunities in the many concerns being voiced. For example, he joined the debates over agricultural practice. Already at Rothamsted he had developed an interest in the interface between natural habitats and agriculture and he promoted the interests of those at Silwood, such as Gordon Conway, interested in sustainable agriculture.[39] One of his own research projects entailed a comparison of the insect fauna in cut and uncut grasslands, a topic to be taken up later by John Lawton. He also initiated research on the ecology of game birds. Organochloro insecticide use had led to a decline in their numbers, notably a decline in the number of partridges and wood pigeons. Southwood conducted a few studies on this problem.[40] He understood the dangers of pesticide overuse and saw the need for refuges around farmed fields so that natural control agents could thrive.

As to the fear of overpopulation, Southwood engaged there too. He was invited to address the Parliamentary and Scientific Committee on the topic in 1969.[41] During the 1970s he gave several public lectures and, clearly a strategic synthesiser, connected overpopulation and other environmental anxieties to the kind of ecological work being carried out at Silwood. In 1979 he claimed that there is 'only a 50/50 chance of the world surviving the dangers of overpopulation'.[42] In 1972, along with 136 others, he signed a letter to *The Times* supporting the 'Blueprint for Survival' document and worked on a proposal for an environmental think tank. It never materialized.[43] Nonetheless, together with one of his new appointees, Gordon Conway, and with the enthusiastic backing of the Imperial College Rector, Brian Flowers, he moved his department in the direction of environmental as well as ecological science.[44]

4.3 The early years of the Silwood Circle

Like Southwood, Conway was a child naturalist. He joined the London Natural History Society at the age of fourteen and went on some of the Society's field trips, including outings to Bookham Common in Surrey for the collecting of insects. Also of interest to him were the long-term plant and animal succession studies that the society was conducting on bomb sites in and around London. Conway was a pupil at Kingston Grammar School where, with the help of an excellent teacher, he did well in mathematics; for botany and zoology he had to attend the local technical college. He was interested in studying ecology and his father looked into where would be a good place to go. They decided on the University College of North Wales at Bangor.[45] Conway entered with a scholarship in mathematics and biological sciences in 1956. He told me that his experience at Bangor was good and that he learned much from his teachers F. W. Rogers Brambell, T. B. Reynoldson, and Paul W. Richards.[46] He also mentioned reading H. G. Andrewartha and L. C. Birch's *Distribution and Abundance of Animals* (1954), a book mentioned by all those in the Silwood Circle with whom I have spoken. It was one of the principal texts of its day and is well remembered both because it taught much and because it was a foil, something to argue against.

On graduating, Conway was offered both an ARC grant to work on a PhD with O. W. Richards at Silwood, and a Colonial Office job to work on insect pests in North Borneo. He must have had a sense of adventure since he chose the latter — perhaps Richards' brother, Paul, the tropical forest specialist, was an influence. Vincent Wigglesworth who was responsible for ARC grants thought that Conway had made a poor decision. But things turned out well and, after further training in agricultural science at Cambridge and in tropical agriculture at the University of the West Indies in Trinidad, Conway moved to Tawau on the southeast coast of North Borneo, close to the border with Indonesia.

Conway told me that his five years as a research officer with the North Borneo (later Sabah) Department of Agriculture were among the happiest of his life.[47] For a start he was, as he put it, in 'entomology heaven'. This despite the fact that plantations of the cocoa tree (*Theobroma cacao*) had been decimated by insect pests and the farmers, using heavy insecticide sprays, were killing not only the pests but also their natural predators. From Tom Reynoldson, his teacher in Bangor, Conway had learned about natural control. Since the plantations were surrounded by tropical forest from which the pests invaded, Conway understood that they would probably be controlled by their natural enemies if the latter were allowed to survive. He recommended that spraying stop. Two of the major pests soon came under the control of parasitic wasps. Another pest was controlled by the removal of some nearby trees that were its natural host.[48] By integrating biological and chemical control the cocoa crop improved dramatically.

In 1963, mid-way through his stay in Borneo, Conway returned briefly to England where he spoke about his work at a conference. His paper made a good impression on Michael Way, the insect pest control specialist at Silwood. But Conway realized that to further his career he would need to learn more both about ecology and insect pests, and he would need a PhD.[49] On Way's advice he flew back to Borneo via California so as to consult with Southwood who was spending a few months at Berkeley as a visiting professor — on a Fulbright educational and cultural exchange programme. Southwood advised Conway to study for a PhD with Kenneth Watt at the University of California at Davis. Watt, a leading figure in mathematical and systems ecology, was interested also in resource management. Southwood must have discussed his own ideas for a book on ecological methods with Conway since, shortly after their meeting, he wrote, 'your kind of enthusiasm for the project encouraged me to regard this as something more concrete than I had hitherto. Any criticisms or comments would be gratefully received.'[50] Conway did indeed comment on a draft of the book *Ecological Methods* which Southwood was to publish two years later.[51]

Conway took Southwood's advice and enrolled as a PhD student at Davis. Watt had a grant from the Ford Foundation and was able to support his research. While a student at Davis, Conway sent Southwood lively letters describing his activities. He wrote about his studies in mathematics and agricultural economics, and how he was learning to programme computers.

He also studied Chinese as a 'scientific language' and described the deciphering of the characters as 'rather like insect taxonomy'. And he explained how he had arrived at his research proposal. After briefly considering working first in genetics, and then on a laboratory study in ecology, he decided instead on computer simulations:

> laboratory experimentation, it seem[s] to me, should be confined to finding the form of functions and the initial parameter values which can then be used in computer simulation. The logic of [C.S.] Holling's approach appeared very convincing and I became interested in the techniques of building mathematical and computer models.[52]

Watt and Holling had worked together earlier at one of the Canadian Forest Service laboratories and were continuing to do simulation work in their new positions.[53] Conway continued:

> Ideally one should wait until a dozen Hollings have worked out every important population process in great detail. However my reading of the literature on locusts and mosquitoes in the preceding months suggested that there really was a great deal known about these two groups and with some luck and insight one could build meaningful models of populations of these ... using 'jig-saw fashion', the best information from as many sources as possible.

Such models would then show where further empirical research was needed. No one, he believed, had attempted a complex and realistic simulation of insect populations. He wrote that while coming up with a perfect model would not be possible within two years, something meaningful could be done. Clearly Conway was impressed by simulation modelling which he called 'a powerful research and teaching tool ... something well known by the chemists and physicists and engineers, but not, so far, by more than a few biologists'.[54] Conway also wrote that he had been using ideas from Southwood's typescript of 'Ecological Methods' when directing some students in an analysis of a limpet population in Bodega Bay. From this experience he was able to suggest some additions to the book. But there were a few problems at Davis. For one, Conway did not receive the National Science Foundation grant that he had been hoping for and, worried about his financial situation, considered dropping out of the PhD programme. Southwood advised him to stay even if funds were limited. He wrote to him that he expected to succeed Richards as the head of department at Imperial College, and that while he could make no promises he would 'strain *every endeavour* to make it possible for you to come and work in this department in a couple of years or so, after you have completed your time at Davis'.[55]

As it happened, Conway made a short but important detour before moving to Silwood. In 1968 he accompanied Watt on a trip north to visit Holling at the University of British Columbia. While there Holling asked Conway whether he would like a temporary job with the Ford Foundation. Conway was interested and flew to New York to be interviewed by Gordon Harrison. Harrison was director of the natural resources and environment work at the Foundation, and in charge of grants in those areas.[56] Harrison offered Conway a short-term consulting job. Like Holling had done earlier, he was to accompany Harrison on a working trip; this one to observe agricultural practices in Thailand, Indonesia and India. The purpose was to figure out ways of best managing agricultural environments, and to determine what kind of programmes the Ford Foundation should support.

After his return to Davis in 1969 Conway successfully defended his doctoral thesis. He must have impressed not only his examiners but also Harrison. He was retained as a consultant on the programmes that followed from their Asian prospecting trip and was awarded a Ford Foundation grant amounting to $500,000.[57] This large grant, similar to the ones that Holling and Watt had received earlier for the founding of their research institutes, reflected the Ford Foundation's new direction of supporting research in global resources, environmental science, ecology and agriculture. It was a good time to be an ecologist. Conway was free to use the grant money wherever he wanted, giving Southwood an added incentive to appoint him at Silwood. On arriving there in the Autumn of 1969 Conway set up the Environmental Resources Management Research Unit. A press release stated that members of the unit would apply 'computer simulation and system analysis methods to the study of natural resources, especially [in relation] to pest management'. It also stated that the unit would encourage overseas students to come to Silwood, and to make use of Imperial College's considerable resources in tropical agriculture.[58]

One year after Conway's appointment, Southwood brought Michael Hassell to Silwood from Oxford. Hassell, too, had been a child naturalist and serious butterfly collector. He studied zoology at Cambridge University with the intention of becoming a marine biologist. However, he was 'converted' to insects by one of his teachers, George Salt.[59] Hassell moved to Oxford for a PhD where he studied under the Hope Professor, George Varley — though on a day-to-day level he received much help from Varley's long-term collaborator, George Gradwell.[60] Varley and Gradwell had made population dynamics a major field in the Hope department. In 1967, on gaining his doctorate, Hassell was awarded a three-year postdoctoral fellowship, one year of which was spent in C. B. (Ben) Huffaker's laboratory at the University of California at Berkeley. In 1970 Huffaker, a leader in biological control methods, became director of the International Centre of Integrated and Biological Control. The centre received major international funding for pest management research. After working in California on some laboratory ecosystems, and focussing on insect parasites, Hassell returned to Oxford determined to continue to experiment, but to take a mathematical and theoretical approach in his work. He told me that on his return he became close to John Lawton and David Rogers. Both said that they learned much from Hassell at the time. Hassell was also an occasional visitor to Silwood and, when a position opened up in 1970, he moved there permanently. Southwood had only a small team at Silwood, but his department now included two highly able and ambitious young lecturers.[61] Their work will be discussed further in chapter seven.

Michael Crawley came to Silwood in 1971 as a doctoral student to work under Conway. Crawley, who grew up in rural Northumberland, was yet another child naturalist. Indeed, he still sees himself as much a naturalist as an ecologist. His father was an Anglican minister but his parents were also sheep farmers and keen gardeners and he learned much from them. Like Conway earlier, he looked for a university where he could study ecology as an undergraduate. As it happened, by the mid-1960s several forestry departments, lacking students and sensing the direction of the political wind, were converting themselves into ecology/environmental science departments, or into such units within other departments. One such was at the University of Edinburgh where Crawley decided to go. But the foresters had not yet fully figured out how to become ecologists and the curriculum was rather vague. There were only four ecology students in Crawley's year and they came to an agreement with their teachers that, instead of attending lectures, they be allowed to read about their subject in the library. Crawley told me that he was fortunate in not having to go through much formal training at that stage. I asked him what he read, and he mentioned the older work of A. J. Nicholson and V. A. Bailey, as well as the above mentioned book by Andrewartha and Birch.[62] Crawley also kept up with contemporary work, much of it American. For example, he read papers and books by Robert MacArthur, Lawrence Slobodkin, Nelson G. Hairston and F. E. Smith.[63] He did not read much of the work coming out of Oxford, and told me that while David Lack was still revered, Charles Elton and E. B. Ford were by then seen as rather old fashioned.

At the end of his undergraduate studies in 1969, Crawley took a three-month summer job working with C. S. Holling at the University of British Columbia. He said that Holling had a huge influence on his later work. Crawley was offered a place at Silwood for his PhD, but decided to delay a year and to do some work instead with the International Biology Program (IBP).[64] He was appointed a computer programmer on a desert biome project, working under David Goodall at the University of Utah. While he found this work useful Crawley claimed that, overall, the IBP was not well thought out. On arriving at Silwood in 1971 he discovered there had been a mix up. Notice of his delay in taking up his NERC grant had not been received and since it had only another two years to run he had to carry out his doctoral research in a hurry. A quick way out was to focus on computing for which he already had some skills; Crawley was one of the few biologists to have them at that time. In the early 1970s one could not buy software. People had to write their own or rely on people like Crawley to help them out. Crawley's doctoral contribution was to write software for his cohort of PhD students and for some others at Silwood, enabling them to carry out statistical analyses of their data. He was awarded a PhD in 1973. His first job after graduating was as a lecturer in ecology at the University of Bradford, but he returned to Silwood in 1979 and has worked there ever since.

1971 was also the year that Robert May made his first appearance at Silwood. May told me that his was an accidental migration into biology. An Australian, he was a pupil at the Sydney Boys' High School, an excellent state grammar school in the Eastern suburbs of Sydney.[65] There he shone academically, especially in mathematics. He was also a keen debater and was advised to enter the legal profession or, if not that, to study medicine. But May decided that neither was for him and enrolled at the University of Sydney in a chemical engineering programme. By his third year he had decided on physics and mathematics rather than engineering. In this he was encouraged by Harry Messel, a Canadian who was building a fine physics department at the university and who had brought in some world-class theoreticians. For his PhD May studied theoretical physics under one of them, Robert Schafroth.[66] May was the first doctoral graduate in that field in Australia. He then took up a two-year visiting lectureship in applied mathematics at Harvard University. While there he met his future wife, Judith Feiner, a young student at Brandeis University who was to play an important supporting role in his life.

May returned to a physics lectureship at Sydney and, at the age of thirty-three, was given the first personal chair awarded by the University. But he also had an interest in environmental matters, and was active in the 'the social responsibility of science' movement.[67] One of its leaders in Australia was Charles Birch, professor of ecology at Sydney. May shared Birch's enviro-political concerns and increasingly also his ecological interests. (Birch was co-author, with fellow-Australian Herbert Andrewartha, of the influential ecology text already mentioned.) In 1971, while May was on sabbatical leave in the United Kingdom and working at the Culham Plasma Physics Laboratory, Birch wrote to Southwood telling him that May had some mathematical ideas that were applicable to ecology. They stemmed from May's having read Kenneth Watt's *Ecology and Resource Management* (1968) and were the start of his challenge to the idea, supported by Watt and G. Evelyn Hutchinson, that complex ecological communities are more stable than simple ones. Birch wrote to Southwood suggesting that he invite May to give a talk on this at Silwood.[68]

May was invited. It was a major turning point in his life. He found common cause with Southwood, Conway and Hassell at Silwood and was confirmed in his decision to give up theoretical physics. He used the remainder of his fifteen-month sabbatical leave to turn himself into a theoretical ecologist. May is lively and charismatic, true even now that he is in his seventies. I imagine he would have been hard to resist when, after his 1971 conversion, he went seeking talent elsewhere. His next stop was at Oxford where three different groups had some ecology interests. These groups will be discussed in the next chapter, but one of the people that May met while in Oxford was John Lawton. Lawton told me that his first impression of May was of a somewhat 'manic Australian', but in retrospect he acknowledged that May 'is the cleverest person I have ever met'. Later Lawton and May would become scientific collaborators, and May would return to Oxford as a professor. Another person whom May met while in Oxford was the professor of mathematical biology, James Murray. Murray put May in touch with George Oster who would soon become a working colleague.[69] May also visited John Maynard Smith, then Dean of the School of Biological Sciences at the University of Sussex. Maynard Smith, a major evolution theorist, was beginning to take an interest also in ecological modelling. His work will be discussed briefly in chapter six.

May had intended spending seven months at Culham and the rest of his fifteen-month sabbatical at the Institute for Advanced Study in Princeton. At Princeton the plan was to work on some astrophysical problems with John Bahcall with whom he had shared a house during his last summer at Harvard.[70] But these plans did not materialize. May consulted further with Charles Birch on switching to theoretical ecology. While not keen on some of the new mathematical approaches to ecology, Birch nonetheless suggested that while at the Institute May talk with Robert MacArthur. The resulting conversations confirmed May in his decision. MacArthur showed him how to frame questions in ecology so that they could be approached using the tools of mathematics and theoretical physics, tools that May had in abundance. They published a joint paper and MacArthur encouraged May to write a book on his ideas in the time remaining on his sabbatical leave. After just four months in Princeton, May returned to Sydney to write his *Stability and Complexity in Model Ecosystems*. He acknowledges his great intellectual debt to MacArthur who was inspirational in at least two ways: first in his conceptual ability, his way of envisaging ecological problems, and in his ability to arrive at simple mathematical models with fairly straightforward, albeit not quantitative, predictions; second, as someone who understood the importance of working closely with serious naturalists and laboratory scientists. Both lessons were well learned by May. They were foundational to his own collective engagement, and to his considerable success.

Sadly, MacArthur already knew that he was terminally ill with renal cancer when May visited him in 1971. Recognizing May's brilliance, MacArthur recommended to the authorities at Princeton that May be his successor. May was offered a chair at Princeton before MacArthur died but, reluctant to leave Australia, he did not accept. However, with the encouragement of his wife, he changed his mind. He returned to Princeton for a brief visit in late 1972 both to visit MacArthur who was then close to death, and to make arrangements to take up the chair in the following year.[71] But Silwood remained a magnet. For May it was 'an inspirational place' and he was delighted when, in 1973, Southwood arranged for a visiting professorship which enabled him to work there for a few weeks each summer.[72]

The summer visits allowed May to work with people at Silwood, and for the exchange of ideas with a community of like-minded theoreticians at the two-day workshops in mathematical ecology which took place during the 1970s. The workshops came about after Gordon Conway gave a paper at the University of York in 1971. The head of the biology department, Mark Williamson, had similar interests and, with Conway's encouragement, decided to set up a summer workshop with the aim of 'bring[ing] people together who are working on mathematical problems in ecology and who use quite different sorts of organisms and mathematical techniques'.[73] Attendance at the 1972 workshop was by invitation only and about thirty people showed up. A second workshop was held in York in the following year. In 1974 and 1975 the workshops moved to Silwood, and a final one was held in York in 1976.[74] The 1973 workshop was devoted to the subject of ecological stability, the topic that May had been working on. In introducing the collection of published papers that followed from the meeting, May stated that he was optimistic about the state of ecological theory, even while recognizing that it lacked the 'crisp determinacy that characterizes the physical sciences'. He also saw the scientists who gathered at York as working predominately on 'the way biotic interactions between and within populations act as forces moulding community structure'.[75] This focus reflected the interests not only of those working at Silwood, but of ecologists working in the United Kingdom more generally. British ecologists were heavily biased toward agricultural insect pests: those found in Britain and those found in former British colonies. They were also interested in human and animal disease parasites and in the role of parasitism in the biological control of pests. Not surprisingly there was a focus on predator-prey and host-parasite interactions; and on the population dynamics of organisms involved in such interactions. May pointed out that North American ecologists had wider interests and that there was more work on competition within trophic levels, and more interest also in birds and other vertebrates. As will be shown in the next chapter this had much to do with the interests of some of G. E. Hutchinson's students, including

Robert MacArthur. Also to be discussed in later chapters is the way in which May, having a foot in both camps, was a major conduit of ideas across the Atlantic.

People at the workshops came from a number of different British universities and research stations. Included were some fisheries biologists from the research station at Lowestoft, a small group of ecologists from Oxford, a fairly large contingent from Silwood and an even larger group from York. In addition to May and Conway, those present included Michael Hassell, Roy Anderson, David Rogers, John Beddington and John Lawton, all people belonging to the Silwood Circle.[76] In 1974 Conway and Southwood organized the workshop at Silwood and, together with Hassell and May, chose whom to invite. Aside from those already working at Silwood,[77] other attendees included Mark Williamson and the people just mentioned, as well as William W. Murdoch, Richard Weigert, John Calaprice, Christine Shoemaker, Patricia Rosenfield, and John Maynard Smith.[78]

In 1973 Murdoch joined Southwood, Conway, Hassell and May at Silwood for an extended working period over the summer. Given the future direction of those in the Silwood Circle, the parallels in his and their careers are interesting. Murdoch had been a student of Charles Elton at Oxford before moving to the United States. One of the foremost ecologists of his generation, like those in the Silwood group he specialized in population dynamics, predator-prey dynamics and biological control. He was later to head the National Centre for Ecological Analysis and Synthesis, located at the University of California at Santa Barbara. And, like Southwood, he early joined the debate on human overpopulation. That these parallel interests were to be found in California (and elsewhere) is illustrative of a more general feature of science, namely that there is, as was discussed in chapter one, a situational logic to much that goes on. Not surprisingly the interests of the Silwood Circle were mirrored in other ecology centres around the world.

Much later, when giving the Nathan Lecture on the Environment, Conway described what happened at Silwood during that summer of 1973:

> It was a long, hot and extraordinarily productive summer; we worked in intensive sessions, interspersed with vicious games of croquet and dips in the Director's pool, producing a number of papers which became the basis for a book entitled *Theoretical Ecology*.

Southwood, who introduced Conway at the Nathan Lecture, noted the presence of Lord Flowers, the former Rector of Imperial College, in the audience. He stated that he thought Flowers would agree that 'I used Gordon as my special weapon in taking forward the environmental agenda'.[79] Southwood did, indeed, use Conway to forward his agenda. And, as Conway mentioned in his speech, Southwood's agenda was helped also by the work they all published in May's edited collection, *Theoretical Ecology*.[80] Conway also helped Southwood by using some of the Ford Foundation money to invite a young Australian theoretical physicist, Hugh Comins, to Silwood in 1974. The invitation came about because May saw that many interesting problems were being generated at Silwood and that the people working there needed help in framing them analytically. He suggested that Comins, one of his former students at the University of Sydney, be invited to Silwood. Conway described Comins as 'an amazing theoretician able to apply his mathematical skills to ecological problems'.[81] In his few years at Silwood, Comins helped several people, including Southwood, to move their work forward. Bill Hamilton wrote of Comins that, 'like May he came a Viking into the quiet hall of Silwood, brandishing an analytical skill against which we entomologists and ecologists had no defence'.[82]

Conway also promoted Southwood's environmental agenda in an undergraduate course in environmental studies given at South Kensington. It was extremely popular with students who crowded into the Great Hall, Imperial College's largest lecture theatre, to hear him. Thus encouraged, Conway and Southwood sought support for a postgraduate research centre in environmental studies at the college. As mentioned, the environment was very much on the political agenda during the 1970s. Many universities introduced courses in environmental science in the hope of attracting both students and grants. Imperial wanted to find a suitable niche from which it could compete. Brian Flowers, understood that were Imperial to have a place within the growing environmental science movement, it would need to include more than just the ecologists. He saw a role for chemists and engineers, and also wanted to help the Royal School of Mines which was then in serious decline. Its staff were looking for new areas of work, including the environmentally-friendly reclamation of old mine sites. Flowers asked the head of the geology department, John Sutton, to look into what people at the college were already doing in the environmental area and to consider what would be a good way forward. The result was the founding of the Imperial College Centre for Environmental Technology (ICCET), a multidisciplinary centre that offered MSc courses in a range of specialties. People from across the college were associated with it, and Conway was appointed its first director. The centre was officially opened in 1976 by Peter Shore, Secretary of State for the Environment. Much government money flowed in including, in 1978, a large grant of £450,000 for work in ecology.[83] The centre's location in South Kensington meant that Conway moved there from Silwood, but connections between the two campuses were strong.

Sites such as Silwood, where scientists come together to discuss their work in an atmosphere that is socially and intellectually engaging, challenging yet also relaxing, are important for understanding much of what happened in twentieth-century science. Silwood was a place where people with young families could feel comfortable. Children had companions of their own age and could explore the large grounds together.[84] They enjoyed watching the live butterflies kept in the conservatory, and those interested were able to study also the collections of dead specimens. There was a tennis court, and there were invitations to use the Southwoods' swimming pool. There are many other such places where regular summer gatherings occur, some of them more notable than Silwood. In the biological sciences Cold Spring Harbour, New York, and Woods Hole, Massachusetts, come to mind.[85] Whether this type of summer social workspace, afforded by major institutions, will continue to be important in the twenty-first century is an open question. People need to network in a personal way so as to share ideas of what is important, and to know how to conduct their own work. But, as electronic communication changes our world, will scientists find extended face to face visits worthwhile in the future? Whatever the case, when looking at twentieth-century science historians need to pay attention to sites such as Silwood, places where new scientific ideas were discussed and nurtured.

People who were at Silwood during the summers of the early to mid 1970s remember it well. Conway was not alone in reminiscing on the intense scientific discussions, the parties held around Southwood's swimming pool, and the 'fierce battles' on the croquet lawn where Bob May is said to have 'held court'. I was told that the outcome of a croquet game could determine the order of authorship on a scientific paper.[86] Important intellectual exchange occurred also in the conservatory during coffee/tea breaks. The summers, remembered as having been highly stimulating intellectually, allowed people to keep up with what others were doing and, perhaps more importantly, with what others thought was worth doing next. For those with ambition in theoretical ecology Silwood was an important place.

As mentioned, one of the first places that May visited after meeting Southwood and others at Silwood was the zoology department at Oxford. It was there he first met John Lawton. Like others in the Silwood Circle, Lawton was a child naturalist. He grew up in Leyland, Lancashire, where his father had a law practice, and was free to roam the countryside on foot or by bicycle. His parents encouraged his growing interest in birds, an interest that he pursued also as a zoology student at Durham where one of his teachers was the ornithologist John Coulson. Lawton was something of an ornithological prodigy but Coulson persuaded him to widen his interests and suggested that he study insects for his doctorate. Lawton told me that he was grateful for this advice since it allowed him to become knowledgeable in two major areas. After moving to Oxford he exchanged ideas with other young ecologists, communication that occurred despite the tension between the various departmental divisions in which they worked. Not without justification Lawton described the zoology department of that time as 'tribal', a point to be taken up in the next chapter.

As will also be discussed, after a period of great activity, ecology at Oxford was in decline by the 1970s. It regained some of its former strength after Southwood's appointment as Linacre professor in 1979. By then the discipline was gathering strength also elsewhere, including at the new University of York where Mark Williamson, a former Oxford research graduate, and the principal organizer of the aforementioned mathematical biology workshops, was head of the biology department. He appointed Lawton to a lectureship in 1971.[87] At the same time he appointed John Beddington to a temporary (later permanent) lectureship. Beddington had studied economics at the London School of Economics. Interested in Karl Popper's lectures, he decided to take a one-year MSc course in the philosophy of science and theory of statistics. Finding the course rather easy, and being interested also in ecology, Beddington spent much time reading in the library at the London Zoo.[88] Like many others, he found Kenneth Watt's *Ecology and Resource Management* (1968) to be of especial interest. He told me that he knew of MacArthur's work but was not much influenced by it at the time. Rather it was his undergraduate education in economics, and especially Paul Samuelson's *Foundations of Economic Analysis* (1948), that informed his ideas. Beddington recognized that some of Samuelson's mathematical models could be applied to ecological systems. He was influenced also by P. H. Leslie's work on age-structure demography and interacting populations.[89] After moving to Edinburgh for a PhD, he was able to use his demographic and mathematical skills in a study being conducted on Red Deer populations. He also heard Conway give a talk at the Merlewood Nature Conservancy Station in the Lake District and, impressed by it, made contact with people at Silwood. Both Lawton and Beddington would later move from York to become professors at Imperial College.

Shortly after the summer workshop held in York in 1973, Lawton suggested that he and Beddington meet with some of Lawton's former Oxford colleagues and with some of the people working at Silwood for occasional, but regular, dinners. The dinners took place about three times a year, often after meetings of the British Ecological Society. The group would meet for a beer and then go for a meal at an Indian restaurant. The idea was to exchange ideas and to keep in touch. Sometimes short informal papers were presented and discussed. Included at these gatherings were Hassell and Crawley from Silwood, Rogers from Oxford, and John Krebs who was in Bangor for a brief time before returning to Oxford. Occasionally they were joined by Southwood, May, or Conway. They were also joined by former Imperial College student, Roy Anderson, whom May, Lawton and Beddington first met at the 1973 summer workshop in York. Later, after Crawley moved to Bradford, the town with its many Indian restaurants and its situation near walking country in the Yorkshire Dales, became another meeting place.

Anderson told me that, while a keen naturalist as a child, he became seriously interested in ecology only in his third year as an undergraduate at Imperial College.[90] He was drawn to the work of Neil Croll, an expert in parasite ecology. He attended a field course that Croll ran at Slapton Ley in Devon which he much enjoyed and found the people he met there, including two other Imperial College parasitologists, June Mahon and Elizabeth Canning, interesting. Mahon was to supervise Anderson's doctoral work on fish parasites and Canning would later hold a chair in protozoology at the college.[91] Anderson, skilled in mathematics, was interested also in the work of Gareth Davies and George Murdie. Murdie was a statistician and biomathematician and Davies was using computer simulations in his biological work; he later held a chair in the biology department.[92] Both men encouraged Anderson to learn more about mathematical biology and to read work by MacArthur, Holling, Murdoch and Watt.

Another outcome of the York workshop of 1973 was the organization of a summer hike, something that proved so successful that collective hiking became an annual, sometimes biannual, event. Among the regular participants in the early years were Lawton, Beddington, May, Hassell, Anderson and Crawley.[93] In 1974 they ambitiously tackled the Three Peaks (Whernside, Ingleborough and Pen-y-Ghent) in the Yorkshire Dales National Park. In subsequent years, often with Bob May doing the organizing, peaks in the Lake District, Snowdonia, the Scottish Highlands, the Swiss Alps and the Pyrenees were tackled. 'Bob's walk', as it came to be affectionately called, has occurred every year since 1973. At first they drove to their destinations in a large van and, according to Crawley, had much fun en route. Later, with less time to spare and with more disposable income, they usually flew. From talking with those who took part in the early walks I have the impression that there was a degree of physical as well as intellectual competition among the participants. Wives and girl friends were largely absent, though some occasionally joined in.[94] In later years the occasional woman scientist was invited to join but few did so. The hikes took place over two to three days during the Easter and/or summer breaks. Like the dinners, they helped to build a clubbish homosocial ethos. The hikes, although remaining exclusive, were soon to include also younger associates.[95] As with the summers at Silwood and the dinners in Indian restaurants, they were important in solidifying both group and individual identities, and in the furthering of careers. They also forwarded the cause of ecological science by providing a further intimate venue for the exchange of ideas.

4.4 Southwood's later years at Silwood

Southwood did not join the hikes but he encouraged his growing circle in their work, and he helped them to forward their careers.[96] He was determined to see the science of ecology advance and for it to occupy a more central place in the scientific life of the country. His excellent administrative and communication skills were demonstrated not only in the running of his department, but in the wider scientific world. He was good with people, remembered names (indeed he was renowned for that) and details of people's academic and private lives. He enjoyed teaching undergraduates as well as research students and became a major voice at Imperial College. He was elected Dean of the Royal College of Science and was appointed the first chairman of the Division of Life Sciences. His department expanded under his leadership. By 1978, shortly before his departure to Oxford, thirty-three of its staff members were funded by the University Grants Committee, something that was then widely seen as remarkable. By the late 1970s the department's main administrative concerns were who should replace Southwood as head of department, and whether it should unite with the botanists into a department of biology, something wanted by the college Rector, Brian Flowers.[97]

One of Southwood's ambitions was to become a Fellow of the Royal Society and to move ecology to a position of importance within the Society, a topic to be picked up in chapter six. His younger associates at Imperial College, those who shared his ambition, were keen to remain close to him. This, and closeness to their generational colleagues, was key to their own advancement. As will be discussed, Southwood helped Conway, Hassell and Anderson to rapid promotion. All became professors at Imperial College at relatively young ages.[98] Also to be discussed, is how Southwood's voice was heard well beyond the college; and how, as May and Hassell have pointed out, he worked 'at the intersection between three distinct and different cultures: academic researchers, civil servants and Ministers.'[99] I would add business culture to this mix. By the time Southwood moved to Oxford in 1979, ecological science was growing in esteem not only within the scientific community, but also in the eyes of the public. Government and industry took notice. For this, Southwood could take some of the credit.

Perhaps a good window on Southwood's field of entomological ecology as it stood in the late 1960s is a volume of papers he edited for the Royal Entomological Society. The papers were delivered at a conference held at Imperial College just before Richards's retirement in 1967.[100] The discipline, still largely descriptive, was showing signs of change. In order to understand the changes that followed, and the increased activity that has been discussed in this chapter, it is necessary to look more closely at how ecology moved from being a largely descriptive science to one with more theoretical depth. For that we have to look beyond Silwood at what was happening elsewhere.

Endnotes

1. 'You played the game as to the manor born and I hope that one day from some elevated Chair you will again have a chance of displaying the same administrative talents'. This prophetic letter was written to Southwood from J. H. Corin, Imperial College Secretary, 8 September, 1958. It is interesting both because it contains some tips for Southwood on the art of administration, and because Corin early recognized that Southwood had talent in that area. As it happens Southwood was, literally, to the manor born. The family lived in the Parrock Manor House, near Gravesend, Kent. Judging from photographs in the Southwood papers the house was large and fairly imposing, though perhaps not in the best of repair. Letter in Bodleian Library, Southwood Papers, B3; photographs in section A.

2. Sir Thomas Richard Edmund (Dick) Southwood FRS (1931–2005).

3. Southwood applied unsuccessfully for a post as demonstrator in agricultural entomology at Cambridge. Richards then invited him back to Silwood, but Southwood only began work in January 1956 since he suffered complications after an appendectomy and needed a few months to convalesce.

4. Southwood gave talks to his school's Nature Study Club and published some early observations in the British Empire Naturalists' Association's journal (the first, in November 1946, was among five papers published while Southwood was still at school) and in the *Entomologist's Monthly Magazine* (he served on its editorial board from 1962). A talk on 'Bug-Hunting' was given to the Gravesend Rotary Club in May 1953. Southwood became a FRES before the minimum age was set at eighteen.

5. (May and Hassell, 2008) 349. A copy of Southwood's memoir is at the Bodleian Library, University of Oxford; Southwood papers, Section A.

6. This is a conclusion I came to after hearing much gossip. It is difficult to pin down. It implies neither that the women he supported had affairs with him, nor that they were academically unworthy. In some letters of recommendation that he wrote for women he mentions their physical attributes, albeit in connection with their ability to do field work — something I did not see in any letters written for young men. Further, he sometimes qualified his praise of women by writing, 'one of my best girl students' or 'one of the most accomplished female ecologists' and so on. In his attitude he was not unusual and mirrored his time. He left an interesting note (undated, but from the early 1970s) to the Rector of Imperial College on dealing with discipline at Silwood in 'the world of feminist politics'. At this stage of his career he was less sure-footed in dealing with the demands of academic women and women students

than he was in other areas of administration. Bodleian Library, Southwood papers, A6.

7. The 'dinners in hall' were first introduced at South Kensington after World War II by the new Rector, Richard Southwell. He had come to the college after a career spent at both Cambridge and Oxford. The dinners were an attempt — only partially successful — to introduce a type of Oxbridge college conviviality at Imperial College. Munro instituted similar dinners at Silwood and they were continued by Richards and Southwood.

8. Conway, personal communication.

9. Southwood's work at this stage had much in common with Munro's, though his doctoral work was mainly on agricultural rather than stored food pests. Later Southwood carried out some consulting work that Munro might well have engaged in. For example, he examined the damage caused by bacon beetles to goatskins during their transit from India, worked on infestations of dried fruit for Foster Clark Ltd., Maidstone, Kent, and on spider beetles in tinned potatoes. Bodleian Library, Southwood Papers B6, B7 and B15.

10. Carrington Bonsor Williams FRS (1889–1981) was educated at the University of Cambridge, and at the John Innes Horticultural Institution. While at John Innes he also studied with H. M. Lefroy at Imperial College. After a period in Trinidad he moved to Rothamsted where he became head of the entomology department in 1932. He was a specialist in insect migration (especially *Lepidoptera*), an interest he passed on to Southwood.

11. Kenneth Mellanby (1908–1993) was a medical entomologist. During World War II, he was an outspoken pacifist. Instead of conventional war service he organized a group of about 40 conscientious objectors who shared a house in Sheffield, close to the university where Mellanby held the Royal Society's Sorby Research Fellowship. The group submitted themselves to various experiments. These included ones on DDT and Mellanby's own experiments on the control of the scabies mite. At the suggestion of nutrition specialists, Robert McCance and Elsie Widdowson, Mellanby was also a guinea pig for Hans Krebs and his work on the digestion of 'national wheatmeal' flour. Krebs determined that, with a calcium supplement, high extraction (85%) flour could safely be used, so saving the country from having to import wheat during the war. After the war Mellanby became Principal of the new University College of Ibadan, Nigeria. On returning to England he worked at Rothamsted before becoming director of the Monk's Wood Experimental Station in 1961. His book *Pesticides and Pollution* (1967) was, in part, a response to Rachel Carson's *Silent Spring* (1962). Mellanby, while warning of the environmental dangers of pesticide use, was more measured in tone than Carson. From letters in

Southwood's papers one can infer that Mellanby was of considerable help to him at the start of his career. For the Athenaeum see Bodleian Library, Southwood Papers A.60. For more on Mellanby, (Weindling, 2004).

12. The work was supported by the ARC. See, for example, (Southwood and Jepson, 1962).

13. Reader was appointed an experimental officer when Southwood took the chair. They worked together for many years. See, for example, (Southwood and Reader, 1976; Southwood, Hassell, Reader and Rogers, 1989).

14. Together with Dennis Leston he published *Land and Water Bugs of the British Isles* (1959), a volume in the popular *Wayside and Woodland* series. Southwood was interested especially in a sub-order, the Heteroptera. The standard authority when Southwood was a boy was E. A. Butler. Interestingly G. Evelyn Hutchinson, earlier, and also as a child, developed an interest in these insects in part because of Butler's mentorship. (Hutchinson, 1979; Butler, 1923).

15. Ten years later I, too, went to Flatford Mill with a school party on a one-week field study course. The mill had once been owned (but not lived in) by the family of the artist John Constable whose paintings made the site and its environs famous. Interestingly its start as a field studies centre was made possible by an endowment from an earlier Imperial College professor, the entomologist Frank Balfour-Browne (see chapter 3). This allowed the Field Studies Council to lease Flatford Mill from the National Trust, starting in 1946.

16. See, for example, (Southwood, 1961).

17. Data was collected from areas of the Silwood estate at different stages of development — from bare fields to woodland. See, for example, (Southwood, Brown and Reader, 1979). After Southwood moved to Oxford, Brown led this continuing research program, and a team of about twenty people. She became a reader at Imperial before leaving in 1993 to become Director of the Commonwealth Agricultural Bureaux' Institute of Entomology. The Institute was located in South Kensington before its later move to Silwood. Brown ended her career as a professor at Reading University, and as Director of the Centre for Agri-Environmental Research.

18. (Southwood, 1977). According to May and Hassell (2008), 351, the address became a 'citation classic'.

19. If not appointed to the chair and directorship at Silwood, Southwood was planning to take a chair at the University of Melbourne where his close friend, Robert Blackith, worked. See correspondence with Blackith, Bodleian Library, Southwood papers, R14. Richards continued to be productive in retirement as a senior research fellow. He published several papers and a major book on American wasps; (Richards, 1978).

20. One very public disagreement was over some fallen elm trees at Silwood. The trees had died from Dutch Elm Disease and Hamilton thought they should be allowed to decay naturally, left in place to be invaded by insects and other life forms. Southwood wanted the trees cleared away — and they were.

21. ICL archives, Zoology department correspondence; Southwood to Flowers, 6 August, 1974. Southwood appears not to have fully understood Hamilton's work, though some others in the Silwood Circle did and likely would have tried to convey its importance to him. Hamilton formalized Haldane's ideas on altruistic behaviour. His model of inclusive fitness, published in 1964, connects genetic relatedness to the costs and benefits of altruistic behaviour. While his equation was not strictly predictive it had enormous heuristic power.

22. (Hamilton, 1996–2005). For Hamilton's remarks on Silwood see vol. 1, chapter 4. These volumes are unusual not only because of Hamilton's interesting scientific papers, but for the short autobiographical pieces with which he introduces many of them. One wishes that more scientists would do the same and not distance themselves from the context of their working lives. In Hamilton's case some passive aggression is apparent, but nonetheless a remarkable human being shines through. He did not, as could perhaps be implied from his memoir, shut himself away in his lab. He was a convivial presence at Silwood. (The third volume of Hamilton's papers, edited by Mark Ridley and published after Hamilton's death, contains reminiscences by some of Hamilton's co-authors). In his obituary (*The Independent*, 3 October, 2000), Richard Dawkins states that Hamilton 'is a good candidate for the title of most distinguished Darwinian since Darwin'.

23. According to a recent paper, the tide has again turned, this time slightly against Hamilton; E. O. Wilson is said to be deviating from his earlier position on altruism; (Okasha, 2010). See also comments on Martin Nowak in chapter 9.

24. In my view the evidence does not support the account of Southwood luring Kennedy and Lees to Silwood as described in (May and Hassell, 2008), 355. However, he may well have been a behind-the-scenes supporter of the move and will no doubt have viewed Kennedy as an ally in his quest to become an FRS.

25. Cox, a specialist in x-ray crystallography, first approached the physicist Clifford Butler to make some informal enquiries: ICA, Zoology Department correspondence KZ/Sir Gordon Cox to C. C. Butler, 21 March, 1967. This was followed by correspondence with Saunders, See, for example, Sir Owen Saunders to Cox, 14 April, 1967. The correspondence included discussion on

how many staff Kennedy and Lees could bring with them, and on whether they could receive conferred chairs while still being paid by the ARC. Charles Percival Whittingham (1922–), a plant physiologist, had earlier worked on new herbicides at Imperial under G. E. Blackman, before leaving with Blackman for Oxford.

26. Bodleian Library, Southwood Papers, R. 153. Letters from 1960 show detailed comments by Kennedy on Southwood's work and ideas. See, for example, letter dated 13 September 1960.

27. For more on Kennedy, (Brady, 1995).

28. Comments by W. P. Jepson dated 18 October, 1969. Bodleian Library, Southwood papers, B14. By 'integrationists' Jepson meant people promoting integrated pest management (IPM), an approach using chemical and biological control methods along with attempts to change farming methods in other respects. In using the term Carsonites he was referring to the admirers of Rachel Carson.

29. Way succeeded Southwood as Director of the Silwood Park campus (but not as head of the new joint department of pure and applied biology created in 1981). The Spraying Machinery Centre became the International Pesticide Applied Research Centre (IPARC) under the direction of Graham A. Matthews, a specialist in the control of cotton pests. He carried out much work in Malawi.

30. (Carson, 1962 and 1963). The Duke of Edinburgh bought several copies of the UK edition and distributed them to people he thought needed to read the book, including the Minister of Agriculture. For details of some British problems, (Sheail, 2002).

31. Hardin (1968). The 'population crisis' was a topic of discussion throughout the 1950s and 60s. It made the cover of *Time*, 11 January, 1960. Another best seller in this vein, (Paddock and Paddock, 1967), written by two brothers, one an agronomist and the other an international development specialist, failed to appreciate the merits of the Green Revolution. They had a Malthusian view of things, predicted the collapse of India, and saw America's role as one of triage specialist, making decisions on which countries might survive famine and were therefore worth helping with food aid.

32. The Torrey Canyon was a Liberian-registered oil tanker that was wrecked off the coast of Cornwall, 18 March, 1967. The oil released caused considerable environmental damage. That, and the overuse of detergent to dissipate oil on the Cornish beaches, was a tipping point for the environmental movement in Britain.

33. For more on this and other groups, (Lowe and Goyder, 1983).
34. Earth Day was established by U.S. Senator, Gaylord Nelson. Among other things it saw the introduction of the highly successful Möbius-loop recycling logo. Earth Day is now celebrated in about 175 countries.
35. H. G. Sanders was professor of agriculture at Reading University. For research study group, (Sheail, 1987).
36. Edward R. D. Goldsmith (1928–2009) was the older brother of the financier Sir James Goldsmith. Goldsmith's basically conservative political ideas later put him at odds with some co-founders of the Green movement in Britain.
37. The origin of the use of the term 'green' to describe environmental movements is obscure. The first political Green Party was founded in Tasmania, but 'green' ideas were everywhere in the early 1970s and several Green Parties sprung up at roughly the same time. The English party was founded in 1973 and was at first called the People Party. It went through a few name changes becoming the Green Party of England and Wales in 1985.
38. UNEP had an endowment of £20 million. Its first director was the Canadian businessman, Maurice Strong, the force behind-the-scenes and the Secretary-General of the Stockholm conference. Strong's business interests were combined with an active concern for the global environment. As the young chairman of Power Corporation, he was mentor to the future Canadian prime minister, Paul Martin. Later, after his term in Nairobi, he became chief executive officer, first at Petro Canada, then at Ontario Hydro. He chaired the 1992 UN Conference on Environment and Development held in Rio de Janeiro, and was a moving force behind the Kyoto Protocol — coincident with his growing financial interest in the Chinese car industry.
39. For a retrospective account of this period, (Southwood, 1997).
40. Bodleian Library, Southwood Papers, A.8 and E.79. The second of these is a lecture given on his bird/insect studies to the Edward Grey Institute in Oxford in January, 1973. Southwood's research on partridges began on Lord Rank's estate in Hampshire in the early 1960s — before the publication of Carson's *Silent Spring*. For some later work in this vein, (Southwood and Cross, 1969). Cereal seeds were being treated with organochlorine compounds such as aldrin and dieldrin, then manufactured by Shell. The chemicals were used both as anti-fungal seed dressings and as insecticides for growing crops. Game birds ate seeds left on the ground and were therefore especially vulnerable to poisoning, and their egg shells to thinning. The egg shells of raptors that preyed on the game birds also thinned.
41. Bodleian, Southwood Papers, E. 58; see notes and correspondence on 'man's population', 1969–70. Southwood began thinking about how to estimate

population more generally. See also E.72, for a paper given at the WHO Conference on Vector Ecology, December 1971; 'The principles of population estimates'.

42. Imperial College London, *Topic*. 11 June, 1979; (copy in ICA).

43. Bodleian Library; Southwood papers, B16. *The Times*, January 25, 1972, 15. Southwood's was the first signature, followed by those of J. S. Kennedy and Gordon Conway. See also (Nelkin, 1976). This article gives a good description of the situation in the United States and how ecologists there were drawn into public debate but claims that many soon became disillusioned when their voices were not heard.

44. Brian Hilton (Lord) Flowers FRS (1924–2010) was Chairman of the Royal Commission on Environmental Pollution, 1973–6. A theoretical physicist, he became Rector of Imperial College in 1973 at a time when bringing money to the college was very difficult. He saw the environmental sciences as a possible area for expansion.

45. As with Southwood, Conway's parents — his engineer father and geographer mother — took much interest in his activities and supported his naturalist leanings.

46. As mentioned above, Paul W. Richards (1908–1995) was the younger brother of O. W. Richards. See his study of a tropical rainforest; (Richards, 1996).

47. North Borneo gained independence from Britain in 1963 while Conway was working there. It became Sabah, a provincial state within Malaysia.

48. (Conway, 1998), 205–206 and *passim*.

49. His replacement in Sabah was a Sabahan native, Tay Eng Book, who had gained a PhD under Nadia Waloff at Silwood.

50. Bodleian, Southwood papers, R.50; Southwood to Conway, 13 Jan 1964, Berkeley, California.

51. (Southwood, 1966). This book was largely based on the material used in teaching the MSc course in applied entomology given at Silwood. It included also methods of trapping insects that Southwood had learned at Flatford and at Rothamsted as well as W. P. Jepson's flotation method for trapping insects that lived in the soil.

52. Bodleian, Southwood Papers, R.50; Conway to Southwood, 27 July, 1967, Davis, California. C. S. (Buzz) Holling was by then working at the University of British Columbia. Some of his ideas will be discussed in chapter six.

53. They worked together at Sault Ste. Marie, Ontario, and were among the first to use computers in ecological modelling (in the early 1960s). Holling told me that they used Fortran programming language. See the highly influential (Watt, 1968).

54. Letter cited in note 52.

55. Bodleian, Southwood Papers, R.50; Southwood to Conway, 10 August, 1967, Silwood. Emphasis in original.

56. Holling's Institute of Resource Ecology was kick-started by a large grant from the Ford Foundation, as was Watt's department at Davis. Harrison, a Harvard history graduate who contributed to the official U.S. history of World War II, also wrote a book on the environment titled *Earthkeeping: The war with nature and a proposal for peace* (1971) — well written and prescient, it is still worth reading today. Later Southwood invited Harrison to Silwood where he wrote another book, *Malaria, Mosquitoes and Man* (1978).

57. Conway remained a consultant to the Ford Foundation for twenty years before joining it full-time as head of the New Delhi office.

58. Report in *IC News*, 31 October, 1969.

59. George Salt FRS (1903–2003) was born in England but grew up in Canada. After gaining his BSc from the University of Alberta, he worked on parasitism in bees and wasps for his Harvard PhD. Later he worked on parasitism in the control of insect pests attacking Canadian wheat crops. He will have introduced Hassell to his ideas on insect parasitism. Salt's papers are held at King's College, Cambridge where he was a fellow for over fifty years. His entomology notebooks are beautifully illustrated and his papers include also some fine illuminated books that display his considerable skill in calligraphy. The Salt papers include correspondence with Southwood who helped him with the identification of some insect specimens. I would like to thank the librarians at King's College, Cambridge for their help with the Salt papers.

60. Varley and Gradwell conducted a long-term study of the winter moth (see chapter 7). Like Holling, they were among the first biologists to use computers in the storage and analysis of data and were in the forefront of those using new quantitative techniques in the early 1960s. See, for example, (Varley and Gradwell, 1968). For more on ecology at Oxford up to the 1960s see chapter 5.

61. Conway and Hassell, in addition to Hamilton and to Professors Way, Kennedy and Lees and their teams.

62. A. J. Nicholson (1895–1969) was born in Ireland and educated at the University of Birmingham where he took a degree in zoology. He then moved to the University of Sydney and a lectureship in entomology — a new discipline at the university. He successfully built up entomology at Sydney and carried out work on parasitoids which led him to improve on the earlier models of Lotka and Volterra. Realizing that they were too simplistic, he attempted to capture more of reality in his own. For example, in the case of parasitoids the death of the host is not instantaneous. Unlike the fish prey in Volterra's model, the insect prey is not immediately consumed

by the predator. It is even possible for an infected host to give birth to a new generation. Together with his colleague, V. A. Bailey, professor and head of the department of physics, Nicholson produced models which included factors such as the ones just described. He paid especial attention to prey-predator densities and believed that they were more important in determining population sizes than external, environmental, factors. This was a much debated topic for many years, and is one to which we will return. (Nicholson and Bailey, 1935).

63. For Slobodkin and MacArthur see chapter 5. Hairston and Smith were eminent ecologists working at the University of Michigan. Smith was much involved with the International Biology Program (see note 64 below); In (Hairston, Smith and Slobodkin, 1960) the authors argue that while competition exists among most groups of animals, herbivores do not compete. Rather, their numbers are limited by the abundance of predators, including parasites and other disease organisms. The paper was much cited and argued over. The authors responded with a series of papers clarifying their position. The debate was to influence Crawley and many others.

64. The International Biology Program was set up in the 1960s and lasted approximately ten years. Several governments put large sums of money into it. One of its rationales was to help foster cooperation between scientists in the West and those in the USSR. Ecology was seen as a suitably neutral area of science where cooperation was possible. Large scale ecological systems known as biomes, areas such as prairie grasslands or boreal forests, were studied with a view to collecting much data. The data was put into computers for system theorists to work with.

65. In an interview with Robyn Williams for the Australian Broadcasting Corporation' Science Show, broadcast on 24 December, 2011, May stated that he had an excellent chemistry teacher, Lenny Basser, who taught eight future Fellows of the Royal Society, including one Nobel Laureate (John Cornforth).

66. Schafroth had gained his doctorate under Wolfgang Pauli. Shortly after May graduated, Schafroth intended returning to Switzerland, and to a chair in Geneva, but sadly he was killed in an airplane crash. May had planned on following Schafroth to Geneva but the tragedy intervened and his life moved in a new direction.

67. See chapter 1 for the British Society for Social Responsibility in Science. May was also encouraged by Harry Messel to think of ways of using his mathematical and physical skills in biology.

68. En route to the United Kingdom, May stopped in Nairobi and gave his very first ecology talk at the university there.

69. James Murray later moved to the University of Washington in Seattle. George Oster had a background in engineering but found his way into biology. When May met him he had begun working as a theoretician in the entomology department at the University of California at Berkeley. Their joint work will be briefly mentioned in chapter 6.

70. John Bahcall (1934–2005) was an eminent physicist known for his theoretical work on solar neutrinos.

71. May told me that without the encouragement of his wife, Judith, and without her doing the hard work of planning an intercontinental move, he would never have left Australia for the Class of 1877 zoology professorship at Princeton. And, without similar encouragement later, he would not have left Princeton for England.

72. Personal communication. Expenses associated with May's first summer visit were paid from Conway's Ford Foundation grant; in later years May paid his own way from his NSF grant.

73. The talks from the second workshop were published in (Usher and Williamson, 1964). For attendance by invitation see vii, for quotation, viii.

74. Meetings continued under the auspices of the British Ecological Society and the Biometrics Society. The societies supported a Mathematical Ecology Group 'to cater for biologists with a quantitative interest and biometricians with an ecological interest'. (Sheail, 1987), 222.

75. May in (Usher and Williamson, 1964), 1.

76. Also present was Joel E. Cohen who was visiting Oxford from Harvard. An eminent ecologist and epidemiologist, he was later to hold a chair at The Rockefeller University as head of the Laboratory of Populations.

77. In addition to the four organizers, those from Silwood in attendance that summer were Professors John Kennedy and Michael Way, Nadia Waloff, and Michael Crawley (then still a student). Also present were some other research students and junior staff.

78. Conway, personal communication, and Bodleian, Southwood Papers, B: Silwood Business Files, 1973. R. G. Weigert is a salt marsh ecologist who worked at the University of Georgia. John Calaprice is a Canadian who was working at the University of California at Davis when Conway was a research student. A fish ecologist with genetic interests, he almost persuaded Conway to move into genetic areas of entomology for his PhD. Patrica Rosenfield and Christine Shoemaker were among the very few women who attended the workshops. Rosenfield was then working on schistosomiasis. Her later career was at the Carnegie Corporation in New York. Shoemaker, a specialist in pesticide management, later worked at Cornell University.

79. Gordon Conway, Second Nathan Lecture on the Environment 'Applying Ecology: A Personal Journey', delivered at the Royal Society of Arts, 8 April, 1997. Introduction by T. R. E. Southwood. (The lecture was sponsored by the Denton Hall Environmental Law Group in honour of Lord Nathan and was privately printed; copy in Southwood's papers, Section R, at the Bodleian Library). Conway delivered the lecture when he was Vice Chancellor of the University of Sussex.

80. (May, 1976). There were at least a couple of important joint papers that preceded the book: (Southwood, May, Conway and Hassell, 1974a and 1974b).

81. Conway, personal communication.

82. Hamilton wrote, 'When I had exhausted May's patience with my requests for generalizations of the evolutionary stable strategies (ESSs) concerned with dispersal, or when perhaps he had decided that other fields of ecology or evolution now needed his battleaxe more than mine did, I found myself passing my challenges to Hugh Comins instead'. Both quotations from (Hamilton, vol. 1 1996), chapter 15. One joint paper, stemmed from collaboration at Silwood in the summer of 1975; (Hamilton and May, 1977). Of this work, Hamilton wrote that May found the first solution to the optimal dispersion problem; a general problem of whether an organism would do better staying at home or dispersing. This was in the heyday of the idea of an evolutionary stable strategy, as promulgated by John Maynard Smith. A later paper based on work from the Silwood period is (Comins, Hamilton and May, 1980).

83. The grant was to support research projects over five years. The Centre received also other government monies and was supported by the Ford Foundation and the International Institute for Environment and Development.

84. Susan Conway told me that her son, Simon, was friendly with Robert May's daughter, Naomi, and that they very much enjoyed the Silwood summers. In his novel *A Loyal Spy* (2010), Simon Conway gives a fictionalized account of life at Silwood. His principal character, the son of a biologist working at Silwood, reminisces about the time he spent there as a child. Silwood comes across as a romantic place, but the portrayal is realistic.

85. Physical surroundings are important to intellectual exchange. This point is well illustrated for Cold Spring Harbour in (Watson, 1991). The book also includes some short historical essays by James D. Watson, 'Landmarks in twentieth-century genetics'. An indirect way of illustrating the importance of Woods Hole in the development of modern biology has been the publication of two volumes of Friday evening lectures given there. The first collection is from the 1890s and the second includes some twentieth-century recollections of, and by, scientists and their work at the Marine Biological Laboratory: (Maienschein, 1986; Barlow, Dowling and Weissman, 1993).

86. This must still have been the case in the late 1980s and early 1990s. After winning a series of croquet matches Hassell became the lead author of the following paper: (Hassell, Comins and May, 1991). May and Hassell also note the many lunchtime gatherings around Southwood's pool, 'often followed by an informal afternoon workshop and more plans for future work', (May and Hassell, 2008), 357. Tennis parties are also mentioned in Southwood's correspondence. Another person, outside the circle, who remembers the swimming parties and the social life at Silwood during the early 1970s is Southwood's co-researcher, Patricia Reader. See http://www.imperial.ac.uk/centenary/memories/reader.

87. One student who gained a BSc and PhD at York was Paul Harvey. Harvey visited Silwood during the 1980s when a lecturer in ecology at the University of Sussex. He later moved to Oxford where, in 1998, he became head of the department of zoology. He can be viewed as belonging to a wider Silwood Circle.

88. Hamilton wrote of his difficulties at the 'very private and patrician' Library of the Zoological Society of London, (Hamilton, 1996), 140.

89. P. H. Leslie, working in the 1940s, is remembered for the Leslie Matrix, an age-structured model of population growth. A demographer, his model was later used also by ecologists.

90. Anderson was a pupil at Hertford Grammar School, the school once attended by Alfred Russel Wallace.

91. Sadly both Croll and Mahon died prematurely. June Mahon (1928–1978) was a lecturer in parasitology (helminthology) at Imperial College. She became unwell while Anderson was her student, and Anderson received much help from R. G. Davies and Croll. Neil A. Croll (1941–1981) was a South African who had been a student at Imperial College. A lecturer in parasitology with an interest in nematodes, he fared less well than perhaps he should have done under Southwood's headship. While Croll was promoted to a readership he saw some of the Silwood Circle favourites gaining, or being likely to gain, chairs before he did. (The professors of parasitology were B. G. Peters followed by J. D. Smyth in 1968). Croll left to take up the directorship of the Institute of Parasitology at McGill University. A role model for Anderson, he moved into the area of human ecology and infectious diseases. Indeed, to further his knowledge (and to be taken seriously as an epidemiologist) he studied medicine (taking the crash eighteen-month program at the University of Miami), and gained an MD in 1980. See his posthumous book co-authored with John H. Cross (Croll and Cross, 1983). Croll invited Anderson to McGill as a visiting professor and Anderson spent a couple of summers working in

Montreal with Croll and with Eugene Meerovitch, Croll's successor as director of the institute. Elizabeth Canning was one of Southwood's contemporaries; they had been students together at Imperial College. She is a world expert on microsporidia parasites.

92. I have spoken to several people who were in the zoology and applied entomology department at Imperial in the 1970s. All agree that Davies was both exceedingly clever and extremely helpful. But he is also remembered as retiring and not competitive. Being highly gifted intellectually is rarely enough to reach the very top. Davies wrote an interesting history of entomology at Imperial College (typescript in ICA).

93. May told me that just five people went on the first hike in the Lake District; May, Hassell, Lawton, Beddington, and a biologist colleague from York, John Currey. Currey knew the area well.

94. Dorothy (Dot) Lawton was an occasional, and Judith May a fairly regular, participant.

95. Over time, younger people such as Jeff Waage and Charles Godfray were invited to join the walks. There were insiders and outsiders. For example, Michael Way, the professor of applied ecology, and a highly accomplished mountaineer, was not invited. In the summer of 2008 the group went to the Julian Alps in Slovenia and, in 2009, to Grindelwald.

96. Perhaps Southwood did not join the hikes because his health was poor. He suffered a minor heart attack in 1967. But he also liked to stand a little above, and apart from, the others in the Silwood Circle. All recognized him as the leader.

97. Brian Flowers was determined to unite zoology and botany and create a department of biology. There was much heated discussion over this and over who should become head of the new department. In the event a Department of Pure and Applied Biology was created in 1981 with botanist R. K. S. Wood as its head. (Wood had recommended that John Harper be invited from Bangor to head the new department but he was not interested.) Wood was succeeded by Roy Anderson in 1984 and, in 1989, the department was renamed the Department of Biology.

98. All were deserving but so, too, were others; for example, Hamilton and Croll as noted above.

99. (May and Hassell, 2008), 363.

100. (Southwood, 1968).

chapter five

Some Important Antecedents to the Silwood Circle

Ecology at Oxford and at Some North American Centres

5.1 Ecology at Oxford University: from the 1920s to the 1960s

During their university education members of the Silwood Circle will have been introduced to the work of two major ecologists who worked at Oxford in the mid-twentieth century, namely Charles Sutherland Elton (1900–1991) and David Lambert Lack (1910–1973). Both Elton and Lack complained about their own undergraduate experiences as zoology students at Oxford and Cambridge respectively. They were not alone in being bored by lectures in comparative anatomy, physiology, descriptive morphology and embryology. The dominance of these fields in the academic curriculum owed much to the centrality of systematics and to questions concerning the path of evolution. Ecology was widely seen as the domain of amateur naturalists, and not something worthy of serious academic study. But the pedagogical norms were already rather stale and were gradually to change.

Like many gifted people Elton was not cut out for examinations. The story of his surprise at receiving a first-class degree, despite his desultory examination record, is amusingly told by Southwood and Clarke in their biographical memoir.[1] Elton was fortunate in being tutored and mentored by Julian Huxley who, rare among his Oxford peers, saw the importance of field work and encouraged Elton in his interests. Of especial importance was that Huxley invited Elton to be his assistant on a university expedition to the island of Spitsbergen in the Svalbard archipelago in 1921.[2] The main thrust of the expedition was ornithological, but Elton was allowed to go his

own way and study arctic mammals. His excellent field work under difficult conditions, and his associated report, won him his First. He owed
much to Huxley who, recognizing his talent, gave him this opportunity.
Making the most of it allowed Elton to stay on at Oxford where he built
his career.[3]

Elton returned to Spitsbergen two more times in 1923 and 1924 and
completed his survey of mammals in the area.[4] In 1930 he studied
another arctic region in Norwegian Lapland. Later he was to visit some
northern regions of Canada. Two books that he had read earlier helped to
focus his thinking. Robert Collett's *Norges Pattedyr* (1911), described the
periodic lemming migrations, and the other, C. Gordon Hewitt's,
Conservation of the wild life of Canada (1921), included discussion of
cycles in the populations of arctic foxes, hares, lynxes and lemmings.[5] The
leader of the Spitsbergen expeditions, George Burney, was soon to join
the Hudson's Bay Company where he appointed Elton a biological consultant. As it happened, the year 1930 was a low point for the populations
of a number of marine and terrestrial animals in the Gulf of St. Lawrence.
The shortage of salmon, cod and lobster, as well as of many fur-bearing
land animals, was causing considerable hardship among the local inhabitants, including the trappers on whom the Hudson's Bay Company
depended. In order to learn more about what was happening, the
Company sponsored the Matamek Conference on Biological Cycles. It
was held in 1931 on a large estate on the Matamek River on the north
side of the Gulf. The estate, site of a large fish-processing plant, was
owned by a wealthy American, Copley Amory. Amory, the principal
organizer and financier of the conference, appointed Elton as conference
secretary.[6] An interesting mix of people attended. Aldo Leopold was there,
representing an arm of the sport-hunting industry. Also present were
other scientists, representatives of the aboriginal community, the fishing
industry, and several trappers. The cyclical low point was a situation
where local knowledge, especially that of fishers, trappers and indigenous
people, was seen as important.[7]

Elton's early experiences laid the foundation for his later work in which
food chains, cycles and fluctuations in animal populations were to be major
themes.[8] In 1921, after returning from the first Spitsbergen expedition, he
was appointed to a junior demonstratorship. As a research student under

Huxley he carried out field studies together with J. R. Baker and E. B. Ford. Each later went their separate ways while remaining colleagues in the zoology department at Oxford.[9] Together they studied small mammals such as voles and field mice living in Bagley Wood, recognizing that these underwent population fluctuations similar to those long known for arctic animals. By 1925 Huxley had moved to a chair at King's College London and was the editor of a series of texts on animal biology intended for undergraduates.[10] He asked Elton to write a book on animal ecology for the series. The result, *Animal Ecology* (1927), was published when Elton was twenty-six years old. It is a wonderfully detailed book in which Elton discussed his arctic and Bagley Wood studies, as well as work carried out elsewhere. For example, the book owed much to the ideas of Victor Shelford of the Illinois Laboratory of Natural History. Shelford had discussed animal and plant communities but saw them as more natural entities than did Elton. Thinking primarily of the northern animals he had observed, Elton saw communities as forming largely as a result of stochastic processes. No enthusiast of the 'balance of nature' idea, he noted that northern animals were very mobile, able to survive in a wide range of climatic zones, and were able to live alongside different plants and animals in their various habitats. This insight into community variance figured in much of his later work. However, Elton also believed that there was probably some underlying principle, as yet unknown, to the number of species forming an animal community. Echoing Linnaeus, he claimed that his ecology text was about 'the sociology and economics of animals'. He emphasized the idea of community and, in particular, the acquisition of food; 'food is the burning question in animal society'.[11] He used the term 'trophic' in his discussion of food chains. Later it was to have a wider meaning encompassing also the more general exchange of energy in ecosystems.

Elton's early work was carried out in roughly the same period as Lotka and Volterra were developing their mathematical models. Like Lotka, Elton believed that populations were controlled by disease, predation, parasites, migration, changes in food sources and supply, and by general instability in the environment. However Elton, the more grounded naturalist, was better able to communicate his ideas to others. Using only minimal mathematics he introduced some concepts that were to be of great use to other ecologists: food chains (later food webs), food cycles (later trophic structure of community), food size, the pyramid of numbers, and the niche. The pyramid of numbers was not entirely new. The idea that there are more small than large animals, and that they reproduce more quickly than larger animals for whom many of them form prey, was widely understood. Elton saw the logical conclusion as one in which large predators, few in number, sit at the top of pyramids consisting of layers of predator/prey animals decreasing in size but increasing in number (and collective mass) as one moves toward the base.[12] The niche was not a new term either, but Elton defined it in a new way in terms of an animal's place in the biotic environment, with an especial focus on food and natural enemies.[13] As others have been, I was struck by one of the analogies he used to get his idea across. Indeed, Elton appears to have been a master of the apt analogy, perhaps something he owed to his father, a professor of English literature. In the case of the niche he stated that when an ecologist says 'there goes a badger' he should think along the lines 'there goes the vicar'.[14] The niche included the role that an animal plays within its larger ecological community. For badger and vicar it was not just the sett in the wood or the vicarage in the village that defined their niches. Elton's ideas were well expressed in some lectures given at the University of London and later published as *Animal Ecology and Evolution* (1930). It appears that he recognized some problems with his conceptualization, including a tension between his belief in the importance of chance in the formation of community, and the idea of adaptive design, a tension that has engaged theorists to this day. There was a problem also with his vicar-badger analogy since he was comparing a type of individual and his role within a population with the role of a species within a community. While recognizing that the vicar has a role to play within human society, Elton did not spell out the roles played by different badgers in badger society. Badgers were just badgers. Ignoring variation in the animal world, *contra* Darwin, and viewing members of animal species as identical occupants of their niche is one way of thinking about ecological systems, but it can be misleading.[15]

By the time Elton attended the Matamek conference he had accomplished much. He impressed other attendees and was encouraged in his ambition to set up his own ecological research unit at Oxford. With a small start-up grant from the New York Zoological Society and the support of the head of the zoology department, the Linacre professor, E. S. Goodrich, Elton founded the Bureau of Animal Population in 1932. Later he had the support of Goodrich's successor, Alister Hardy. Another Oxford scientist who was very supportive was Howard Florey.[16] Florey had been the team doctor on the 1924 Spitsbergen expedition and had helped Elton with his field work. He and Elton remained good friends. For the Bureau's work to be seen as scientific and professional, however, entailed publication. Elton was one of a group within the British Ecological Society pushing for a new journal. *The Journal of Animal Ecology* was founded in 1932, the same year as the Bureau, and Elton was appointed its first editor. Other small grants followed the one from America, but the Bureau of Animal Population was never financially secure.

Elton appears to have planned the Bureau's research with pragmatic intelligence. First, the name of the unit is significant; tracking animal populations, census taking, was something widely understood as being useful. Second, knowing that he would be unable to attract many students or academic scientists to an area still seen as the domain of amateurs, Elton enlisted several good amateurs willing to track animal numbers. Third, he chose to focus on animals that had some economic significance such as game birds, red and grey squirrels and other rodents. This focus brought in funding from the Agricultural Research Council (ARC). Along with some other small grants, it enabled Elton to appoint a few members of staff including George Leslie. Leslie, a biomathematician, developed a matrix for dealing with various population parameters, and thought of ways of organizing the data delivered by amateur and student naturalists.[17] During World War II the Bureau was able to expand because of its expertise in rodents which were a major threat to scarce food supplies. Elton was not a self-promoter, but his popular talks on the BBC brought further attention to the Bureau.

The expanded Bureau survived the war and its associates were then able to work at a new site, Wytham Woods, on land given to the university.[18] Elton began what was to become a major ecological survey of the area. The long-term studies he initiated were carried out by a mix of amateur and academic ecologists, including students who were becoming increasingly interested in his work.[19] In 1958 he published *The Ecology of Invasions by Animals and Plants* which prompted work on biological invasion, and contains a chapter on the kind of debate over stability and diversity within ecological communities that was taken up later by Robert May and others in the Silwood Circle. Just before retiring Elton published a further book, *The Patterns of Animal Communities* (1966), but it failed to capture the imagination of young scientists in the way that his earlier *Animal Ecology* had done. Interestingly the book includes discussion of a study linking Wytham with Silwood, and work (at Silwood by Richards and Waloff) on the similarities of grasshopper species found in both locations. But Elton's principal legacy lies in the work stimulated worldwide by his first book, along with the studies he initiated at Wytham Woods. As Thomas Park, an ecologist associated with the Chicago School of Ecology in the mid-twentieth century, wrote:

> Very occasionally a unit emerges within a larger academic structure and, through time, endows a field of inquiry with new orientations, new meaning and an expanded understanding of its own conceptual significance. ... Usually, also, there is one key figure...

Park rightly saw the Bureau, and its key figure, Charles Elton, as having significantly shaped the history of ecology in the mid-twentieth century.[20] Elton had an enormous influence on ecologists of his and Park's generation. His theoretical ideas, especially those having to do with the regulation of animal numbers, were highly suggestive and provided a framework for much field and experimental work of the 1930s, 40s and 50s.

However, by the time Elton retired in 1967 his approach was becoming old-fashioned and his still wavering over the role of natural selection in evolution suspect.[21] At about the same time the incoming Linacre professor, J. W. S. Pringle, successfully lobbied for the construction of a large new building for the department of zoology. He wanted to bring various outliers such as Elton's Bureau under a single roof together with more mainstream zoologists. The rationale was that zoologists in different fields needed to exchange ideas, and that having isolated units such as the Bureau of Animal Population made little sense. Pringle claimed, probably correctly, that for ecologists specializing largely in mammals, cross fertilization with other Oxford ecologists, as well as with other zoologists would be a good thing. Elton resisted the move, but to no avail. The Bureau was closed in 1967 and a new Animal Ecology Research Group was founded within the zoology department under the direction of the entomologist John Phillipson, Elton's successor as reader (later professor) in animal ecology. The new research group included some other entomologists, some mammal and other animal specialists, but many of the ecologically oriented ornithologists remained within a separate group.

With a few exceptions (see below) the ornithologists worked at the Edward Grey Institute of Field Ornithology under the direction of David Lack. Lack and Elton did not get along and communication between people at the Bureau and those at the Institute was minimal despite their having occupied neighbouring accommodation before the move.[22] Since Elton continued working in the newly unified department after his retirement, the tension between ecologists working in the new animal ecology research group and the ornithologists at the Edward Grey Institute, also corralled within the new building, persisted.[23] According to John Lawton, then a demonstrator under Phillipson, entomologists in the new animal ecology research group found themselves in an awkward situation. Not only were they resented by mammalian specialists from the Bureau, unhappy with their new association, they were also distanced from the mainstream entomologists working under George Varley in the Hope department which was then located at the Museum of Natural History.[24] It seems that it took some time for Pringle's vision of interdisciplinarity to take hold — if it ever fully did. In Lawton's case he got on well with Lack's group, helped by his great knowledge of birds. He also met Michael Hassell from the Hope department and John Krebs from Niko Tinbergen's unit and so was able, as a young person, to bridge some of the tribal divides. That these three found common ground was important also to future developments within the Silwood Circle.

Lack was a child ornithologist, and very focussed. So focussed that when he read Elton's *Animal Ecology* in his teens it made little impression on him.[25] He learned more about ecology after becoming a student at Cambridge at the age of nineteen. There he was exposed to population biology, though the new theoretical ideas of R. A. Fisher and J. B. S. Haldane were only beginning to take form at that time and were not yet on the curriculum.[26] After taking his final examinations Lack, like Elton, was given an opportunity to see something of the Arctic. He joined a sailing vessel on an exploration of Scoresbysund on the east coast of Greenland. On his return he took a job as a biology teacher at the progressive school, Dartington Hall in Devon. He continued to pursue his ornithological interests and included the schoolchildren in some of his activities. He also kept in contact with other ornithologists. One who took an interest in him was Elton's former mentor, Julian Huxley. Huxley encouraged Lack and helped him to travel to Tanganyika to study birds there. He also helped in finding some grants to support a small expedition to the Galapagos islands in 1938. Lack returned from the Galapagos via the United States where he met Ernst Mayr. Mayr's determined Darwinism had a major influence on Lack.[27] It can be seen in his book *Darwin's Finches* (1947), a work that brought Lack to the attention of other zoologists. In it he showed that the various species of finch found on the islands are isolated in the niche sense as understood by Elton. Lack thought seriously about the niche also in relation to Gause's earlier competitive exclusion principle and saw that in occupying different niches there was no direct competition between the different finches.[28]

During World War II Lack left Dartington Hall and was recruited, as a civilian, into operations research and radar work. For a short time he worked in the Orkneys which allowed for some new ornithological opportunities. The group of scientists with which he worked included the entomologist and future Hope Professor, George Varley. Together they interpreted some mysterious radar echoes as coming from migrating birds. Meanwhile, at Oxford there were some new developments in the field of ornithology. Already in the late 1920s the Oxford Ornithological Society had found people willing to conduct a bird census for the region. The census was part of a national effort sponsored by the British Trust for Ornithology (BTO). Among the enthusiasts was the former politician Lord Grey of Fallodon, chancellor of the university. He died shortly before the war and money was raised to found an institute in his memory, and to have the census work continue there. The result was the Edward Grey Institute.[29] Early work at the Institute was carried out under Wilfrid B. Alexander who also built up a good ornithology library. Alexander retired at the end of the war, coincident with the appointment of Alister Hardy as Linacre Professor and head of the zoology department. On Hardy's advice, the university decided to expand the operations of the Institute and to rename it, more descriptively, the Edward Grey Institute of

Field Ornithology. David Lack was appointed Director in September 1945. The BTO continued in its support of the new institute. As at the Bureau of Animal Population, work at the Edward Grey Institute was carried out by a mix of academic and amateur scientists.

While the Bureau was rather more community focussed, the new Institute focussed almost exclusively on population problems. As Lack put it, 'the object of this Institute is to find out why birds are as numerous as they are'.[30] The idea was to study bird populations over the long term and work began on this in the recently acquired Wytham Woods. At first Lack wanted to study robins, the subject of his book *The Life of the Robin* (1943), based on observations made while at Dartington Hall. But robin nests proved too elusive so he switched to great tits which obligingly took to nesting boxes.[31] The nesting box was a rare artifact for Lack who tended to shun experiment and artifact for direct field observation. Thinking always about numbers, Lack looked at reproduction rates but also, as did Varley in his insect studies, at population control. Like others, he came to the Malthusian conclusion that the regulation of bird numbers depended mainly on checks to the population from disease, parasites, predation and food shortages; and that the severity of these factors increased with population density. The idea of the density dependence of population seems highly plausible, but how it works in practice is still unclear. It was a problem that interested Lack whose ideas were well expressed in his *The Natural Regulation of Animal Numbers* (1954).[32] Interestingly 1954 was also the year in which Andrewartha and Birch published their influential *Distribution and Abundance of Animals*. They took a slightly different line in assuming that population densities were rarely high enough for density dependent control to take effect.[33] It would also appear that Andrewartha and Birch took a more empirically inductive approach in their work and Lack a more hypothetico-deductive one. The practical consequence of this was that Lack looked at rather simple ecological situations to see whether or not they backed up his ideas. He was criticized for not dealing with nature in all its complexity, but his was the way of the future and ensured him a following among young ecologists. One of Lack's major hypotheses was that the clutch sizes of birds were optimal (in the sense of maximising the number of surviving offspring) for any given set of conditions and that this tendency was the result of natural selection. His 1954 book was a major synthesis. It included also ideas from other ornithologists,

from Vito Volterra, from his entomologist contemporary at Oxford, George
Varley, as well as from earlier entomologists such as A. J. Nicholson, a proponent of density dependence arguments, and W. R. Thompson an
opponent.

The book was not without its critics and competing ideas soon appeared;
for example, those expressed by V. C. Wynne-Edwards in his book, *Animal
Dispersion in Relation to Social Behaviour* (1962).[34] Wynne-Edwards had been
a student at Oxford where one of his teachers was the young Charles Elton.
At the time, Elton was much taken by A. M. Carr-Saunders' book, *The
Population Problem* (1922) and he lent Wynne-Edwards his copy. Carr-
Saunders wanted to show that Malthus was wrong and that, historically,
human populations consciously adjusted their reproduction rates in light of
circumstance, and that they did so cooperatively. The book filled a need for
those who were unable to accept that for humans, too, nature was 'red in
tooth and claw'. Both the ideas of Charles Darwin and those of Pyotr
Kropotkin, whose *Mutual Aid* was republished in 1914, were powerfully
present in the period. Like many others, Carr-Saunders drew from both.[35]
Wynne-Edwards appears to have taken two ideas from Carr-Saunders. First,
that individuals within a community competed for status and thus only indirectly for food and other resources. Hierarchy, Wynne-Edwards claimed, was
the principal way of limiting population and of preserving the health of the
group. Second, that for natural selection, the unit being selected was the
group rather than the individual. This was a belief that fitted well his conviction, following Kropotkin, that cooperation was as natural as competition.[36]
While debate over this still continues, Wynne-Edwards' ideas were shortly to
be undermined by, among others, William Hamilton working quietly in
London and then at Silwood. Lack responded to Wynne-Edwards's book
with one of his own, *Population Studies of Birds* (1966), which included an
appendix on the then existing theoretical controversies concerning animal
populations. While largely directed at Wynne-Edwards, the appendix is a
good summary of the theoretical situation facing the young scientists who
were to form the Silwood Circle.

Another important zoological scientist working at Oxford during
the post-war period was the Nobel Laureate, Nikolaas (Niko) Tinbergen
(1907–1988).[37] Lack met Tinbergen in Holland at the end of the war and
was delighted when Hardy invited him to Oxford a few years later. While

animal behaviour had long been a topic of interest among naturalists, Tinbergen was a pioneer ethologist and thus a forerunner also of behavioural ecology. Like others in my study, Tinbergen was a keen child naturalist. He claimed that as a young man he was more keen on outdoor activities, as naturalist and sportsman, than on academic studies. But, like Elton, with some good breaks he managed to get through university. Also like Elton he spent an extended period in the arctic, two summers and a winter in Angmagssalik, a small town on the east coast of Greenland. At Leiden he pursued his interests in both birds and insects and was encouraged in this by Konrad Lorenz who came to the university as a visiting scientist. Tinbergen also spent time with Lorenz at his home near Vienna. After a difficult war for both they met again in Cambridge once the war was over.[38] They were invited there to a conference organized by W. H. Thorpe, Lack's early friend and mentor, and later professor of ethology at Cambridge.[39] Before moving to Oxford Tinbergen had published important work on both herring gulls and on the homing ability of digger wasps.[40] At Oxford he built a small research group in animal behaviour, helped by funds from the Nuffield Foundation and the Nature Conservancy.[41] Unlike Lack, Tinbergen was an experimentalist as well as a field naturalist and his work covered a wide range of topics. After World War II interdisciplinary projects were specially favoured by granting agencies; departmental heads needed to take notice.[42] Hardy requested that Tinbergen join together with some neurophysiologists to found the Oxford School of Human Sciences. He agreed and began to apply his ethological skills in a study of early childhood autism.

While Tinbergen and his wife Lies are known for their autism studies, more important from the perspective of ecology was his work on instinct and on natural stimuli in animal behaviour. Like Lorenz, Tinbergen carried out experiments using enhanced, artificial, stimuli to provoke animal behaviours especially as they related to communication in mating rituals, and in the signalling of danger.[43] John Krebs has written about Tinbergen's work and on the ways in which it has been superceded.[44] Tinbergen just missed the revolution in Darwinian theory begun by Hamilton and others. Today's behavioural ecologists look at topics such as courtship and the signalling of danger differently than Tinbergen did. Impressed also with ideas from information theory, they place much emphasis on assessment. Courtship displays, for example, are no longer seen simply as mechanisms for attracting mates but as genetic signals of the health, mating, and future parenting ability of the animal performer. According to more recent work on animal communication there is said to be co-evolution in the behaviour of both actor and reactor (assessor). One wonders whether there is any historical significance to the fact that obsession with assessment and survival in the academic world was very much on the rise when these new ideas emerged. I won't pursue this idea, but another point of cultural context is that universities, with so many young people present, are places where sex, as well as competition and assessment, looms large. New sexual freedoms of the 1960s and 70s could be relevant. Throwing Cold War obsessions and game theoretic ideas into the mix, one could argue that the academic world of the 1970s was the ideal habitat for behavioural ecologists to develop the kind of theory that they did.

5.2 Ecology in North America: G. E. Hutchinson and his students

Oxford and Silwood were not the only places where ecological work was carried out during this period. There were a number of other important centres including several in the United States and Canada. Like many young British scientists of their generation those in the Silwood Circle spent short periods of time working in North America and made many useful contacts while there. Here just a few ecologists will be discussed, those who were important intellectual forerunners of the Silwood Circle. Principal among them was George Evelyn Hutchinson (1903–1991), Sterling Professor of Zoology at Yale University. Together with his students he had a profound influence on the direction ecology was to take.[45] Hutchinson grew up in an academic family in Cambridge (England) where he took a degree in natural sciences. He wrote about his early life in a memoir and, like Lack ten years later, was critical of the teaching in the zoology department.[46] Hutchinson was a prodigious child naturalist. It would appear that, as with musical performance, ecology is a discipline where there is considerable advantage in developing skills at an early age. In Hutchinson's case his childhood interest in aquatic insects presaged his future as a limnologist. After gaining his Cambridge degree he spent a brief period at the Stazione Zoologica in Naples, and then as a lecturer at the University of the Witwatersrand, before moving to Yale University where he remained for the rest of his career.[47] Despite his complaints about

Cambridge zoologists, Hutchinson received a well-rounded scientific education. He also read widely. As he noted in his memoir, he was a 'hunter gatherer of information' and some who knew him have testified to his extraordinary breadth of knowledge.[48] In his memoir he mentions having read Elton's *Animal Ecology* while travelling to New Haven from South Africa.

Hutchinson was the author of a four-volume work on limnology.[49] This major synthesis included material on his own research and that of scientists worldwide. Much of his own work was carried out on dry lakes in South Africa, the Himalayas (he was chief biologist for Yale University's North India Expedition in 1932), and Nevada. He also worked with his students on lakes in Connecticut, especially Linsley Pond which is close to the university. Hutchinson was open to ideas from geology and chemistry and will have learned something also from his mineralogist father. His early work on the biogeochemistry of Linsley Pond was stimulated in part by the ideas of the Russian geochemist V. I. Vernadsky.[50] Vernadsky's son George, an eminent historian of Russia and a colleague at Yale, helped Hutchinson in translating some of his father's writings. The older Vernadsky was something of a visionary who believed that living things not only affected the history of the planet in a chemical and geological sense but that there was some purpose to this global regulation. A proponent of the biosphere concept, in some respects he anticipated James Lovelock and the Gaia hypothesis.

Linsley Pond was in a eutrophic phase.[51] Hutchinson's early student Edward Deevey, together with research assistant, Anne Wollack, studied core samples from the lake bed, allowing for some understanding of its history over the past few thousand years.[52] The samples showed increasing productivity in the lake from its glacial origin to its more recent trophic balance. As with the growth of animal populations, lake productivity followed a sigmoid curve toward equilibrium. This early work attracted Raymond Lindeman, a graduate of the University of Minnesota, who went to Yale in the early 1940s on a postdoctoral fellowship. Lindeman had just completed a five year doctoral study on the ecology of the Cedar Creek bog in central Minnesota, identifying the many animals living in the bog and thinking about possible food chains. At around the time he went to Yale, cybernetics was the big new idea so it is perhaps not surprising that Lindeman thought about the many feedback loops, chemical, physical and biological, that kept places such as the Minnesota bog and Linsley Pond in eutrophic equilibrium.[53] In his paper

'The trophic-dynamic aspect of ecology', based on a chapter from his doctoral thesis, Lindeman used the term 'ecosystem'. He had counted and measured many of the life forms found in the bog and must have had Elton's idea
of the pyramid of numbers in mind when thinking about how to organize
his findings. His paper is in part a reformulation of Elton's idea. It gave the
term 'trophic' new meaning, and was suggestive of further work that others
would take up.[54] Lindeman was interested in the capture of energy from the
sun and its flow through the plant and animal life of an ecosystem until
released back into the environment — possibly for further biological use. He
considered both the living and non-living parts of the ecosystem, studied the
efficiency of production at different levels of the food chain, and related
these to patterns of succession within the bog community. Lindeman's study
lent itself to abstract mathematical analysis and he attempted a spatial analysis of matter and energy in continual rearrangement. In this his ideas were
not unlike those expressed earlier by Lotka.

Hutchinson, thinking along similar lines, was influenced by Lindeman.
Related to this line of work was another of his many interests, namely radioec-
ology. In a study initiated just after World War II, Hutchinson introduced
radioactive phosphorus into Linsley Pond. The idea was to trace the element's
route through the biogeochemical system.[55] He first discussed this work in a
1946 paper in which he encouraged his readers to think along ecosystem
lines.[56] His student H. T. (Tom) Odum and Odum's brother, Eugene Odum,
were to bring ecosystem ideas to the forefront of ecology.[57] In one sense the
Odums' approach was reductionist and traceable through Lindeman back to
Lotka's desire to have chemical thermodynamics at the root of all biological
explanation. But Eugene Odum, a self-identified holist, would not have
accepted the label 'reductionist'.[58] He applied the new and fashionable cybernetics, and associated system theory, to ecosystems as wholes.[59]

The term 'physics envy' has sometimes been applied to biologists of the
1950s and 60s implying their wanting to acquire the same high status that
physicists then enjoyed not only within the scientific community but in the
wider world.[60] According to some biologists the way to forward their discipline
was to make it more quantitative and to find some useful mathematical models with predictive power. I do not think the expression 'physics envy' applies
to Hutchinson who was a devoted naturalist. But he saw value in physico-
chemical ideas and brought people with ability in the physical sciences and

mathematics into his research group. Given the socio-cultural climate of the 1950s it would be just as easy to make a case for physics guilt as for physics envy. While many people moved from the physical sciences and mathematics toward biology, was it because they saw some novel opportunities and wanted to help mathematize the discipline, or was it because they wanted to work as far as possible from disciplines that had produced the atomic bomb? These are difficult questions to answer. Another difficult question is whether it is sensible to seek analogies from the physical sciences and apply them in biology. Up to a point it appears to be a useful way of proceeding and this book follows the careers of people who believed that it was. However, it is also the case that naturalists and some others, for example indigenous peoples, have great stores of local observational knowledge. They can have an understanding of ecological communities of equal, if not greater, value in the making of predictions — and thus decisions on environmental and resource matters — as those in possession of mathematical models. This is not to belittle theory but to recognize that theoretical ecology is still in its infancy.

Hutchinson was a polymathic naturalist attracted to the physical sciences. He was interested in the biosocial not only in system terms but also along a line of thought that came more or less directly from Darwin. While primarily interested in the ecological community, he understood also the need to think of populations as composed of competing individuals. Covering a number of aspects, he had students working on the levels of population and community, as well as in the energeticist and biogeochemical areas mentioned above. He was also aware that as mathematicians entered biology during the late 1950s and early 1960s, the mood was not only for more quantitative approaches, but also for the integration of different subdisciplines. For ecologists this meant fusing their discipline with genetics and evolution theory. One could imagine a different result, but what in fact ensued, at least in the shorter term, was the dominance of population and community-based approaches in ecology. This, as we shall see, was in part because of the persuasiveness of some of Hutchinson's later graduate students.

Complex disciplines such as ecology occasionally need someone to clarify exactly what is going on, to state clearly what are the problems to be solved, and to come up with new conceptual approaches. It would appear that Hutchinson played that role in the late 1950s. His own attempt at mathematization was with the niche, a specialized topic within the fields of community and population studies. Elton was an important forerunner but Hutchinson saw himself as clearing the way forward by giving an old concept clarity.[61] He metaphorically described the niche, as he formulated it in 1957, as a 'vacuum cleaner' that would remove much detritus and allow ecologists to look afresh at old questions. Two major addresses that he gave to symposia in the years 1957 and 1959 did indeed set people running. Of course there were many important problems that he did not address, but in clearing the ground Hutchinson was to influence a wide range of research. His two addresses drew also on some ideas from one of his mathematically trained students, Robert MacArthur. MacArthur, to whom we will return below, gained his zoology PhD at Yale in 1957 and then spent a postdoctoral year with David Lack at Oxford. He returned not only with some new ornithological ideas but also having had his own methodological inclinations bolstered by Lack's hypothetico-deductive approach.

Hutchinson's vision of the niche was introduced to a large audience in his concluding address to a conference on population biology held in 1957.[62] The audience was, as he put it, a 'heterogeneous unstable population', consisting as it did of demographers, animal population theorists, field biologists of various stripes, and laboratory scientists attempting experiments under strict environmental controls. Thinking abstractly, Hutchinson saw the niche as including all environmental factors that impacted an animal — from food sources to places of shelter; from temperature to seasonal change. All variables, physical, chemical and biological were to be considered. From the set-theoretic point of view he adopted, placing an animal in its niche meant placing it somewhere in an n-dimensional hyperspace. Each niche was seen as occupying a region of that space, namely a hypervolume, the dimensions of which represented the animal's various needs. A community was seen as a collection of niches occupied by different species.

The newly mathematicized niche was highly suggestive. It could not answer all the interesting questions then being discussed, but it did help to reframe many of them. For Hutchinson it meant a new emphasis in his own way of thinking — one moving away from competitive exclusion and toward coexistence. To what extent was coexistence possible? Niches may well overlap — a mathematical possibility for sure. But to what extent was overlap biologically possible? And what about more or less optimal habitats? Surely there are better and worse niches. Further, how does organization within the hyperspace change over time? And, given an available habitat to what extent was it, and the resources within it, used? Do given habitats support the maximum possible population sizes, or the maximum possible number of species? How important was predation, as compared to competition, in community maintenance? These were old questions but Hutchinson's model was a guide to thinking them anew. It was also attractive to field naturalists seeking ways to organize their data. Good models should have heuristic value, something just as important in furthering scientific activity as true descriptive power. Of course descriptive and predictive power are the ultimate goals for theoretical constructs in science, but fully achieving such would put an end to one's line of inquiry. Judging from the flurry of work following Hutchinson's 'concluding remarks' it appears that he gave ecology a considerable heuristic boost.

Hutchinson's second address in 1959 added yet more for people to think about.[63] One thing to note about these and several other of his addresses, including the one in 1946 (see note 56) where he promoted his earlier system ideas, is that Hutchinson appears to have delivered his more important papers as plenary lectures in front of large audiences at major conferences. It effectively drew attention to his ideas which could later be read in published form. His 'Homage to Santa Rosalia' was prompted by some observations made of life in an artificial pond, part of a shrine in Palermo devoted to S. Rosalia, Sicily's patron saint. Hutchinson saw a large number of *Corixidae* (water boatmen) in the water, but belonging to only two species. Why were only two species present? This led Hutchinson to a more general discussion of why there are so many animals in the world, why more are found on land than in water, why about three-quarters of known species are insects, and why there are more small animals than large. Thinking first along food chain lines he dismissed the food chain, as others had done earlier, as an explanation for biodiversity. Lotka had already shown that more than five links from the bottom to the top of a food chain was highly unlikely on energeticist grounds and that this low number of links was insufficient to account for the number of different animals observed. But perhaps the explanation for animal diversity could be pushed back and become a problem of plant diversity. Because there are so many plants, Hutchinson argued, there are many possible niches for small herbivores and therefore for many millions of possible species, especially insects, to prosper. Thus many millions of food chains could be considered — also food webs since predators could have many different food sources.

Robert MacArthur had earlier produced a mathematical model which implied that communities with more complex food webs were more stable than those with less complex webs.[64] Communities should therefore be strengthened by newcomers. Hutchinson agreed with his former student on this point. As was later shown, however, MacArthur's model was faulty. For a start the concept 'stability' is itself semantically far from stable. Does it mean that a community remains static with regard to the number or type of species it carries? Does it mean the ability to withstand perturbation, or does it mean that the community remains healthily stable in some other way? MacArthur appears to have drawn his ideas on stability from Eugene Odum; namely that stability is a function of the number of ways that energy can flow through a system, through food webs and so on. If one species drops out of the web, in complex systems there will always be other routes for the energy to take and for stability to be maintained within the system. Clearly there are greater numbers of species in tropical ecological communities than in northern or southern ones. But whether tropical communities are fundamentally more or less stable than others was something still to be determined. Complexity, ecosystem area, and the environment all had some role to play in stability, but what? These were problems to which ecologists were to turn in the 1970s.

More than in his earlier address, Hutchinson's 'Homage to S. Rosalia' invoked the theory of evolution in thinking about the colonization of different habitats. He brought in ideas similar to those Lack had expressed earlier in his book on the Galapagos finches, ideas relating to the evolutionary advantages of the partitioning of food niches and the adaptive changes that result. How finely the hypervolume could be partitioned became an interesting question. How many species are possible? As to the Corixidae in the pond in Palermo, Hutchinson speculated that even though they were different in size, it was not simply their different food requirements that allowed their coexistence. He speculated that the fact that they reproduced at different times of the year was also significant. The problem of coexistence was discussed further in a later paper, 'The paradox of the plankton'.[65] In this Hutchinson raised the question of why so many plankton were able to coexist in environments with limited resources — with little light and few nutrients. The idea of competitive exclusion would lead one to assume that only very few species should survive. Hutchinson drew attention to two possible ways of explaining the paradox. First, that the environment could be very fine grained (the hyperspace finely partitioned) and that light gradients, small differences in water turbulence, and so on, could be important in the creation of niches. Second, it was possible that such ecosystems were not in equilibrium.

This kind of thinking aloud was very stimulating. It suggested all kinds of research possibilities but the overall direction of research that followed from these talks was toward trying to understand ecological diversity and stability.[66] This fitted well also the environmentalist agenda mentioned in chapter four. Ecologists could feel that their science was not simply 'blue sky', but that it possessed utility. It would help in thinking how to protect ecosystems, manage resources, better conserve natural habitats, preserve biodiversity, and so on. These became the concerns of many, including those in the Silwood Circle.

Looking back one can see Hutchinson's students as having moved in two general directions. Those following in the line of Hutchinson's earlier interests went in the same direction as H. T. Odum, adopting the energeticist and so-called holistic systems approach. Later work carried out by the Odums and others was heavily focussed on large environmental regions or biomes, and on the interactions between the living and non-living components in such environments. On the other hand, those inspired by Hutchinson's niche ideas of the later 1950s focussed more on the population problems of diversity and species packing in different ecosystems. The two areas will perhaps need to come together if Hutchinson's question of why are there so many species is to receive a satisfactory answer. However, since this book is focussed on the Silwood Circle, and because its members were largely interested in problems of the second sort, I will follow the Hutchinson trail only in that direction.[67] Here I will limit discussion to the ideas of just two of his students, Lawrence Slobodkin and Robert MacArthur.

Slobodkin entered graduate school at Yale in 1947 at the young age of nineteen. He claimed that when he first met Hutchinson he didn't even know that the field of ecology existed.[68] He was far from alone. However, by the time he gained a PhD in 1951 he had become a convert to ecological work. After a few years on the faculty at the University of Michigan he set up a fine new department of ecology and evolution at the University of New York at Stony Brook. It was the first of its kind in the United States. Aside from founding an important department, he is best remembered for one of his papers and for a book that followed a year later.[69] Slobodkin's book, *Growth and Regulation of Animal Populations* (1961), revised and enlarged in 1980, helped to promote mathematical modelling as central to ecological science. However, Slobodkin's view of what might one day be accomplished was overweeningly ambitious. In concluding his book he wrote:

> We may reasonably expect to have eventually a complete theory for ecology that will not only provide a guide for the practical solution of land utilization, pest eradication, and exploitation problems but will also permit us to start with an initial set of conditions on the earth's surface (derived from geological data) and construct a model that will incorporate genetics and ecology in such a way as to explain the past and also predict the future of evolution on earth.

This quotation neatly pulls together many of Hutchinson's lines of thought, though to the best of my knowledge Hutchinson never made such extravagant claims. The book was reviewed by MacArthur who used the opportunity to set out his own position:

> We can divide ecologists into two camps. One so aware of the complexities of nature that it is critical of simplifying theory, is content to document observations at endless length. The second, primarily interested in making a science of ecology, arranges ecological data as examples testing the proposed theories and spends most of its time patching up the theories to account for as many of the data as possible. Slobodkin's little (184 pages) book is the best account yet written from this second viewpoint.[70]

As it turned out Slobodkin's enthusiasm for mathematical modelling in ecology was infectious. His book provided an introduction to the field for those with interests in the ideas of Lotka and Volterra, in niche theory, in competition, and in prey-predator interaction.[71] In the above passage MacArthur set out the credo for many, including Robert May and others in the Silwood Circle. Another person drawn to Slobodkin's book was E. O. Wilson who, as he wrote in his memoir, *Naturalist* (1994), was having some problems of his own in the early 1960s. As an associate professor of zoology at Harvard, Wilson had to struggle to have his kind of field work taken seriously at a time when molecular biology was in the ascendent. He wrote of the 'molecular wars' and of the period in the late 1950s when James

Watson joined Harvard — as a golden boy highly confident after his recent discovery of the structure of DNA. According to Wilson, 'Watson radiated contempt in all directions', but especially toward ecologists, and he reported Watson as stating that 'anyone who would hire an ecologist is out of their mind' — not exactly helping the ecology cause at Harvard.[72] With some of the best biological brains draining in the molecular direction, Wilson realized ecology needed stronger leadership and, like Hutchinson, saw a way forward in the mathematization of the discipline. Mathematics was not Wilson's forte but Slobodkin's book was highly suggestive of approaches that ecologists could take. Wilson got in touch with Slobodkin, then a rising star at the University of Michigan, and proposed some joint work. Slobodkin introduced Wilson to Robert MacArthur. With him Wilson did move forward and was able to reignite his own career.[73]

As mentioned above, Hutchinson early saw the need to bring his discipline in line with developments in evolutionary theory and genetics. With respect to the former he was to be guided in part by his student, Robert Helmer MacArthur (1930–1972). MacArthur joined Hutchinson at Yale after gaining a masters degree in mathematics from Brown University. Like many others in this story MacArthur was a child naturalist. He had a strong interest in birds and his Yale PhD thesis (1957), based on some bird observations, was designed to examine the limits to Gause's principle of competitive exclusion. After returning from a postdoctoral year with Lack, MacArthur moved to the University of Pennsylvania where he quickly climbed the academic ladder. He was appointed to a chair at Princeton in 1965 and accomplished much before his premature death in 1972. In their obituary Hutchinson and Wilson wrote that MacArthur was the principal founder of a type of evolutionary ecology that emerged in the 1960s and 70s. It was, they claimed, MacArthur who 'reformul[ated] many of the parameters of ecology, biogeography and genetics into a common framework', and thus helped to 'set the stage for the unification of population biology'.[74]

MacArthur began publishing some interesting papers while a doctoral student. They gained almost immediate attention. His first paper, noticed by Wilson, included, as mentioned above, the claim that diversity in an ecosystem enhances its intrinsic stability. Another paper, published in 1957, included some ideas that were similarly later discarded. However it is worth saying a little about the 1957 paper since it encapsulates a methodological approach that was to be widely emulated, especially by Robert May and others in the Silwood Circle.[75] Before its publication, most people who studied the abundances of bird species did so by observation and by graphically plotting their results. This allowed for some comparisons of different populations, some speculation on species packing, but not much more. The approach was typical not only of ornithology, it was an inductivist empiricist approach typical of ecology more generally. MacArthur had been trained in mathematics where theorems are deduced from axioms. He had studied under Hutchinson, a bit of a dreamer, who came up with hypotheses that he liked having his students test. Why not, Hutchinson suggested, turn tables on the inductivists. Why not begin with some simple hypotheses and devise mathematical models from which one could deduce, at least approximately, the same type of graphical patterns being produced by field ecologists?

In his 1957 paper MacArthur came up with three such models and credited Hutchinson with having suggested both problem and approach to him. MacArthur relied on some observational data, and on a claim made by Lack and others that most of the bird populations studied were fairly stable and thus probably in some near-equilibrium state. He then came up with three models that could, perhaps, lead to such equilibrium states. To do so, he first postulated an abstract space. He then posited a simple metaphor for a single dimension of that space, namely a stick. The mathematics used to describe the one-dimensional stick could apply also to a model of many dimensions — to Hutchinson's hyperspace for example. At first MacArthur considered a randomly broken stick seeing the pieces as regions of the abstract space. The stick pieces and associated regions were metaphors for non-overlapping niches. The lengths of the pieces were said to be proportional to the abundances of species in the respective regions. In this way MacArthur was able to roughly reproduce some of the graphical results arrived at empirically by naturalists; for example his model roughly agreed with the population results of a bird census in the Quaker Run Valley in Pennsylvania. In his second model he assumed that the regions (or sticks) overlapped at different points. In his third he abandoned the stick metaphor which was not useful in thinking about non-overlapping, non-continuous, niches. In order to predict species populations in such cases he imagined throwing particulate units at random into different urns (representing species); but with an infinite number of tosses all species become equally abundant so the model had its limitations. Luckily for the future of this way of doing things, his model of the first hypothesis for non-overlapping niches looked promising. It appeared to have some support and was worth testing further. However, what the underlying mechanism was for real-life distributions remained an unanswered question.

The fact that MacArthur's three models were soon abandoned is beside the point being made here. What he showed was that mathematical modelling can play a role in ecology. His paper suggested a new analytical way in which old ecological problems could be approached. Though new to ecology, MacArthur's method was in a very old scientific tradition. It puts one in mind of Ptolemy's approach to astronomy. Ptolemy's use of epicycles and eccentrics was a way of conceptualizing what, similarly, had been only partially observed; in his case the movements of the heavenly bodies. Ptolemy's belief in the superiority of circular motion may have been misguided, but his models were good in the predictive, if not the descriptive, sense — something that was true also of the later Copernican model. His approach represented the future direction of mainstream science (where prediction, and the saving of appearances, is deemed more important than true description) and is why he is remembered as a major historical figure. Most sciences have progressed through mathematization; the science of ecology is likely to do the same. As with circular planetary motion it is possible that the niche concept will one day be totally overturned, but as a mathematical device it has shown its worth. Since there are many possible ways of modelling the niche, competition for resources, and Darwinian adaptation as a way of minimizing competition, space was opened up for discussion of behavioural strategy among species within ecosystems. MacArthur, himself, famously worked on a community of five insectivorous warbler species of similar size, observing that competition between the species was limited because of different microbehaviours. The different populations did, indeed, occupy slightly different niches. Some warblers fed early in the day, others later; some high in the canopy and some lower down — and so on.[76] It was a start at what became the new behavioural ecology. MacArthur followed his early work with a series of collaborative studies, notably with Richard Levins. They focussed especially on biodiversity and packing among bird species.[77]

As mentioned, one of MacArthur's admirers was E. O. Wilson. In retrospect one can see that Wilson has a gift for finding the right people to help with his various projects — perhaps one of the marks of a great scientist.[78] He also has an uncanny ability for large scale conceptualization and for thinking up big projects. One of his ambitions was to do something big for biogeography and, for this, he thought that islands might be the key. He recognized that MacArthur shared his vision of going beyond description, and that what ecology needed was a foundation of simple premises from which explanations could be deduced, developed, tested and then refined. Wilson persuaded MacArthur to cooperate in constructing theoretical models that could be tested by biogeographers in the field, and thought that modelling island populations would be a good way forward. The turnover in the fauna of small islands had long been recognized by biologists. Ernst Mayr, for example, had noted that invasion and extinction were common features of island life. These were issues that Wilson had already pondered in connection to his biogeographical work on ant populations. Perhaps new models could direct further thinking on islands, on species succession, packing and extinction.

While working with MacArthur, and to test some of their ideas, Wilson organized a field experiment on some small Mangrove islets in the Florida Keys.[79] This work was largely carried out by his student, Daniel Simberloff.[80] The four islets chosen were small and relatively newly formed and so were unlikely to be home to many different animals. Each was about 15 m across, and at varying distances from a larger island. The closest was about 2 m away, the farthest about 533 m. The size and newness of the islets made census taking relatively easy. Wilson, Simberloff, and some assistants scoured the islands from 'mud bottom to tree tops' making lists of all the arthropod species (largely insects), though they needed systematists to tell them exactly what they had found. Wilson then gained permission to have a pest control company fumigate the islands to kill off the small fauna (but not the plants). The result was not exactly new born islands, but presumably they were devoid of arthropods. Recolonization began almost immediately and it took no longer than a year for equilibrium to be reestablished. Spiders arrived early after the extermination but were prone to extinctions; mites which arrived relatively late persisted with less turnover. On the islet closest to the large island 44 species established themselves where earlier there had been 43. On the island farthest from the large island 22 species established themselves where earlier there had been 25. In both cases the species were not identical to those that had been there at the start of the experiment and the population densities were lower. The equilibrium, which held for a second year, was dynamic with species dying off, immigrating and emigrating. At mid-point less than half the species were among those noted in the pre-fumigation census.

A later experiment was carried out in Brazil by one of Hutchinson's students, Thomas Lovejoy. The 'islands' were created within a rain forest and were surrounded by clear-cuts made by logging. Here the emphasis was on what was left behind in the forest islands. Lovejoy wanted to determine what was the minimum size of rainforest that could still sustain the animals and plants of the region. Determining biodiversity in the forest islets proved difficult. Not until the 1990s were the ways that diversity declines with area well understood; that small populations are especially vulnerable, and that scarcity carries the risk of both inbreeding and extinction.[81]

But how to approach these problems from the theoretical end? In the preface to their book, *The Theory of Island Biogeography* (1967), MacArthur and Wilson note that at first they struggled to understand each other since the language of the zoogeographer and that of the mathematically trained ecologist were very different. But their coming together resulted in a fertile exchange of ideas. Taking off from Darwin's comment made on leaving the Galapagos, that the 'zoology of archipelagoes will be well worth examination', the book sets out to prove him right.[82] For a start the authors expanded the notion of 'island' to include areas such as caves, tidal pools, small streams, islands of taiga within tundra and the reverse, small woods within agricultural landscapes, forest areas surrounded by clear cuts and so on. Could theorizing about such places throw light on old questions: historical questions relating to the origin and extinction of different species in different places, questions relating to species packing, dispersal and adaption, and questions related to the theory of evolution more generally? It turned out that indeed it could.

The book begins quite simply with the claim that larger islands are likely to support more species than are smaller ones.[83] By and large the evidence supports this claim but it is far from being the whole story. Numbers depend on variables other than just area. For example, they depend also on the taxon, the geographical region, historical events, and the distance of an island from other areas of habitation. Questions relating to species abundance were being asked also by others, among them Frank W. Preston. He is interesting as someone who entered the debate from outside academic circles and, while he is remembered, the memories of what he accomplished are selective.[84] Like Wilson and MacArthur later, Preston focussed on the immigration and emigration of species into different habitats and recognized that species diversity could be understood in terms of dynamic equilibria.[85] He wanted to understand the relative abundances of species within any given habitat, and was especially interested in rareness and commonness. This was no easy problem. Figuring out whether observed abundances are the result of biological processes such as interspecies competition, or simply what one should expect statistically, was something that had exercised the minds of others, including Ronald Fisher in the 1930s. Preston had some further complicating thoughts and, taking a long-term view, thought that many ecological communities could still be in a transitory state as a result of the last ice age. He was less confident than MacArthur that even long established ecosystems were in equilibrium. Preston was an engineer and schooled in statistical mechanics. He tried to model the distributions of populations along Boltzmannian lines, and then to test his models against several bodies of evidence — mainly, but not exclusively, from the bird studies that he knew best.[86] Like others, he noted that commonness and rarity are distributed roughly in lognormal fashion but that a variety of distributions have some empirical justification, including a version of the lognormal, the canonical.[87] As is often the case with enthusiasts, Preston introduced masses of supporting data, impossible for an outsider to evaluate. However, while difficult to understand, it seems he was thinking along roughly the same lines as MacArthur and Wilson. Like them he believed that islands could be the key to understanding community structure, and like them he began by looking at islands of increasing size. He understood that the size of an island, and its distance from a larger land base, would determine roughly how many species were present, but not what those species would be. He also included a thought experiment in one of his papers not unlike the actual experiment carried out in Florida. He imagined denuding one of the Comoros islands of its birds. He predicted how soon the bird population would re-equilibrate, how many species would be present and how rarity and commonness among the species would be distributed.[88] And he was not alone. Another person to come up with an equilibrium theory of island biogeography was Eugene Munroe, in his Cornell University doctoral dissertation of 1948.[89] Munroe's case, together with some others, is discussed in a paper that illustrates well the situational logic that gave rise to MacArthur and Wilson's work.[90] Together, Preston and Munroe's work raises the interesting question of authority, and of reception — of when and how ideas come to be taken seriously.

The Theory of Island Biogeography is a short book full of equations and graphs. I will do my best to do it justice without using either of those mathematical descriptors. The book is undoubtedly important in the history of late-twentieth theoretical ecology, not so much for any good predictive models but rather because it showed how interesting questions could be framed and approached. This is true in two senses, the first expressed well by the authors themselves; namely 'insularity is ... a universal feature of biogeography' and that what applies to islands applies to a lesser or greater extent to all natural habitats.[91] Second, by simplifying, by looking at community problems at just the species level, and by relating species distributions to concepts in population ecology, they came up with theories suggestive of new relationships. These could then be explored and tested in the real world. The view of community assembly expressed in the book differed from Hutchinson's niche assembly ideas in certain respects. But like Hutchinson's two lectures, the book had enormous heuristic value, both in stimulating other theorists and in showing the need for much new biological data; in other words for suggesting to theoretical, field and laboratory ecologists what to do next.

The book begins, as did Preston, with some known species-area curves. Some of Wilson's data on ant populations in various island habitats were used alongside other data, much of it on bird populations.[92] Preston's views on the balance between immigration and emigration and the idea that diversity could be expressed in terms of a dynamic equilibrium were acknowledged. But area alone while a fairly good predictor of species diversity (possibly the best there is) is by no means perfect. Even adding distance from a mainland source of species helps only a little; so where to go from there? MacArthur and Wilson slowly added different assumptions to their model and looked to see where that led them. They tinkered with variables; for example, changing the configurations and ages of islands, adding possible stepping stones leading to the islands — indeed, adding any kind of geographical perturbation that seemed plausible. Although it seems likely that which colonizer arrives on an empty island first is largely a matter of chance, and that what follows is partially determined by what came before, testing such ideas is no easy matter.

The authors claimed that islands usually have lower biodiversity than mainland areas of similar size and that this could be explained by hypothesizing higher extinction rates for smaller populations as well as greater problems with re/colonization. This point was later taken up by conservation biologists interested in the optimal size for national parks and reservations. MacArthur and Wilson looked to all kinds of data and much of it pointed to equilibria being reached — for numbers of species, if not for any given species — shortly after new territory is opened up for settlement. One of their examples was data on the repopulation of birds on the island of Krakatau. The explosive volcanic eruption of 1883 blew much of the island away and killed almost everything on it. But life soon returned and bird sightings were made from 1908 onwards. From about 1919 onward the sightings did not change much, and by the 1930s the birds observed were the same as those sighted a decade earlier.[93] This suggests some kind of species settlement was arrived at, but it leaves open the question of whether a different stable grouping would have been just as likely. Since there was considerable turnover before equilibrium was reached, chance could have played a major role in what happened. Evidence from elsewhere seemed in rough agreement with the Krakatau pattern of resettlement. This picture of some turmoil before equilibrium is reached (if it ever truly is), and no obvious pattern to the eventual mix of species, presented something of a challenge to existing ideas of ecological communities as collectivities of co-adapted species. However MacArthur, while bringing something new to the table, was not ready to fully embrace the idea that communities are arrived at wholly by chance, nor was he ready to abandon earlier ideas, including the niche ideas he had developed with Hutchinson.

So, what factors should one consider beyond the geographical? What makes for a successful colonist? In looking at this question MacArthur and Wilson brought different reproductive strategies into play. It was assumed that the best colonists will be those of reproductive age, but how should they behave to maximize population? Here another body of theory was introduced. In some situations, it was claimed, maximizing reproductive rates pays off, in others a lower reproductive rate and the maximization of parental care could be the better strategy. The theoretical discussion of so-called r and K strategies proved to be of enormous interest and fed into a number of future subdisciplines, notably into sociobiology one of Wilson's developing interests.[94] Hutchinson's niche ideas were also important in suggesting when, or if, invasion was possible. Could a newcomer change its niche requirements and so succeed as a colonist? Perhaps it could change its diet or, if its traditional food source is readily available in the new habitat, perhaps it could extend its range. But, if food is plentiful, perhaps a better strategy would be to restrict travelling, stay put and save energy. Competition is clearly a major but not exclusive factor in all these possible strategies. And, as the authors demonstrate, much lends itself to mathematical modelling — a revelation to many ecologists.

Finally the authors considered what kinds of evolutionary changes are likely to follow colonization and how natural selection might work within an island community. Climate could be a factor here. For example, very cold winters leading to booms and busts in food supplies could favour r selection, while a more uniformly benign climate might favour K selection. Another factor could be the temporal point of colonization. For the earliest inhabitants it is a toss up whether r or K selection is best. For the next set of invaders it seems that r selection might be the better strategy and for the late colonizers K selection might prove best. But nothing is straightforward and all kinds of complicating factors enter the picture — for example, the presence or absence of predators. All in all, and despite the fact that the book does not consider the evolution of new species, the argument is well made that islands are good places to study evolution — just as Darwin had surmised.[95]

I have touched on only a few of the many points made in the book. On the merits of the mathematical models I have said nothing. Data was introduced in partial support of some of the models but my point is not to make a scientific argument, but a more philosophical one. All I wish to claim is that much can be systematized in mathematical form and that models can provide ideas not only for further modelling, but also for many and varied empirical studies.[96] MacArthur had learned something from Hutchinson, namely that in science it is as important to show the way forward as it is to come up with any definitive answers. MacArthur and Wilson's book succeeded in pointing out a direction that looked promising. It has been discussed here because of its importance to the intellectual development of members of the Silwood Circle. But MacArthur's legacy lies in far more than this one book. In his short academic career he published two other books and numerous papers, all of which helped in the better understanding of biodiversity, in the developing of ecological theory from both community and evolutionary perspectives, and in showing how an analytical turn of mind can be brought to bear on biological questions.

However, the successful advance of MacArthur's approach did not rest solely on its intellectual merits, considerable though they were.[97] As will be shown, success for his modelling approach entailed also a rhetorical onslaught by like-minded ecologists, including some among the Silwood Circle, on what was labelled an anti-theoretical establishment. Many claims were made during the 1960s and 70s: claims for what constitutes 'good scientific method', claims for what the future direction of ecology should be, claims for the need for mathematical modelling and claims that a general theory of ecology was, as MacArthur believed, a realizable possibility.[98] In my view this onslaught worked, but only because MacArthur's work, and work in a similar vein, had great heuristic power, and because it appeared at a time when environmentalism was on the upswing.

Endnotes

1. (Southwood and Clarke, 1999). Elton is quoted on his 'marked inability to pass examinations' (134). See also (Hardy, 1968).

2. Several people from Imperial College joined this expedition, an early zoological link between the two institutions. Ownership of Spitsbergen and Bear Island, also visited during the expedition, was fiercely contested by Sweden and Norway. The Spitsbergen Treaty of 1920, part of the post-war Versailles negotiations, gave Spitsbergen to Norway. Britain had mineral interests in the territory which is one reason why the expedition received financial backing.

3. Peder Anker writes about Elton in the context of Imperial science, seeing his work as reflecting Britain's commercial interests and seeing ecology at Oxford in the period before World War II as a tool for the management of nature and empire (Anker, 2001, chapter 3). My aim in writing about Elton and his colleagues at Oxford is to provide a historical context for the later concerns of the Silwood Circle.

4. The 1923 expedition, with financial support from the resource sector, was sponsored by Merton College and Elton was appointed its chief scientist. Further supporting Anker's thesis (see note 3), the 1924 expedition was sponsored by the university with major grants from the government and the Anglo-Persian Oil Company (British Petroleum from 1954).

5. Elton translated Collet's book (*Norwegian mammals*) into English but his translation was not published.

6. (Crowcroft, 1991), 11.
7. Elton also studied the records of the Hudson's Bay Company held in London which included data on past cycles in the populations of fur-bearing animals.
8. See, for example, (Elton, 1942).
9. Edmund Brisco (Henry) Ford FRS (1901–1988) studied genetics with Julian Huxley and was to work closely with R. A. Fisher. Ford was a specialist in *Lepidoptera* and, with others, demonstrated microevolution in the peppered moth. He later became professor of ecological genetics. His popular *Mendelism and Evolution* (1931) went into eight editions and he later published *Ecological Genetics* (1964). John R. Baker (1900–1984) was another of Huxley's research students. As with Elton, and later Lack, Huxley helped Baker to go on an overseas field expedition. Baker went to the New Hebrides (now Vanuatu) and turned himself into something of an anthropologist. Together with Michael Polanyi and Arthur Tansley, he founded the Society for Freedom in Science in 1940. Theirs was a classic liberal response to the view that science should serve society in practical ways, a view held by, among others, J. D. Bernal and expressed in his *The Social Function of Science* (1939). Baker, Polanyi and Tansley believed there was a clear distinction between pure and applied science and stated that the former should be free of social function, and a strictly intellectual endeavour. A lecturer in the department of zoology at Oxford, Baker's principal interests were in microscopy (he developed some stains for looking at living tissue) and eugenics. For Baker and eugenics, (Kenny, 2004).
10. Huxley resigned his chair a few years later to become a roaming man of science, much engaged with conservation projects in Africa, and with a number of public issues in the UK. After the war he was appointed the first Director-General of UNESCO. He wrote much and is best remembered for his contributions to modern evolution theory. See note 27 below.
11. (Elton, 1927), chapter 5.
12. Later, in 1959, G. Evelyn Hutchinson and Robert MacArthur suggested that the number of species increases directly with a decrease in the surface area of the body of the animals or inversely with the square of their weight. Both of these cannot be true and I am not sure what was intended. Whichever is the case it can be true only up to a point. In very small niches, with very small animals, there will be departures from such rules.
13. For 'niche' as a contested term with many meanings, (Odum, 1971), chapter 6. For a long discussion of the niche see also (Hutchinson, 1978).
14. (Elton, 1927), chapter 5.

15. The debate over whether it is better to develop minimalist models in which individual differences within populations are ignored or whether to include detail such as size, age or behavioural difference was debated already in Elton's time. There are good arguments for both. Needless to say many behaviourists will depart from Elton in seeing difference almost everywhere. This is not to say that animals have cultures, or that any given species exhibits the types of regional cultural difference one sees among human beings — though perhaps they do.

16. Edwin Stephen Goodrich FRS (1868–1946) was an accidental zoologist. He was a student at the Slade School of Fine Arts when he befriended E. Ray Lankester who piqued his interest in zoology. When Lankester was appointed to the Linacre chair he took Goodrich along as his assistant. Goodrich enrolled as a student at Oxford and was awarded a degree in zoology. He was an outstanding comparative anatomist with a strong, but not exclusive, interest in fishes. Not surprisingly he was also a fine zoological illustrator. Sir Alister Clavering Hardy (1896–1985) was also educated at Oxford. In the 1920s he joined the Royal Research Ship *Discovery* on its trip to the whaling grounds in the oceans around Antarctica, after which he was appointed chief zoologist at the Colonial Office. He later held chairs at the universities of Hull and Aberdeen, returning to Oxford and the Linacre chair in 1945. His area of expertise was marine plankton. He was very supportive of field studies at the university. Howard Walter (Lord) Florey (1898–1968) was a young Australian medical graduate when he went to Oxford on a Rhodes Scholarship. A physiology student under Charles Sherrington, he later held posts in Cambridge, Sheffield, and London before returning to Oxford and the chair in pathology in 1934. At Oxford he brought together the team, including chemists and biochemists, that led to the isolation of penicillin. Together with Alexander Fleming and Ernst Chain, Florey was awarded the Nobel Prize for medicine in 1945.

17. He is not to be confused with the demographer P. H. Leslie, also of matrix fame.

18. Raymond ffenell, who made a fortune in South African gold and diamond mining, purchased the Wytham estate and surrounding land in the early 20th century. He owned about 1250 hectares situated about 5 km northwest of the City of Oxford. A little more than half of the land was farmed, the rest was woodland. ffenell sold the farming estate to the university and, in 1942, donated the woodland with the provision it be conserved for posterity and used for instruction and research. The woods lie within a loop of the Thames and rise from marshy land to hills reaching about 164 m above the river; (Savill, Perrins, Kirby and Fisher, 2010).

19. These long-term studies were various and included work on small mammals as well as on the arthropod communities that formed on fallen oak trees. However, the short-tailed field vole became Elton's principal animal of study.

20. Park, quoted in (Crowcroft, 1991), preface. Thomas Park (1908–1992) was a member of the Chicago School during a period well covered in (Mitman, 1992). Mitman's main focus is on Warder Clyde Allee. He also shows how the ideas of Darwin and Kropotkin competed for attention in the period after World War I. While animal ecology at Chicago was being built as a science based on laboratory and field investigation, and on the existing science of physiology, there was much emphasis on group behaviour and on collaboration. Mitman relates this to anti-German sentiment fuelled by the war, and the association of Germany's aggression with Darwinian competition — 'nature red in tooth and claw'. Interestingly the paleontologist George Gaylord Simpson was to attack the 'aggregationist ethics' of the Chicago School, seeing it as totalitarian (Mitman, p.6) as opposed, presumably, to the classically liberal individualism of Darwinism.

21. Elton had questioned the role of natural selection in some adaptations such as fur colourations seen in northern animals. He held on to the idea of non-adaptive, neutral, mutation even though most of his Oxford colleagues, for example E. B. Ford and David Lack, were strict adaptationists.

22. One story has it that Elton barred the door joining the premises of the two groups. Both Lack and Elton appear to have been sympathetic but retiring people. However, being retiring does not account for the overall lack of communication.

23. Pringle may well have wanted a nameless ornithology unit in his department, but the terms of reference in the original endowment precluded that. The Institute also had some administrative independence from the zoology department.

24. The Hope department moved to join the other zoologists only in 1979 when Southwood moved to Oxford as the Linacre professor.

25. For Lack, (Thorpe, 1974). See also a collective obituary, which includes a memoir written earlier by Lack himself, as well as brief memoirs by a number of his friends and colleagues; (Lack, 1973).

26. Haldane spent ten years at Cambridge (1922–1932) where he taught a course on enzymes.

27. Both (Mayr, 1942) and (Huxley, 1942) were to be important influences on Lack's thinking. Like Darwin, both authors believed that different parts of a population could be under different selection pressures, leading to slightly different modes of life. Before the war Mayr had been on an expedition, organized

by L. W. Rothschild's Tring Museum, to collect bird specimens in New Guinea and the Solomon Islands. His Darwinian ideas crystallized later when working as a curator at the Museum of Natural History in New York — which is where Lack met him. There Mayr brought order to the vast Whitney collection of birds from the South Seas. He studied small variations in the various species, and matched them to geographical placement, and to displacement from what he thought was the principal and original source of many of the birds, namely the main islands of New Guinea (the Solomon Island birds were a different group). He was able to provide solid evidence for Darwin's idea that new species arise when populations disperse, and when pockets become isolated from the rest — so called 'allopatric' speciation.

28. (Gause, 1934).
29. Oxford was the hub of the national bird census, in part because of the earlier presence of Julian Huxley, his student B. W. Tucker, and E. M. (Max) Nicholson, a historian (later civil servant) and keen amateur ornithologist. As David Allen put it, Nicholson was 'destined to rival De la Beche in the bountifulness of a career of natural history institution-building'. Like Huxley he was a practical visionary. Later he helped to found the Nature Conservancy and the World Wildlife Fund. The British Trust for Ornithology was founded in 1932 and the movers and shakers, including Nicholson, who set it up raised enough funds which, when combined with monies already in the Edward Grey Memorial Fund, enabled the university to found the Edward Grey Institute; see (Allen, 1976), chapter 14; quotation, 252.
30. Lack, quoted in (Thorpe, 1974), 277.
31. Lack heard about great tits taking to nesting boxes from the ornithologist H. N. Kluyver whom he visited in Holland at the end of the war. Kluyver was also the inspiration for Lack's long-term studies. On the same visit Lack met Niko Tinbergen. They were to meet again later in the United States when Lack made a second visit to Ernst Mayr in New York. Tinbergen was to move to Oxford in 1948 (see below). Lack's successor in Oxford, Christopher Perrins, increased the number of nesting boxes to about 1000 and population studies on great tits and blue tits continue to this day.
32. For yet earlier work on density dependence see, for example, (Nicholson, 1933). However, the idea goes back even further, at least to the ideas of biological control discussed in (Howard and Fiske, 1911) if not to Darwin — and Malthus. Lack's colleague at Oxford, the entomologist and Hope Professor, George Varley, had noted that it was unclear how or when a density dependent process would occur after a change in the population. For example, thinking along the same lines as Volterra, he claimed that parasite populations could peak some

time after peaks in the populations of their hosts. See, for example, (Varley, 1953). For more on Varley see chapter seven.

33. Andrewartha and Birch studied especially local populations in small scale environments. Such populations are especially prone to extinctions — not the result of dense populations but of factors such as climate and environmental hazard; extinction was often followed by recolonization of the space. See further discussion below.

34. V. C. Wynne-Edwards (1906–1997) taught at McGill University for many years before returning to Britain and a chair in Aberdeen.

35. Kropotkin was an anarchist and did not believe in governmentally directed communitarianism. Carr-Saunders (1922) combined some of the ideas of Malthus, Darwin, Mendel and Kropotkin in an essay on human fertility and population. He had studied zoology at Oxford and biometrics under Karl Pearson in London. While arguing against Malthus, Carr-Saunders was a keen eugenicist. Late in his career he became Director of the London School of Economics. Another zoologist who drew ideas from both Darwin and Kropotkin was Peter Chalmers Mitchell FRS (1864–1945), biographer of T. H. Huxley, and Secretary of the Zoological Society of London from 1903; (Mitchell, 1915).

36. These ideas were around also at the Chicago school of ecology. Wynne-Edwards was inspired also by Elton's work in the Arctic and was later to carry out work there of his own. He was part of the Baird Expedition to Baffin Island (1950 and 1951). He believed bird colonies regulated population according to environmental circumstance; (Wynne-Edwards, 1952). For a major critique of Wynne-Edwards, (Williams, 1966).

37. Tinbergen shared the 1973 Nobel Prize for physiology/medicine with Konrad Lorenz and Karl von Frisch. His brother Jan Tinbergen won the Nobel Prize for economics in 1969. For more on Tinbergen, (Kruuk, 2003).

38. Tinbergen spent the war in occupied Holland. Lorenz joined the Nazi Party before the war. While this led to some preferment within the academy (in Germany), he was later drafted into the army. He was captured and spent time in a Russian prison camp. Later he directed the Max Planck Institute for Behavioural Physiology in Seewiesen, Bavaria, and became an internationally renowned figure. Lorenz was a group selectionist in that he believed that ritual fighting (not going all the way) was the norm in animal competition and that this served the interests of the larger group through avoidance of wholesale mutilation. Like Wynne-Edwards, Lorenz had his critics. Notable among them was J. B. S. Haldane.

39. William Homan Thorpe FRS (1902–1986) was a specialist in bird song and a friend and associate of both Tinbergen and Lorenz.

40. While at Oxford Tinbergen published some popular books: among them (Tinbergen 1953 and 1958). The second of these has the appealing title, *Curious Naturalists*, and tells of Tinbergen's various field studies on the wasp parasites of bees, and of his studies of birds and other animals carried out in Holland, Britain and the Arctic.

41. Sir Peter Medawar (1915–1987), who persuaded the Nuffield Foundation to support the group, and 'institution builder' E. M. Nicholson (see note 29) at the Nature Conservancy, were both important in helping to nurse the new research group in its early years.

42. This academic behaviour is perhaps an example of what ecologists term co-evolution. The question of ecologists responding to political and cultural developments is a topic to which I will return in my concluding chapter. The move toward interdisciplinarity may have some connection to the success of large wartime projects such as the Manhattan Project. For a discussion of this point, (Etkowitz and Kemelgor, 1998).

43. (Tinbergen, 1951).

44. (Krebs, 1991). Krebs attended Tinbergen's lectures as an undergraduate and later became a doctoral student in Tinbergen's unit (see chapter 8).

45. (Slack, 2010). I am indebted to Nancy Slack for letting me read two chapters of her book prior to publication. See also (Kohn, 1971).

46. (Hutchinson, 1979), 76–78 and *passim*.

47. Like many of his generation Hutchinson never gained an advanced degree though he was awarded several honorary doctorates in later years. While in South Africa he became friendly with Lancelot Hogben, professor of zoology at the University of Cape Town. After the termination of Hutchinson's lectureship at Witwatersrand, Hogben invited him to work for a while in Cape Town and persuaded him to apply for a position at Yale. Earlier Hogben had worked in the zoology department at Imperial College and, while there, had befriended the mathematician Hyman Levy. Levy taught him mathematics and Hogben then became a staunch advocate of the use of mathematics in biological work. Perhaps Hutchinson was influenced by Hogben in this. Later Hogben and Levy were members of the 'Visible College' and Tots and Quots (see chapter one, note 34).

48. For example, (Slobodkin and Slack, 1999); especially p. 29.

49. (Hutchinson, 1957–1993). Hutchinson had intended a fifth volume but was unable to complete it. The first volume includes much of his own biogeochemical work. The second volume contains ideas and data on lake biology, focussing especially on ecological ideas. The third volume on plant life in lakes is based largely on other people's work, the fourth volume on animal life was published posthumously.

50. Vladimir Ivanovich Vernadsky (1863–1945) was born in St. Petersburg but later moved to Ukraine for health reasons. He was a founder of the Ukrainian Academy of Sciences and its first president. He knew of Lotka's work (see chapter 2).

51. Today ecologists at Yale are trying to reverse the recent history of Linsley Pond. During the industrializing nineteenth and twentieth centuries dams were built on some of the rivers in the water system of which Linsley Pond is a part. This hindered the movement of fish and other wildlife. The idea is to re-engineer the system to allow fish species that once lived in the lake to return. By introducing keystone predators into the lake it is hoped that kingfishers and some mammals, including otters, will return to the area.

52. (Hutchinson, Deevey and Wollack, 1939; Hutchinson and Wollack, 1940).

53. While the older logistic equation with its sigmoid curve represented systems approaching equilibrium, cybernetics had something to say about instability and Hutchinson saw its possibilities as a theoretical tool in ecology. Wartime cyberneticists were working on problems such as aiming guns at moving targets. Getting guns to move back and forth smoothly while tracking the target was not easy; violent oscillations were not uncommon. Hutchinson saw a connection to the maintenance of stability in living organisms, and in ecosystems. Relatedly, it was known that an infectious disease outbreak can reach epidemic proportions from a relatively stable base. For some of the wider influences of cybernetic thinking on biologists see (Haraway, 1981–82). Haraway was one of Hutchinson's doctoral students.

54. Lindeman had difficulty in getting his work published and sadly, while knowing of the paper's final acceptance, did not live to see it published and see its future influence within the discipline; (Lindeman, 1942). His paper owes something to Arthur Tansley's use of the term 'ecosystem'. For a general history of ecosystem ecology see (Hagen, 1992; Golley, 1993). Golley's book shows some interesting connections between ecosystem ecology and the military. After World War II much research was funded by the military, especially the tracing of radioactive isotopes through natural systems. The Odum brothers were major beneficiaries.

55. Vernadsky had done something similar earlier. ^{32}P and ^{33}P have half-lives of 14 and 25 days respectively, long enough to carry out some studies (given that the element is quickly taken up) and not long enough to be a serious environmental hazard (provided the isotopes are used in small quantities). At the time only small quantities were available; Hutchinson obtained his isotopes from the Oak Ridge National Laboratory, Tennessee. See, for example, (Hutchinson and Bowen, 1950).

56. (Hutchinson, 1948). This paper was presented at a 1946 conference of the New York Academy of Science in front of a large audience. The conference drew people interested in a wide range of cybernetic issues. Another attendee was Hutchinson's friend Gregory Bateson who was interested in cybernetics and social behaviour. Hutchinson's paper, which followed one given by Norbert Wiener, included a discussion of his radiochemical/biogeochemical work from a cybernetic perspective, as well as some vague comments on populations being governed by feedback loops. The conference was supported by the Josiah Macy Foundation. For the role of this foundation in supporting cybernetic thinking, (Heims, 1991).

57. For an interesting discussion of the connections in conceptual thinking between Hutchinson and his student H. T. Odum, (Taylor, 2005), chapter three. At Yale Odum carried out a study of oceanic calcium and strontium levels, claiming that a steady state had prevailed in the oceans for at least 40 million years. After graduating, and while working at the University of Florida, he carried out studies at Silver Springs not unlike Lindeman's earlier studies in Minnesota. For a source of some ecosystem ideas of the period, (Odum, 1953). Eugene Odum was the founder of the Institute of Ecology at the University of Georgia. His influential textbook was organized around the idea of the ecosystem as understood by his brother and by some other early Hutchinson students. Just as Hutchinson was influenced by his mineralogist father, the Odums appear to have been influenced by their sociologist father who saw human society as strictly part of the natural world, (Hagen, 1992). As with Lindeman, the Odum's naturalism was premised on the flow of energy. In 1954 they conducted a study of energy flow through a coral reef ecosystem on the Pacific island of Eniwetok, one of a group of islands captured from the Japanese and then used by the United States as a testing site for nuclear weapons. A neighbouring island had been totally obliterated during an H-bomb test. The reef had been in a stable state before the explosion. Ironically, the evolution of stabilizing mechanisms needed to maintain natural systems was to become the subject of much discussion. Also in the 1950s, Eugene Odum carried out some research on ant hills in Wytham Woods.

58. Jan Smuts's philosophy of holism had drawn something from an earlier generation of American ecologists, notably from the ideas of Frederick E. Clements. Smuts was trained as a lawyer but envisaged a plan for ecological work in South Africa. However, this intellectual thread is beyond what can be discussed here. See (McIntosh, 1980), especially 208.

59. The classic figure in biological system theory is Ludwig von Bertalanffy who applied it in his work in physiology. His emphasis was holistic and organismic.

The Odum brothers were influenced by him, as well as by Claude E. Shannon and Warren Weaver's work on communication and information theory, and by the types of computer simulations carried out by Norbert Wiener and other people working in operational research during and after World War II. John Von Neumann's game theoretic ideas were to inspire both the biological system theorists and the population theorists.

60. For use of the term 'physics envy' see, for example, (Cohen, 1971). Today the tables have turned and the biological sciences, especially the biomedical sciences, are in the ascendent. Much is a consequence of Crick and Watson's discovery of the structure of DNA, the use of mathematics in biology, and the rise of computer technology. But the full story is hugely complex.

61. Another forerunner not already mentioned was Joseph Grinnell (1877–1939). While Elton thought of the niche in terms of an animal's role in its community, Grinnell thought of it more strictly as a small region within the larger environment. Hutchinson incorporated both ideas within his expanded definition. Grinnell, Director of the Museum of Vertebrate Zoology at the University of California, Berkeley, a mammal specialist and a pioneer in Californian ecology, may have been the first to use the term 'niche' in ecology, (Grinnell, 1917).

62. (Hutchinson, 1957). In developing his mathematized niche idea, Hutchinson was indebted to the earlier ideas of the Russians, V. A. Kostitzin and G. F. Gause. (Kingsland, 1995), 144 and 181.

63. 'Homage to S. Rosalia', (Hutchinson, 1959).

64. (MacArthur, 1955).

65. (Hutchinson, 1961).

66. A major conference on diversity and stability was held in 1968 at Brookhaven National Laboratory. See chapter 6.

67. Tribalism is a fact of scientific life and Hutchinson can be seen as the parent of at least two tribes in theoretical ecology, and possibly others in biogeochemistry and limnology. Members of the Silwood Circle belong to the MacArthur tribe and were early keen to distance themselves from the Odum tribe of systems ecologists.

68. Slobodkin obituary; (Anon, 2009).

69. (Hairston, Smith and Slobodkin, 1960). The authors argue that while competition existed among most groups of animals, herbivores do not compete. Rather, their numbers are limited by the abundance of predators, including parasites. This much cited paper came to be known simply as HSS.

70. (MacArthur, 1962). Interestingly, in the same journal issue, Frank Preston published the second of his papers on the rarity and commonness of species (see note 84 below).

71. Later Slobodkin became more wary and stated that mathematicians needed to know some biology before entering the field. Borrowing ideas from statistical mechanics or economics was all well and good, but knowledge of what was being modelled was essential; (Slobodkin, 1975).

72. (Wilson, 1994), 219–220.

73. In his memoir (1994), Wilson stated that MacArthur first came to his attention with his 1955 paper. Understanding community stability was the dream of just a few ecologists in that period, a dream shared by many more in the following decades. Interestingly MacArthur had his own small coterie which met at his summer place in Vermont during the early 1960s. Its focus was on creating a new population biology based on modelling. After having been introduced by Slobodkin, Wilson became a member of the group. It consisted of about six people and included also Richard Levins. For Wilson's recollection of the group, (Segestrale, 2000), 43–44.

74. (Wilson and Hutchinson, 1989), quotation, 319.

75. (MacArthur, 1955 and 1957).

76. (MacArthur, 1958). MacArthur was not the first to observe such microbehaviours. For example, G. F. Gause mentions work carried out in 1923 by the Russian ornithologist A. N. Formosov on four closely related species of tern living together in a colony on the island of Jorilgatch in the Black Sea. 'It appears their interests do not clash at all, as each species hunts in perfectly determined conditions differing from those of another.' (One species feeds on land, another fishes in the open sea, another fishes fairly close to land and the fourth species fishes in swampy sea marshes.) See (Gause, 1934), 19–20.

77. Levins was someone who also brought considerable mathematical skills to bear on biological problems. See Levins (1968). The collaboration helped MacArthur to refine his approach. Levins is an interesting figure but this is not the place for any in-depth discussion of his ideas. Some from that time can be seen in (Levins, 1966). Like MacArthur, Levins enjoyed thinking about complex systems in abstract terms and he, too, had the mathematical ability to construct models applicable to evolutionary ecology. Like MacArthur, he recognized that it was worth sacrificing realism for generality, that general models are usually false but can be suggestive. Levins is also a social theorist, and during the time he collaborated with MacArthur was a committed Marxist. He had spent time in Puerto Rico as a fruit farmer before his doctoral studies in zoology at Columbia University. He therefore had some first-hand experience of pest problems and had interests not unlike those of the people working at Silwood during the late 1960s and early 1970s, and those of C. S. Holling, working in British Columbia on forest insect pests. Like them, Levins was to use population mod-

elling in the study of herbivorous insect pests and their natural enemies. The first of several joint papers, (MacArthur and Levins, 1964), touched on patchy environments and character displacement, something earlier noted by Wilson. Populations of similar species that coexist exhibit more pronounced variation than do populations of the same species living apart. For more on patchy environments see chapters 6, 7 and 8.

78. Great scientists, as a rule, need many helpers, a topic I have written about in other contexts. See, for example, (Gay, 1996).

79. Before settling on the red mangrove islets, Wilson considered working on islands in the Dry Tortugas and waiting for hurricanes to wipe them clean.

80. (Simberloff and Wilson, 1969a and 1969b); (Simberloff, 1969). Reminiscing, Simberloff claimed he was sceptical when Wilson suggested he carry out this large project. He understood the mathematical modelling, but with little evidence that islands truly behaved in the ways suggested by the models, he was not sure he should risk spending two years in the mangrove swamps of Florida working on a PhD project that might not pan out; (Simberloff, 1984).

81. (Wilson, 1994), 226–228.

82. (MacArthur and Wilson, 1967); Darwin quotation, 3.

83. This had long been recognized, explicitly so by J. R. Forster the naturalist on Captain James Cook's second voyage (1772–1775 on the *Resolution*). Johann Reinhold Forster (1729–1798), an Anglo-Prussian, was a teacher at Joseph Priestley's Warrington Academy and later at the University of Halle. He replaced Joseph Banks who withdrew from the voyage at the last moment. See (Thomas, Guest and Dettelbach, 1996) for an edition of Forster's, *Observations made during a voyage round the world*, originally published in 1778. Forster and his eldest son Georg (also on the voyage) were outstanding naturalists. They also provided evidence in support of Buffon's observation that separate geographical regions, even when environmentally very similar, have distinct forms of plant and animal life.

84. Preston's work was cited in (MacArthur and Wilson, 1967), 10–20. Frank W. Preston (1896–1989) was a man with many interests. Born into a poor family in Leicestershire and unable to afford conventional higher education he took external degrees in engineering from the University of London. He gained a PhD while employed at a glassworks in Loughborough. After a peripatetic couple of years in Europe, Africa and Australia, he set up a glass engineering firm in Butler, Pennsylvania, in 1926. The business was highly successful with major contracts coming in from optical firms in Rochester, and from Corning Glass (Preston devised and built the furnace that made Corelle Ware possible). Preston moved to nearby Meridian where he purchased several large tracts of land with conservation in mind. He was a serious naturalist with interests in both geology and

ornithology. He mapped the gravel moraines and streams of the area around Meridian, noting the extent of some old but extinct glacial lakes. He helped to promote a major reclamation scheme by offering to donate a large tract of forest land to the state provided it could be combined with some other lands, including a narrow river valley, into a major conservation area. He also suggested that the valley be reflooded, something that he knew from his geological studies would be relatively easy to do. The valley had once housed a lake, and was largely plugged at its exit by the remains of a late Pleistocene glacier. An abandoned railway ran along the valley. Preston acquired much of the track. By gathering together with others in the Pittsburgh area who were also interested in conservation, a huge tract of land, including Preston's donations, and including some reclaimed oil and gas-well sites, was made into parkland. The valley was indeed flooded to create Lake Arthur, named for an earlier amateur geologist admired by Preston. The lake and the various lands are now part of Moraine State Park, established in 1970. See (Mayfield, 1989); also the website of the Department of Environmental Protection, Pennsylvania — section on environmental leaders. This is all aside from Preston's serious ornithological contributions, and his ways of thinking about ecological communities that drew the attention of Robert MacArthur.

85. Preston began publishing in the 1940s. A later version of his ideas can be found in (Preston, 1962).

86. He borrowed the idea of canonical distribution from Ludwig Boltzmann's statistical mechanics — namely the distribution of collections of molecules of fixed number, volume and temperature. In such distributions the populations are exponential with respect to something (energy, for example), but what exactly that something should be in the ecological context was not clear. The problem for future ecologists was to explain the differing widths of the curves from habitat to habitat.

87. For a later discussion of lognormal, canonical and other distributions, (May, 1975); for the canonical see especially p. 91. See also (Diamond and May, 1981) May told me that this chapter in his *Theoretical Ecology* contains refinements to the earlier paper, especially in the ways of 'estimating species that have come and gone between census intervals and the problems this can create if you do not take account of it'.

88. One might imagine he would have considered rarity and commonness among the species present in nearby land masses when thinking about the commonness of species invading the islands, but I don't think he did.

89. (Brown and Lomolino, 1989). Munroe's PhD thesis was primarily on the distribution of butterflies in the West Indies. His ideas were later summarized in a

couple of papers published while he was working at the Entomology Research Institute of the Canadian Department of Agriculture in Ottawa; (Munroe, 1963). This very general paper includes a brief statement of his earlier ideas along with the claim that four species of West Indian butterflies are the evolutionary result of early, yet different, colonizations by a single species from the mainland. Brown and Lomolino show some interesting parallels between the ideas of Munroe and those of MacArthur and Wilson. They speculate on why Munroe's theory received so little attention yet, when it was independently discovered and newly framed a quarter of a century later, it became the centrepiece for much ecological theory.

90. (Lomolino, Brown and Sax, 2010).

91. (MacArthur and Wilson, 1967), 3.

92. Bird data from the Solomon Islands was used. Similar data, including much of his own, was later used by Jared Diamond in some related work. See, for example, (Diamond, 1975) and (Diamond and Mayr, 1976). Assembly rules, and whether they even exist, became a matter of much debate.

93. Such observations could reflect permanent residence or extinctions followed by repeated recolonization. Turnovers could be random or not. Indeed, there are many possibilities and only careful field studies can determine what actually happens. Good models can guide such studies. The classic work on the recolonization of Krakatau was carried out by Karel Willem Dammerman in the early twentieth century.

94. For r and K, (MacArthur and Wilson, 1967), chapter 4. It was an idea MacArthur had discussed already in 1961 and is implicit in the Verhulst equation (see chapter 2 and appendix 1) from which the letters r and K were taken. The symbols relate to the intrinsic rate of reproduction and to the overall carrying capacity of a particular environment. The authors' modelling was based on the idea that large populations are more likely to survive than small ones. But how to achieve a large, or at least stable, population? This is where so-called r and K strategies enter. As MacArthur and Wilson show, what is best will depend on circumstance. r-selected organisms were said to recognize their environment as unstable and were under evolutionary pressure to act opportunistically in good times. Their populations were kept at a low level with occasional exponential booms followed by busts. K-selected organisms see a more stable environment and their populations remain more steady. But these are caricatures; in the real world organisms lie somewhere on a continuum between the two extremes. Raising questions such as how newcomers to an island avoid extinction; and how later arrivals compete; and by introducing reproductive

strategies as a way to address such questions, proved very useful in the years that followed. Reproductive strategies were only one of several types of survival strategy entertained in this book, and by MacArthur in his other publications.

95. Perhaps the most famous graphical representation in the book is the one showing the crossed curves of colonization and extinction rates on an island as functions of the number of species already present. The curves cross at the equilibrium point indicating the number of species then present. The shape of the curves, though not the particulars, follows a general pattern from island to island.

96. (Schoener, 2010) looks at the citation record for MacArthur and Wilson's book. The number rises sharply from 1967 to 1985, levels off and then rises sharply again after 2000. Schoener is not entirely sure how to interpret the numbers and this caution is warranted. But the book has clearly had a major influence in ecology which is why Schoener and others celebrated its fortieth anniversary.

97. A similar point is made in (Fretwell, 1975). Fretwell gives a good account of MacArthur's major contributions to ecology and of their immediate impact. He sees MacArthur as a genius who had the common touch, and who was much loved by his students and colleagues — among them Fretwell himself. But Fretwell also claims that MacArthur 'bypassed' the usual refereeing problems that young scientists face: first, by having Hutchinson support his publications in the *Proceedings of the National Academy of Sciences*, and second, by his founding of the Princeton Monograph Series where he could publish his own work and the work of others sympathetic to his approach. There may be something to this, but MacArthur's early publications appeared in a fairly wide range of journals where he would not have been protected from criticism. It cannot have been easy bringing new and somewhat abstract mathematical ideas into ecology, but having Hutchinson firmly in his camp will have helped. I will discuss some later championing of MacArthur in chapter six.

98. It is curious that scientists often argue over who is doing science in the 'proper' way and who is not. One wonders what philosophies of science they carry around in their heads. For my own philosophical views see chapter 10. MacArthur and his followers were not the only ones trying to take hold of 'good' scientific method in support of what they were doing. Needless to say they were accused by others of not being properly scientific. For one of their accusers see (Peters, 1991). In my view this book, written by a distinguished limnologist, adopts too narrow a view of what counts as scientific — even narrower than the view of Karl Popper whom the author takes as an authority. Peters thought ecology a lesser science because of its poor record in prediction. But good prediction, while possible in some disciplines such as physics, where

simple mathematical models work well, is not the only criterion for something to count as scientific. For other sciences, including ecology, precise predictions are scarce. MacArthur's models had great heuristic power, even when not highly predictive. Like Popper, Peters was also concerned with tautology in science, and labelled a number of ecological concepts, including 'diversity' and 'competitive exclusion', as tautological and thus meaningless.

Illus. 1. The Manor House at Silwood Park. *(Courtesy ICA)*

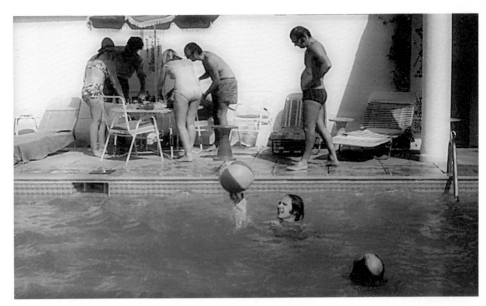

Illus. 2. A lunch party at Dick Southwood's swimming pool in the 1970s. Southwood, wearing blue shorts, is by the table. Gordon Conway is walking toward the table and Mike Hassell is in the pool with ball. *(Courtesy R. M. May)*

Illus. 3. Croquet at Silwood in the early 1980s. From left, Roy Anderson, Bob May, Günther Hasibeder and Mike Hassell. *(Courtesy M. P. Hassell)*

Illus. 4. Hiking in the Lake District in the late 1970s; from left, Bob May, Mick Crawley, Bryan Grenfell, Roy Anderson and Andy Dobson. (*Courtesy M. P. Hassell*)

Illus. 5. Hiking in the Lake District in the early 1980s; from left, Mike Hassell, Jeff Waage and Dan Rubinstein. Although then in fashion, within the Silwood Circle red socks were said to be worn on the hikes by aspirants to fellowship in the Royal Society. *(Courtesy M. J. Crawley)*

Illus. 6. Mick Crawley when a postgraduate student in the 1970s: seen here in his office at Silwood — before the days of desk-top computers. *(Courtesy M. J. Crawley)*

Illus. 7. David Rogers and Sarah Randolph with their children: from left, Emily, Jack and Thea. The photograph was taken in South-West Kenya near the Tanzanian border in December 1988 or January 1989. Rogers and Randolph were on a Royal Society exchange scheme, visiting the International Centre of Insect Physiology and Ecology. John W. S. Pringle FRS, Linacre Professor of Zoology (1961–79), had earlier helped to found the centre for African scientists. Rogers and Randolph were at the Centre's Nguruman Field Station near the Ewaso N'giro River, studying tsetse flies in the area. Here they are at a camp fire breakfast after an early morning drive to spot wild animals. (Courtesy D. J. Rogers)

Illus. 8. Nash's Field, Silwood Park. Long-term experiments are being carried out on the exclusion of rabbits from grassland, as well as on the effects of liming and fertilizing. Insect herbivores and molluscs are excluded from half of the plots. It appears that insects increase plant diversity by reducing grass dominance, while molluscs reduce plant diversity by feeding selectively on the seedlings of herbaceous perennials. *(Courtesy M. J. Crawley)*

Illus. 9. Sir Richard Southwood. This portrait, by Mark Wickham, was commissioned by Merton College when Southwood was Vice-Chancellor of the University of Oxford. The portrait is highly iconic in that Southwood gave instructions that the artist include some important life markers — as, for example, the insects specially chosen from his collection, the hand lens in the left foreground, and a first-edition copy of his book *Ecological Methods* (held in his hand). For Southwood's full instructions see chapter 7, note 112.
(Courtesy the Warden and Fellows of Merton College, Oxford)

Illus. 13. Dick Southwood planting a tree at Silwood to mark the opening of the Southwood Halls of residence in 1983. *(Courtesy ICA)*

Illus. 14. Laboratories, named for Professors Munro, Kennedy and Lees, were opened at Silwood in 1987. *(Courtesy M. J. Crawley)*

Illus. 10. John Lawton standing outside the Centre for Population Biology at Silwood in 1995.   (*Courtesy J. H. Lawton*)

ECOTRON
CONTROLLED ENVIRONMENT FACILITY

Illus. 11–12. Left, one bank of ecotron chambers, 1993. *(Courtesy J. H. Lawton)* Right, one small ecotron community. *(Courtesy ICA)*

Illus. 15. Silwood Manor House, showing the conservatory where during the 1970s many work-related conversations took place among members of the Silwood Circle. *(Courtesy ICA)*

Illus.16. Mertens Acres, Silwood Park. This oak woodland regenerated on a former turnip field during the myxomatosis epidemic of the 1960s when rabbit numbers were very low. (*Courtesy M. J. Crawley*)

Illus. 17. From left, Brian Flowers, Mary Flowers, Gordon Conway, Nigel Bell and John Beddington at the celebration to mark the 25th anniversary of the Imperial College Centre for Environmental Technology in 2002. Conway, Bell and Beddington were successive directors of the centre. Lord Flowers, a former Rector of the college, was instrumental in setting up the centre. (Courtesy ICA)

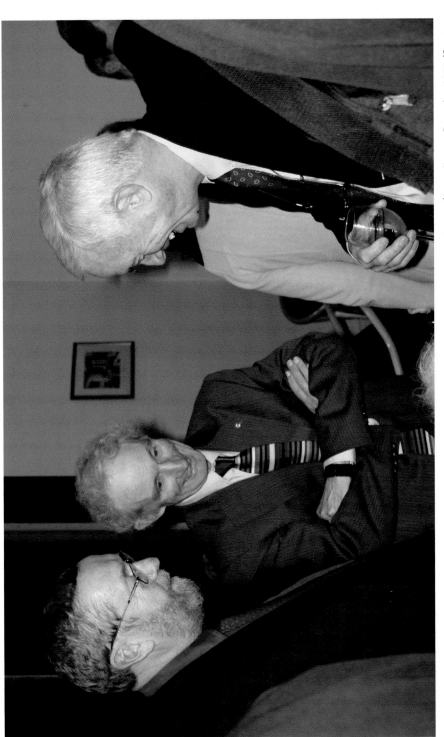

Illus.18. From left, John Beddington, Bob May and Roy Anderson at Silwood on the occasion of Mike Hassell's retirement/farewell party, December 2007. *(Courtesy M. J. Crawley)*

Illus. 19. Bob's walk, Grindelwald, 2009. From left, Zoe and Michael Stumpf, Michael Bonsall, Sebastian Bonhoeffer, Paul Harvey, Bob May, John Lawton, Charles Godfray and John Krebs. *(Courtesy M. P. Hassell)*

chapter six

Hard Work and the Making of Reputations

Robert May and Richard Southwood, 1971–1979

> *Theory supplies landmarks and guideposts*
> *and we begin to know where to observe and where to act.*[1]

The emergence of new ecological theory in the 1960s and 70s is discussed in the first section of this chapter. Its main focus is on some ideas proposed by Robert May in the early 1970s.[2] The second section shows how May's theory influenced field and laboratory work, and focusses especially on work carried out at Silwood. May's work was a guidepost, illustrating the more general claim in the epigraph above. However, not everyone was convinced it pointed in the right direction. The third section looks at the wider reception of the new ideas, and at how they were defended by members of the Silwood Circle. The fourth section includes discussion of Southwood's growing reputation and covers both his and May's election to the Royal Society. (A tip for readers: the more difficult material is discussed in section 1. The other sections are comparatively straightforward.)

6.1 Robert May and *Stability and Complexity in Model Ecosystems* (1973, 1974 and 2001)

> *It should be emphasized that the study of model ecosystems was never more than a corner of a larger canvas, painted by field and laboratory experimenters.*[3]

Robert May's fifteen-month sabbatical leave from the University of Sydney (1971–1972) was extraordinarily productive. He learned much from his visits to Silwood Park and Oxford, his conversations with Robert MacArthur at Princeton, and from further exchanges with others. After returning to Sydney in March 1972, May spent the final three months of his leave writing his influential *Stability and Complexity in Model Ecosystems*. In the late 1960s the theoretical ecology literature was not large. Its major expansion during the 1970s can in part be attributed to May's book. Cheap computation also played a role — not that May himself used computers very much.[4]

What was so special about May's book? I read it for the first time a few years ago, over thirty years after its initial publication. It is easy to imagine its impact. Many of the ideas discussed in the book were already around, but May brought them together with elegance and concision. He introduced models for the dynamics of multi-species communities under a range of conditions, and included also ideas from cybernetics such as the role played by time-delays in feedback mechanisms.[5] Overall the book should probably be read as a generalization of the Lotka and Volterra equations to more than two populations, and to randomly distributed (environmental) fluctuations in the parameters.[6]

Using plausible definitions of stability and complexity May showed that, counter to the view of most naturalists, his models indicated that stability does not necessarily increase with complexity.[7] Complex systems may have stable points and may show oscillatory behaviour around those points. Overall, however, the models suggest that as the complexity of an ecosystem increases its stability decreases. May acknowledged that naturalists could well be correct in their observations, but argued that the moral to be drawn was that they pay attention to how, as it were, ecosystems adopt biological strategies that promote what, from a mathematical point of view, is atypical — namely stability. His point being that ecologists and environmentalists need a better understanding of what engenders, and what endangers, the overall stability we so often observe.

Like MacArthur, May's view was that theoreticians should study simple ecosystems and construct models for the range of dynamical behaviour open to them. However, both men recognized that for the models to be realistic there was a need to take into account the work of laboratory and field ecologists. But how much empirical detail was needed? Another theorist, John Maynard Smith, stated, 'whereas a good simulation should include as much detail as possible, a good model should include as little as possible'.[8] This was a little confusing since the term 'simulation' has more than one meaning. Models that are intended to be as general as possible in their application should, indeed, have little empirical detail built in. Such models are sometimes amenable to analytical solution and sometimes not. When not, computer simulation can be used so as to arrive at numerical solutions. But the term simulation, as used by Maynard Smith, refers to models that have much biological detail built in. Usually such models are intended for a particular ecosystem, such as a species-specific fishery. While fishery simulations may be helpful, for example, in determining safe catches, they are not usually intended to have wide application. I use the term 'model' more generally, recognizing that it encompasses a range — from the highly analytical to what Maynard Smith labelled simulation. Today, as then, most ecologists appear to construct models that are closer to the simulation end of the spectrum. May understood Maynard Smith's point, but he also recognized that existing models of the Lotka-Volterra type were too remote from reality to be useful. There was a need, even for analytical models, to include a few empirically determined generalizations, ones relating to environmental fluctuation, behaviour, life cycles, age demography, or to some other features deemed important. Deciding which are the relevant variables is no easy matter. As May had already shown, models incorporating environmental fluctuation are tricky and not always helpful.[9]

In other areas of science it was well understood that analytical solutions for differential equations of scientific interest were few and far between. Chemists and physicists — economists too — were practised in the ways of inserting empirical data into their models, in the use of numerical approaches, and in the use of difference equations.[10] Non-linear mathematics had long been used in science, including in ecology. But up to the 1970s ecologists had focussed on stable oscillations or dynamic equilibrium points and had assumed that departures from these were due to chance perturbation caused by external factors, not to any intrinsic property of ecosystems themselves. It appeared that most observed perturbations were in any case soon damped. Chemists on the other hand had known about oscillating chemical reactions at least since the 1940s, had used limit cycles to describe such phenomena, and had recognized that the oscillations were inherently unsustainable.[11] Recognizing that this was true also in ecology, May introduced new concepts that stimulated much research. His book included models for communities of one species, and for two and three interacting species (all abstractions, see appendix two) that showed a way to move forward.

The models were of great interest to those working at Silwood in the early 1970s since much of their research focused on limited species interactions, much pertaining to herbivorous insect pests and the parasites and parasitoids that preyed on them. They understood that May's models not only had intrinsic scientific interest, they also had potential application in biological pest control work, and in thinking of ways to reduce agricultural dependence on chemical pesticides. The dynamical properties of the models were exciting. Indeed, some theoretically inclined ecologists became so enamoured with the properties of non-linear equations that they spent more time learning about them than they did about current work in their own fields. Roy Anderson told me that this was true of him for an extended period in the 1970s — though he held the strong belief that the equations had something important to teach ecologists.

One person who, by the early 1970s, had been modelling ecological systems for several years was C. S. (Buzz) Holling, director of the Institute of Resource Ecology at the University of British Columbia.[12] Michael Crawley who spent a summer at Holling's institute was much influenced by his ideas, as was Gordon Conway. Holling worked toward the simulation end of the model spectrum. Like May, he wrote a review of the kinds of problems that interested him. It, too, was published in 1973.[13] There are some parallels as well as some differences in the way these two envisaged ecological theory. Looking back from 1973, some of the parallels can be traced to a symposium on the diversity and stability of ecological systems held at Brookhaven in 1968. Richard Lewontin gave a paper at the symposium titled 'The meaning of stability' in which he suggested that there could be multiple stable states for an ecological community.[14] By the time he heard Lewontin's paper, Holling had been working on prey-predator systems for a few years. In that work he used two concepts, namely numerical and functional response, introduced earlier by M. E. Solomon in connection with the more general idea of predator population response to prey density. Holling looked closely at actual predator-prey behaviour and redefined both the numerical and functional response in light of his observations (see appendix 2).[15] Basically, the numerical response is the change in the abundance of a predator as a function of the abundance of its prey, whereas the functional response relates to prey consumption by a predator as a function of prey population. Advancing some ideas expressed earlier by Hutchinson, Holling claimed that fairly complex behaviours can affect the consumption of prey, and thus also the population numbers of both prey and predator. For example, a predator might well become satiated if its prey were abundant and would not then consume every prey item in its path. Further, it takes time to eat and digest an item of prey, time spent before going on to look for more. The prey species, in turn, may learn how to hide so that more time would then be needed to seek it out. For reasons such as these there could be a range of decelerating rates of predation with a consequent range of outcomes for the two populations.

Hearing Lewontin at the conference prompted Holling to shift his focus slightly to the ways in which species populations can, up to a point, bounce back after near collapse, possibly to even greater numbers than before. He introduced the idea of resilience to describe this or, to put it another way, to describe the amount of disturbance a system can take before it shifts in such a way that it comes under the control of a new set of variables. Holling, like May, was seeking new ways of thinking about stability and was coming to the view that it was more important to focus on the variability of ecological systems than on constancy.[16] His simulation modelling led to conclusions not unlike some reached by May using a more analytical approach. As mentioned, simulations are the models of choice for more empirically minded ecologists, and for those who, like Holling, have an interest in resource management. Simulation was also the methodology of choice for Holling's former colleague Kenneth Watt whose interests in forest insect pests were similar to Holling's, and for Watt's former student, Gordon Conway, when he began working at Silwood in the late 1960s.

Robert MacArthur, too, was prompted to think more about stability after the Brookhaven conference. But he did so from an analytical, hypothetico-deductive, perspective. In this he was to have an important influence on May. With problems relating to dynamic equilibria, complexity, and stability in mind, May published a number of papers that anticipated his 1973 monograph.[17] His book brought many of the then current themes together. It set out some analytical models in which immediate practical utility and empirical detail was sacrificed to get at basic principles. May's models were intended to apply not to any specific ecosystem, but to ecosystems in general. His book remains a fine introduction to this type of modelling. May took into account the kinds of refinements to Lotka and Volterra discussed earlier by Hutchinson and Holling, both of whom understood that there are all sorts of reasons why population growth cannot respond immediately to perturbation. Further, he recognized that the Lotka-Volterra equations are not truly stable.[18] His major contribution, however, was to show that multi-species versions of the equations are no more stable than the two-species version — indeed they are likely to be less so. 'In mathematical models of multi-species communities, complexity tends to beget instability rather than stability'.[19] This claim ran counter to MacArthur's intuitive argument for stability in complex ecosystems.[20]

As mentioned in chapter five, following Elton, E. P. Odum and others, MacArthur believed that food webs with many links would be more stable than those with few. He claimed that if one link were to break, energy could be exchanged through the many other available links allowing the system to move back toward stability. But, as May showed, the mathematics suggested otherwise. May's models also ran counter to much observation. Usually when heightened population oscillations occur, they are damped and stability is restored. But can we rely on that? The fact that complex systems exist all around us and appear relatively stable raises the possibility that complexity may have evolutionary advantages not fully understood. Perhaps instability is the price we have to pay for the evolution of new species. Perhaps there is something peculiar to the complexity observed in nature that allows for a degree of stability as well as for change. Nature, as May remarked, may possess 'devious strategies'. However, as he also pointed out, 'nature only represents a small and special part of parameter space' — perhaps the most stable part.[21] His book concludes with a number of speculations and some advice. Perhaps most important is that:

> complex and stable natural systems are likely to be fragile, tending to crumble and simplify when confronted with disturbances beyond their normal experience [and that] until such time we better understand the principles which govern natural associations of plants and animals we would do well to preserve large chunks of pristine ecosystems.[22]

The properties of the simple non-linear equations May introduced in his book were an eye-opener for many ecologists and raised philosophical questions about the descriptive role of mathematics in nature — not the first time that such questions have been raised about mathematical theories in science. May was sensitive to this issue. He wrote 'my background is in theoretical physics, and I am at least aware of the danger that my interests are liable to be animated too much by elegance and too little by common sense.'[23] Perhaps what especially caught the eye of ecologists were the graphical representations of the various equations which showed clearly the different behaviours that, in principle, were open to non-linear systems.

May was not alone in thinking along these lines, further illustrating the existence of a situational logic of the type discussed in chapter one.[24] I have mentioned Holling, Watt and Maynard Smith but there were others, including Richard Levins and Simon Levin whom May acknowledged in his book. Scientists who wish to move forward need to be aware of the situation they are in, and to have the tools to exploit it. May had the tools and was more aware than most. Like Hutchinson and MacArthur earlier, not only did he weave together some important situational strands, he produced a map suggestive of paths that other ecologists could profitably take. May's book was reprinted with a new introduction in 2001. Looking back at the scene-setting chapter with which the first edition began, May remarked that it 'looks prehistoric. But it does mark the beginning of a seismic shift'. In that he is surely correct. Stability in nature came to be seen in a new light, as something maintained by complex dynamical interactions, and as something fragile. For centuries 'the balance of nature', albeit understood in dynamical terms, had been a comforting fiction, but now it had to be given up — or so it was believed. Even though the fragility of the limit cycle (see appendix 2) and its dependence on feedback was not well understood, ecologists were able to see something important in the new mathematics. They began to look seriously at the properties of difference and differential equations. May, himself, was soon to discover the idea of deterministic chaos in this way. He was not alone. Mathematical chaos was discovered independently by several people at around the same time.[25] May's contribution was to show the properties of simple non-linear difference equations used to describe the growth of biological populations with non-overlapping generations. He showed that one can get chaotic behaviour even in the simplest one-dimensional difference equation.[26]

Leaving aside the exact properties of the equations, it was soon recognized that old debates about density dependence had to be rethought. Density dependence could, perhaps, produce population fluctuations as erratic as anything external to the system. Indeed, the discovery of chaos encouraged some to take up the problem of distinguishing the random from the deterministic — not only in ecology. As May noted it was a problem calling for resolution in many other places; in the stock or currency exchange markets for example. May has played a role in that area too.[27]

For my story the mathematical details are not that important. More so is that the new non-linear dynamical models changed the ecological agenda. They brought recognition that the earth is a complex dynamical system, and that its present state is not necessarily permanent.[28] The models were not fully descriptive, but the ways in which they deviated from reality suggested new experiments and ways for the theory to be improved. While confirmation is sketchy, the heuristic value of the new approach, and its role in changing attitudes, has been considerable.[29]

Although May's book was a major contributor to the changing intellectual climate among ecologists, it far from covered everything of interest, nor did it get everything right. For example, as May acknowledged in the 2001 edition, his 1973 treatment of niche overlap and the limits to similarity among competitors was shown to have problems. It is interesting that the question of species coexistence as framed by Hutchinson, and the subsequent widespread discussion of limits to the similarity between species, dropped out of fashion during the later 1970s. May thinks it an area still worthy of study.[30] However, one area treated only briefly in his book was soon to receive much attention — namely the spatial heterogeneity of ecosystems.[31] Such heterogeneity is important, especially for prey-predator populations where the patchy distribution of prey is a notable feature and may have advantages from the point of view of population stability. As will be seen in the next two chapters, patchiness was of interest to several members of the Silwood Circle.[32]

Some experimental studies prompted by May's book were not conducted entirely in the spirit of the book — not that this matters since interesting information was gained nonetheless. I am referring to studies related to the issue of biodiversity, a topic that has generated much heat in the literature. If it is true that complex ecosystems have the potential to be less stable than simple ones, what exactly are the advantages of diversity? We hear many calls for the preservation of biodiversity but on what grounds is it desirable? Certainly on aesthetic and possibly economic grounds, but surely there are other more pressing reasons? May would like to see an answer to the question of *why* there are more species in some places than in others.[33] While experiments performed since the book's publication appear to show that species-rich systems are more productive of biomass (see chapter 8 and the BIODEPTH project, for example), they have not addressed the question of why this is so. As May put it, it is important from the conservation point of view to know why there are different kinds of ecosystems: 'some natural systems are very productive with few species, whilst others are very species-rich'.[34] Here, as elsewhere, May's environmentalist anxieties shine through. His models have proven useful but they have also bolstered current belief that we are living on the edge.[35]

As mentioned in chapter four, May took up the professorship at Princeton University in 1973. There he joined a small group of ecologists, all inspired by MacArthur's example; it was a good place to be. Two of May's Princeton colleagues were Henry Horn and John Terborgh.[36] Another who joined a little later was Dan Rubenstein, now chair of the much larger Department of Ecology and Evolutionary Biology. I asked May who else in North America he exchanged ideas with at that time. He mentioned Robert Paine at the University of Washington in Seattle, Simon Levin at Cornell, William Murdoch then at the University of Wisconsin at Madison, Jane Lubchenko at Oregon State University, and Jared Diamond at the University of California at Los Angeles.[37] His wider circle included Edward Wilson, Richard Lewontin, Lawrence Slobodkin, Joseph Connell, Nelson Hairston, Richard Levins, Jonathan (later Joan) Roughgarden, Eric Pianka, Joel Cohen, and Kenneth Watt. Within a very short period, May had become known to many of the leading North American figures in ecology, parasitology, and evolution theory. He also made contact with the mathematical biologist, George Oster, working at the University of California at Berkeley. Together with Oster he would soon develop his ideas on chaos.

6.2 T. R. E. Southwood, Robert May, and the Silwood Circle

Given the excitement of the new ideas, and their potential to help with some of ecology's pressing problems, it is not surprising that people working at Silwood were keen to associate with May. They wanted to learn more about his models and the possible application of non-linear difference equations and other new ideas to their own work. Southwood, attuned to the situation, was quick to invite May back to Silwood, this time as a visiting professor. The professorship, beginning in 1973, extended over many summers until May moved permanently to England in 1988.[38] For May, too, the association was important since he needed to work closely with experienced entomologists and ecologists. People at Silwood were working on simple systems open to modelling of the type he favoured. Southwood organized daily meetings at which people brought their specific problems to the table. Gordon Conway told me that they 'had problems coming out of their ears' and that much time was spent both privately and collectively thinking about how to approach them. Silwood, with its relaxing, yet competitive, atmosphere was an important backdrop to this, to the formation of new associations, to successful collaboration, and to the publication of many joint papers. At this early stage, those who appear to have profited most from their association with May were Southwood, Conway and Hassell.

In the early 1970s Southwood was well known as an entomologist and had a growing reputation as an ecologist, the latter stemming largely from his *Ecological Methods* and from the work discussed in chapter four.[39] He was also building a reputation as a man of affairs. He took a major role in college administration and, as will be discussed below, served on several national committees. He also gave a number of public talks on matters related to the environment. But to have a serious voice in national affairs he needed to join the fellowship of the Royal Society. It would appear that the papers he published between 1974 and 1976 with Robert May, Gordon Conway, Michael Hassell and Hugh Comins helped in this endeavour.[40] Southwood's decision to work with May and to attach himself to the MacArthur tribe was beginning to pay off. But, while in awe of May, Southwood was a grounded naturalist and a little wary of theory. He understood its importance and was welcoming to theorists at Silwood. Nonetheless he believed that some of the theoretical ideas being discussed were too remote from the natural world to be useful. This is illustrated in an equivocal review that he wrote of the proceedings of a symposium held in MacArthur's memory.[41] While critical of some of the contributors whom he saw as being carried away by new generalizing models, he was full of praise for those whom he saw as bringing theory and observation together in interesting ways. MacArthur, himself, was held up as exemplary — someone who had a sound basis as a naturalist from which to arrive at useful mathematical models. MacArthur's supporters often stressed his naturalist credentials — that he was a child naturalist, that he was an expert on warblers, and so on. It was a way of validating also his modelling. In his review Southwood especially praised two of the senior contributors, G. E. Hutchinson and J. H. Connell, whom he saw as similarly grounded.[42] Just as MacArthur had done, others, he suggested, could learn from their example. Clearly he believed that people at Silwood, well trained in field or laboratory work were, with May's help, using models in constructive ways.

The Symposium papers are interesting in showing what was perhaps a high-water mark for MacArthurism. John Krebs also reviewed the volume and stated:

> Without any doubt Robert MacArthur was the most influential ecologist of his time. He completely revitalized the study of community ecology, both by brilliant intuitive insights and by developing the tool of mathematical modelling. Through his influence the study of communities and ecosystems was raised from the doldrums of ecological energetics to become an intellectually challenging branch of ecology, and within the space of a few years it had displaced population ecology as the most fashionable area of study. ... Indeed, it is hardly an exaggeration to assert that virtually all the younger leading ecologists in North America are academic children, nephews or nieces of MacArthur.[43]

Krebs's rhetoric, while suited to the reviewing of a memorial volume, was based in genuine conviction. It is interesting both in showing how MacArthur was remembered and in its reflection of the tribal divide in the department at Oxford, briefly discussed in the previous chapter. Clearly Krebs saw community ecology as superior to the energeticist and systems approaches then being introduced at Oxford by John Phillipson. Historically, being in the MacArthur camp turned out well for many.[44] In Southwood's case the joint papers and his promotion of the new ideas were to help his 1977 election to the Royal Society.

Another important way of identifying with the MacArthur tribe was to be a contributor to May's *Theoretical Ecology,* published in 1976.[45] This book, like May's monograph, helped to set the ecological agenda over the next quarter century. In introducing the volume May wrote 'our broad aim in constructing mathematical models for the populations of plants and animals is to understand the way different kinds of biological and physical interactions affect the dynamics of the various species'.[46] As to his own aims, he reiterated what he had stated in his monograph, namely that he wanted to get at general principles rather than at any quantitative predictions. And to do this he would examine very simple systems, including single populations which do not exist naturally but which, like two- and three-species communities, were being studied seriously in laboratories at Silwood and elsewhere.[47] The various chapters in the 1976 book showed how May's principles and models could be applied.

Southwood contributed a chapter in which he looked at single populations, and at what he called their bionomic strategies (eg size, longevity, fecundity, range and migration habits) which he saw as evolving to maximise an organism's fitness to its environment.[48] He discussed some bird and insect populations and made use of May's ideas as well as some from MacArthur and Wilson's *Theory of Island Biogeography*. The ideas were well integrated with results from some empirical studies being carried out at Silwood. In his own chapters, as in his monograph, May discussed how two and more-interacting species populations could be envisaged. In a chapter on two interacting populations he looked to the empirical work of John Lawton and John Beddington; whereas for a more discursive chapter on complex communities he, like Southwood, looked back to the work of MacArthur and Wilson. Similarly retrospective is the chapter that he wrote together with Jared Diamond in which some conclusions are drawn on the optimal design for nature reserves.[49]

While most of the contributors to the volume were working in America, among them Diamond, Edward O. Wilson, Henry S. Horn, Eric R. Pianka, Joel E. Cohen and Stephen J. Gould, there were two other Silwood contributors, Michael Hassell and Gordon Conway.[50] Hassell picked up on Holling's idea of predator search behaviour and how this affects the population numbers of both the predator and its prey. He introduced extensive laboratory data on parasitoids, some of it his own. Conway's chapter, more applied in nature, illustrated the use that the new theoretical ideas could have in the area of pest control.[51] For example, new theory allowed for a clearer understanding of pesticide resistance. When pesticides are overused the result can be a resilient rebound with pest populations reaching even higher levels than before treatment. Bringing Holling's ideas on resilience to bear, as well as the ideas of K and r reproduction (see chapter five) and different pest reproductive strategies, Conway discussed his earlier experiences with pest control in the cocoa plantations of Sabah. At Silwood Conway was assisted by the economist Geoffrey Norton whose appointment, like Comins's was made possible by Conway's Ford Foundation grant. Norton was of considerable help both to Conway and Southwood in thinking through their use of models, and to Conway with his interests in agronomy and sustainable development. Norton also contributed to the teaching of MSc students in the Centre for Environmental Technology. Later he was to move to the University of Queensland.

May's *Theoretical Ecology* helped in bringing the work of the Silwood Circle to the fore, while at the same time linking some of its members to leading ecologists and evolution theorists in the United States. The book was used in colleges and universities worldwide where it helped in introducing students to new ideas in theoretical ecology. It also helped to establish the authors as authorities in the field. The book went into three editions, the content changing with the science. The latest edition was published in 2007.

Southwood's new associations, the work he helped to nurse along at Silwood, and the papers he published together with May and others, brought him much attention. While his earlier book on ecological methods was respected, until the 1970s he was better known as an entomologist. By the mid-1970s, however, he had become associated with serious work in theoretical and experimental ecology. Further, not only had he provided a good working environment for ecologists at Silwood, he continued to push the environmentalist agenda. The opening of the very successful Imperial College Centre for Environmental Technology in 1976 was a major achievement in a difficult economic period. Getting anything up and running in British universities at that time was seen as highly admirable. Southwood had acquired a reputation both as a scientist in touch with new ideas, and as a successful administrator. Many job offers came his way and he was invited to give lectures all over the world.[52]

But Southwood also wanted to advance the careers of his appointees in ecology. Gordon Conway and Michael Hassell soon became readers at Imperial College. And before he left for Oxford Southwood lobbied successfully for them to be awarded chairs and for Roy Anderson to be given a readership.

Southwood also wanted Hassell to succeed him as director at Silwood and discussed the possibility with May who was supportive.[53] But the college Rector, Brian Flowers, had a different idea and appointed Michael Way. Hassell had to wait until 1988 before becoming director at Silwood.[54] Southwood saw Conway, Hassell and Anderson as deserving both in terms of their research and in their having successfully built up teaching and other departmental work in ecology and environmental science. The letters he wrote in their support, first for readerships and then professorships, contrast in their enthusiasm with the letter he wrote for William Hamilton.[55] He also had his eye out for John Lawton. Wanting some friendly faces to join him in Oxford he was thinking of Lawton and Anderson. May wrote to Southwood that he would be unhappy to see Anderson drawn away from Silwood since the two had by then formed a good working relationship. He also wrote that he thought that Lawton would be unwilling to leave York for Oxford.[56] However, letters from Lawton to Southwood during the 1970s suggest he would have been very willing to move to Silwood — as, indeed, he was later to do. In one letter Lawton complained of 'a lack of anybody to talk to about ideas. Sometimes York feels it's a million miles from Silwood'.[57] Southwood appreciated Lawton's research and thought he added liveliness to the 'Silwood mob'. He helped him to gain grants and supported his promotion to reader, then professor, at York. The two also co-authored a book.[58] Lawton's work will be discussed further in chapter eight. Conway, Hassell and Anderson will be discussed further in chapter seven.

6.3 The reception of new mathematical modelling by the ecological community

Not all in the ecological community welcomed May's ideas, or those of other mathematical population/community theorists. For a start the division among the former students of G. E. Hutchinson, discussed in chapter five, continued into the successor generation. Still influential on one side of this divide was E. P. Odum who had written a textbook, *Fundamentals in Ecology* in which systems ideas were promoted.[59] A third edition was published in 1971 and, like May's *Theoretical Ecology*, it was widely read by students. Odum and his brother H. T. Odum believed that the sheer complexity of nature demanded a systems approach if ecology was to move forward. The Odums and their followers saw the new population models as simplistic and reductionist in the extreme. Just as the Cody and Diamond volume of 1976 promoted ideas in the MacArthur vein, the system theorists had their gatherings and publications. One set of symposium papers published in 1975 contains a paper by Eugene Odum whose visionary rhetoric matched anything coming from the MacArthur camp. Odum believed that systems ecology was 'revolutionizing' the field and that in this 'new age of environmental awareness' ecology was no longer 'just a subdivision of biology' but 'a new discipline' integrating the 'biological, physical and social science aspects of man-in-nature interdependence'.[60] Odum, his brother, and those sympathetic to their viewpoint were enticed by new computing power. Huge data bases were brought to bear in the modelling of complex ecosystems. The International Biology Program (IBP) which ended in 1974 saw a number of such projects.[61] One snag was that there was no uniform understanding of what a system in the ecological context meant.[62] Further, the simulations were complex. While less complex than nature itself, they were similarly difficult to interpret. Nonetheless the battle lines were drawn.

On the other side of the divide were some papers delivered in 1978 at the twentieth symposium of the British Ecological Society. The symposium was devoted to population dynamics. The aim of the organizers, one of whom was Roy Anderson, was 'to provide an up-to-date review of past work in this rapidly expanding field'.[63] The focus was on the dynamical aspects of populations and the relevance of new ideas to resource management, the epidemiology of infectious diseases, the use of parasites and parasitoids as biological control agents, and on the effects of human behaviour on ecosystems. Serious system theorists were excluded. Even a cursory glance at the papers show that the new theoretical ideas of MacArthur and May on population and community were catching on, and were influencing the way ecologists thought about their problems. This was true also of those who were not themselves theoreticians, illustrating the considerable heuristic power of the ideas. Spatial heterogeneity, mentioned above, was a major theme, one developed in papers by Anderson and Hassell. Single-species models were prominent, as were studies of simple predator-prey, host-pathogen and host-parasite associations. A few papers looked at multi-species communities and at Holling's ideas on resilience and multiple stable states. But overall, keeping things relatively simple and using simple models was the principal characteristic of contributions to this symposium.[64]

As to chaos ideas, they did not immediately catch on in the way that more general non-linearity did. While May's books were highly suggestive of further work, it seems that there were (still are) limits to empirical ecologists' belief in the descriptive power of difference equations. But the intellectual climate slowly changed in this regard. First, it came to be understood that chaotic behaviour is not limited to difference equations and is a property also of some of the Lotka-Volterra type of differential equations long accepted by ecologists. Second, mathematical theorists began to recognize some of the characteristic markers that one should look for when deterministic chaotic processes are at play. So work in this line continues. Driving it is what drove May earlier; namely the goal of explaining complex behaviours, including the fall into chaos, in terms of simple deterministic rules.

While wary of May's type of mathematical modelling, many ecologists working in the 1970s were also wary of the intrusion of system ideas from engineering and operational research. This can be inferred from papers given at two meetings held in 1974. One was a workshop held at the University of Utah at which Simon Levin noted the many subgroups within ecology, and the divisions between systematists and population theorists.[65] The other meeting, the First International Congress of Ecology held in The Hague, drew far more participants and was an occasion for the proponents and critics of methods used in the IBP to have their say; and for supporters of the MacArthur programme and those critical of May's type of modelling to have theirs. At bottom the debate was about the future direction of ecology: was it to be resource management-engineering focussed, or was it to be science based? And, if the latter, should the approach be as MacArthur and May envisaged or should ecologists seek some other direction?[66] A few years later May showed his frustration with all the methodological debates and wrote about 'naively simple formulations of "The Way to do Science"', especially when put forward by 'doctrinaire vigilantes'.[67] His frustration is understandable, but there was a counterside. It appears that many experimental ecologists turned their backs on the new types of modelling, and on the methodological debates surrounding them, intent rather on clever manipulations in the laboratory and in some natural habitats[68] That there was a certain amount of anarchy when it came to method and approach may well have helped ecology over the longer term.[69]

6.4 T. R. E. Southwood in the Wider World

As May and Hassell noted in their biographical memoir, beginning in the 1970s Southwood's skill on committees and his skill as a committee chairman were increasingly recognized.[70] Entering his forties, Southwood was gaining voice on the national stage. In 1969 he was appointed to the governing board of the Glasshouse Crops Research Institute (GCRI). Established in 1954 at Littlehampton, Sussex, the Institute was formed by merging a number of other institutes, notable among them the horticultural research station at Cheshunt, located in a part of Hertfordshire long known for its greenhouse nursery industry. The GCRI employed about 200 people and was administered by the Agricultural Research Council (ARC). Southwood remained a governor until 1981. One of the reasons for his appointment was increasing interest in biological and integrated pest management in greenhouse production. The GCRI became a leader in this field and some of the work then being carried out at Silwood, notably the nematode work begun earlier under B. G. Peters, contributed to its success.[71] Southwood's involvement with research council business was to increase considerably in the following years.

The origin of the Natural Environment Research Council, founded in 1965, was discussed briefly in chapter three. Southwood's predecessor at Silwood, O. W. Richards, played an important role in its creation. In 1972 Southwood was appointed chairman of its Terrestrial Life Sciences Grants Committee and, in 1977, he joined the Advisory Board on Research Councils. His voice was thus heard on financial matters, and on how research monies in the life sciences should be distributed. As a trustee of the Natural History Museum from 1974, he supported changing attitudes towards the role of museums in society. Southwood saw the need for opening up the museum to a wider public and, while not wishing to see its role as a research institution diminished, recognized that more attention needed to be given to its educational role. Given the economic recession of the 1970s and new policies introduced after the Thatcher government came to power in 1979, museums were being asked to become more financially independent. New sponsors had to be sought and their interests heard. It is perhaps not surprising that Southwood, skilled with people and a good communicator, was appointed chairman of the trustees in 1980. May would later follow him in that role.

In 1974 Southwood joined the Royal Commission on Environmental Pollution then under the chairmanship of Brian Flowers.[72] He actively participated in a number of its reports. The fifth and sixth reports of 1976 were on air pollution and pollution from the nuclear industry respectively. The seventh, in 1979, was on agricultural pollution something Southwood had been thinking about for several years. The eighth, in 1981, was on oil pollution at sea and the ninth report of 1983 was on lead in the environment. Southwood contributed to all of these, but having been appointed chairman of the commission in 1981, he worked especially hard to get the recommendations of its ninth report made law. In this he was successful despite serious opposition from the petroleum industry. Lead additives to petroleum and some other commercial products were banned surprisingly soon after the report's publication. It was a hugely important public health measure. The episode is discussed well in May and Hassell's biographical memoir. They make clear the high regard in which Southwood was held by civil servants, and the admiration they showed for the way in which he pushed the matter through the political maze.[73] Southwood also oversaw the tenth report (1984) concerned with tackling pollution more generally.

During the 1970s the Silwood Circle was an active ginger group within the British Ecological Society (BES). It can be seen as a tightly knit group of young people trying to change attitudes and priorities. Southwood was elected president of the BES in 1976. Because of the kinds of environmental concerns discussed in chapter four, ecologists played an increasing role in the making of public policy decisions. It was worthwhile having a voice in the BES since the society was being consulted on a wide range of issues. Southwood was seen not only as a good scientist and committee man, but as a good science politician, someone who would help to solidify the BES as a political force.

However, as mentioned above, for Southwood to be taken seriously on the national stage he needed to be a Fellow of the Royal Society. This was not easy, in part because ecology was poorly represented in the Society. Support for Southwood was therefore narrow. There were also opponents to his election; one was the Oxford zoologist E. B. Ford. Southwood was first proposed in 1971, but elected only in 1977.[74] After his election he immediately began thinking about bringing in other ecologists to join him.

He also thought about how to make their election easier than his had been. May was Southwood's first choice and he got to work on his behalf.[75] He advised May to make an effort to impress biologists, and to show them that he knew something about their subject — that he was something more than a good mathematical theoretician. With that in mind May delivered a good paper at the ninth symposium of the Royal Entomological Society in September 1977. It was on some of the dynamical and evolutionary factors influencing species diversity. He included competition, predation, food web structure, the numerical abundance of species, and evolutionary genetics. All these factors underscored the interesting and then fashionable question, raised earlier by Hutchinson, of why there are so many more insects than other animals. May speculated also on the relation between the numbers of species of different types of terrestrial animals and their size. His far-reaching paper is interesting in and of itself. But it also shows May consciously demonstrating to his audience that he had thought deeply about problems of interest to them. And, in connecting entomological species numbers to more general questions about species numbers, he showed that he had thought also more generally about evolutionary zoology. Southwood sensibly held that such a demonstration was necessary for someone wishing to be seen as an ecologist and who, only five years earlier, had been a professor of physics.[76]

Some of Southwood's correspondence from that time illustrates his concern over arrangements at the Royal Society. For example, shortly after his election he wrote to James Dodd, professor of zoology at the University of Wales at Bangor, and chairman of the British National Committee on Biology. He stated that he was unhappy with how the nominating sections at the Royal Society were organized, and that in his view the zoology section had too little clout. He thought it overshadowed by the sections on human and animal physiology and he wanted more support for other areas of zoology, notably ecology. In my view he was justified in believing ecology to be under-represented at the Royal Society.[77] After his own election, and worrying about how to get May elected, Southwood wrote out a list of people whom he thought worth seeking out for support. Those who early responded favourably were John Maynard Smith at Sussex, Maurice Bartlett at Oxford, and Southwood's Imperial College colleagues, the zoologist J. S. Kennedy, the statistician D. R. Cox, and the college Rector, Brian Flowers. Also on side was R. O. Slatyer, head of the zoology department at the Australian National University. In addition to having May seen as a worthy ecologist by zoologists, Southwood worried that physicists would judge him on his contributions to physics rather than on those to ecology. In retrospect that worry appears to have been unfounded, but Southwood's concern is telling. He was someone who planned for all eventualities. In this connection he contacted the cosmologist Fred Hoyle (a curious choice) and Roger Blin Stoyle, a physicist and colleague of Maynard Smith's at the University of Sussex. Both wrote that May's contributions to physics were acceptable and would not hold back his election. Hoyle wrote that because others saw him (Hoyle) as an astronomer — neither physicist nor ecologist — it would be inappropriate for him to sign May's certificate, but that he would support his election.

But there was also the problem of which biological theorist should be the next to be elected. Southwood very much wanted it to be May. But others thought William Hamilton more deserving. Interestingly one of those was V. C. Wynne-Edwards whose work had been much criticized by Hamilton. He wrote to Southwood that he would sooner support Hamilton, but that if Southwood thought Hamilton should wait another year, then he would go along with that. O. W. Richards was firmer, stating clearly that he thought Hamilton deserved to be elected before May and that 'I had better leave May alone'. Charles Elton wrote that he could not support someone whose work he did not understand. Clearly getting support was not straightforward and entailed much letter writing.[78] But Southwood's campaign was successful and May was elected in 1979.[79] May wrote to Southwood that he was delighted, and that he was 'particularly aware that the whole happening is due largely to your kind efforts ... only a colonial can really fully appreciate how enjoyable it is to be elected to immortality in this way'.[80]

Southwood went to similar lengths for others that he later wished to see elected — including John Krebs, Roy Anderson, Michael Hassell, and John Lawton in the 1980s, Richard Sykes in the early 1990s, and Michael Crawley, John Beddington and Gordon Conway in the early 2000s.[81] After both he and May were elected, Southwood knew that it would be easier to move his ecology agenda forward. It would also make things easier for others in the Silwood Circle.

Endnotes

1. (Holland, 1995), 5. Holland's book is a good introduction to the more general problem of complexity, not solely as it pertains to ecology.

2. A few mathematical and technical details relevant to the chapter are given in appendix two.

3. (May, 1973, 1974, 2001); Epigraph (2001), xx. In the preface to the 1973 edition, May credited the following with inspiring him: L. C. Birch 'who started it all', J. H. Connell, N. G. Hairston, M. P. Hassell, H. S. Horn, S. A. Levin, R. Levins, J. Roughgarden, L. R. Slobodkin, T. R. E. Southwood, K. E. F. Watt and 'above all', Robert MacArthur. Charles Birch was the professor in Sydney who encouraged May's nascent interest in ecology (see chapter four). The text in the third edition is unchanged from the second (1974) edition but includes a new introduction.

4. May was one of the first two doctoral students in physics at Sydney. He used the Silliac machine (Sydney's version of the University of Illinois' Illiac computer), the first main-frame computer in Australia, to carry out major calculations. May told me that the experience left him with 'an abiding tendency' to delegate computer tasks to graduate students. Early ecologist computer users included Kenneth Watt and Buzz Holling (see below). Early users in the Silwood Circle were Gordon Conway, Roy Anderson and Michael Crawley.

5. There had been some new work on this. See for example, (Rosenzweig and MacArthur, 1963).

6. In his book May used the methods of classical stability analysis, techniques he would have learned when studying engineering and physics.

7. 'Roughly speaking, we here take complexity to be measured by the number and nature of the individual links in the food web, and stability by the tendency for population perturbations to damp out, returning the system to some persistent configuration'; (May, 2001), 3.

8. (Maynard Smith, 1974), 1. John Maynard Smith FRS (1920–2004) was a student of J. B. S. Haldane at University College London and his principal interests were in the area of evolution theory. His early work in that field was on aging. During the 1960s he sought an explanation for the evolution of sexual reproduction, seeing it as a mode of introducing variation into populations, something that could be of advantage in changing environments; but he puzzled over its value to individuals. In the 1970s he became known for applying game theoretic approaches to animal behaviour so as to throw light on natural selection. Together with George Price, using models based on the prisoner's dilemma, he introduced the idea of evolutionarily stable strategies, (Maynard Smith and

Price, 1973). Maynard Smith won the Copley Medal of the Royal Society in 1999. He was the founding Dean of the School of Biological Sciences at the University of Sussex which opened in 1965. I met him during the 1970s when I was a DPhil student at the university, studying the history and philosophy of science. I remember some friendly chats about the history of science and about the new sociobiology; and he gave me some tips on multi-tasking, useful while parenting young children.

9. (May, 1971). In the fifth chapter of his 1973 book (see note 3). May introduced some random environmental fluctuation into his models but concluded that they did not seriously affect the more interesting conclusions reached using strictly deterministic models.

10. May used difference equations since they can recursively define population size at discrete time intervals. A continuous model using differential equations would not be a good fit for insect populations, for salmon, or for other animals with discrete generations. For many insects an entire adult generation can die off during one year with the next one appearing only in the following year.

11. The best known oscillatory behaviour in chemistry is perhaps the Belousov-Zhabotinsky reaction (malonic acid and a bromate; plus a catalyst, usually iron or cerium). It is famous because the different concentrations of reactants can be observed in the form of coloured rings. This reaction and other chemical oscillatory behaviour is discussed in (Higgins, 1967). Increasing the flow rates of the chemicals involved in oscillating reactions can lead to periodic doubling in the oscillations and further loops to the limit cycle (see appendix 2). Such Hopf bifurcation is named for Eberhardt Hopf who recognized the possibility in his mathematical models already in the 1940s. It was also recognized that eventually, as flow rates increase, the chemical systems become chaotic — not that the term chaos was then used or the phenomenon understood. In ecology it is population numbers that oscillate but, as May recognized, under extreme conditions it is possible for the oscillatory pattern to bifurcate, even to the point of chaos and ecosystem breakdown.

12. Holling had worked out some of his ideas on the functional response (see below) when working at the Canadian Forest Services Laboratory at Sault Ste. Marie, Ontario in the late 1950s. Both he and K. E. F. Watt (also at Saul Ste. Marie) used difference equations in their modelling.

13. (Holling, 1973).

14. (Lewontin, 1968). Papers from another important symposium of this period were published in (Lewontin, 1969). Lewontin's introduction to this volume was a little defensive. He claimed that the papers it contained were 'a vindication

in biology of our faith in the Cartesian method as a way of doing science' and that he saw the new molecular biology and the new population biology as progressing by 'breaking down bigger systems and examining the parts'. He wanted a synthesis of molecular and population biology: 'no population ecology without population genetics'. As an aside, (Lewontin, 1961) which uses game theoretical ideas may also have influenced Maynard Smith (see note 8). Lewontin, however, early gave up on game theory as applied to genes and behaviour and became a major critic of sociobiology.

15. (Solomon, 1949). For Holling and his interpretation of the functional response, (Holling, 1961, 1965, 1966). Holling was not alone in rethinking the functional response, further illustration of the semantic instability of commonly used terms.

16. For Holling stability meant that a population returned to an equilibrium point after some brief disturbance; resilience meant the persistence of a population, even after some major disturbance. Resilience was thus consistent with wild fluctuations in population numbers as sometimes observed, notably among the forest insect pests of interest to him. However, as Holling noted, resilience goes only so far and eventually a system may collapse into a qualitatively different state. In this connection he cited the work of K. E. F. Watt on major insect pests in Canadian forests, (Holling, 1973).

17. See, for example (May 1971a, 1972a, 1972b). MacArthur was interested in diversity and species packing in this period, something that can be seen in both what was the lead paper in a new journal, and in another written with May in 1972: (MacArthur, 1970), (MacArthur and May 1972).

18. Technically, stable systems are those that move toward some specific steady state. Unstable systems do not — they blow up in various ways. The Lotka-Volterra equations behave in neither of these ways. They behave like an undamped pendulum, persistent, not truly stable, but what is termed 'neutrally stable'.

19. (May, 2001), 74.

20. In discussing his book in the *Citation Classics* series, May mentions his first coming across the argument that complex ecosystems are more stable and better able to withstand human interference in (Watt, 1968). Interestingly Watt did comment that this was something that ran counter to everyday experience. May read the book before leaving Sydney on his sabbatical and immediately carried out some calculations on n-prey–n-predator models showing they were even less stable than the corresponding simple models; (May, 1988). The issue of diversity and stability as it stood at the time is well covered in a 1975 review by Daniel Goodman. He looked at a wide range of literature and concluded there

was little evidence to support the idea of increasing complexity leading to stability but was nonetheless equivocal. He included May's work in his review but, while sympathetic, was unwilling to fully support it. But he did recognize some of the consequences for environmentalism were May's modelling shown to be correct; (Goodman, 1975). Clearly this was a topic much discussed in the early 1970s. For an early, more general, discussion of the problem, (Gardner and Ashby, 1970), 784. This paper is a forerunner of (May, 1972b) which includes the equation that has become known as the May-Wigner theorem.

21. May (2001); quotations, 174 and 76. Whether the stable species groupings we find in nature actually correspond in some way to the stable solutions of simple models is an open question. However the new ideas proved persuasive and, more to the point, fecund. As to the possible role of evolution in favouring complexity over stability, that raises the spectre of the discarded superorganism ideas from the early twentieth century.

22. (May, 2001), 174.

23. (May, 2001), vi. Many ecologists of the 1960s and 70s were unhappy with what they saw as a too mathematical, deductive, approach. Even those who did use mathematics were unhappy with the overly analytical. This is well illustrated in an exchange that took place between C. S. (Buzz) Holling and Robert MacArthur in 1965. In that year Holling won the George Mercer Award of the Ecological Society of America, an award for young ecologists. Holling was recognized for some computer modelling of predator behaviours that he had recently published (see note 15). MacArthur wrote to Holling that his model contained too much detail to be useful. As he put it, all you need to know about a ball rolling down a hill is the height of the hill and where the bottom is. But Holling did not believe that ecology should mimic physics in this way. As he put it, he was interested in the ruts in the ball's path, and in the many diversions it faced on its way down the hill. The difference between these two scientists was aesthetic; both their approaches are useful. Which of the two comes to the fore at a particular time is contingent on many factors. As it turned out Holling had a major career, but the somewhat more inductivist approach he favoured was overshadowed by MacArthur's and May's hypothetico-deductivism during the 1970s and 80s — for some of the reasons discussed in chapters 4 and 5. I am indebted to Holling for telling me about his exchange with MacArthur.

24. (May, 2001). May's 'Afterthoughts' for the 1974 edition are interesting in that they imply a very active field and his need to acknowledge the many ideas that were around at the time his book was initially published. For example, he discussed Holling's resilience ideas, and some ideas raised by Maynard Smith. In

his introduction to the 2001 edition May had some thoughts stimulated by yet another situation. He stated that networks, such as the internet or www, which have connections with high coefficients of variation, are robust to the random removal of links but not to targeted attacks on the most highly connected nodes (xxi).

25. As May noted in the introduction to the 2001 edition of his book, while there was no discussion of chaos in the 1973 edition it was already implicit. His papers on chaos followed almost immediately. See (May, 1974, 1976a; May and Oster, 1976). May also met the mathematician Mitchell Feigenbaum in Los Alamos in the early 1970s. Feigenbaum was then working on formalizing chaos theory. While May's work led to the discovery of chaos in the properties of equations with relevance to ecology, most famous at the time — possibly in part because of its catchy title — was Edward Lorenz's paper delivered at a meeting of the American Academy for the Advancement of Science in Washington DC in 1972, 'Predictability: Does the flap of a butterfly's wings in Brazil cause a tornado in Texas?' It was long known that some equations are highly sensitive to initial conditions but, around 1970, when people routinely used computers in their numerical approaches and were able to follow the behaviour of equations over long periods, it was only a matter of time before someone took seriously the bizarre solutions that computers sometimes threw out. Lorenz did take notice — in his case, as meteorologist — that very different weather predictions resulted from only minute changes in initial conditions. The term 'chaos' was coined by James Yorke and soon used in his paper (Li and Yorke, 1975). And, as Gleick was to put it, 'now that science is looking, chaos seems to be everywhere': (Gleick, 1987), 5. Robert Merton may well have been pleased to see a further case of simultaneous discovery (see chapter one). Unlike Lorenz, May explored the properties of his equations without using a computer. As he noted in introducing the 2001 edition of his book, Lorenz used three deterministic differential equations in his work, while what he and other theoretical ecologists found interesting was 'the striking simplicity of the 1-dimensional difference equations'. Once this was understood, May's work became highly influential in the emergence of chaos ideas. (May, 2001), xv.

26. The main features of non-linear phenomena have been discovered and independently rediscovered several times. James Yorke noted that he was not the first to demonstrate chaos but he was the last (i.e. to have done so definitively). In the same way May can be said to have been the last to demonstrate chaotic behaviour in one-dimensional difference equations.

27. See chapter 9.

28. We may, for example, be heading into a new dynamic equilibrium where there is no longer any permanent ice in the arctic and where global temperatures are much increased.

29. In the 2001 edition of his book, May cites the following work as supporting his theory, but there was further support: (Constantino, Cushing, Dennis, and Desharnais, 1995).

30. He may well be right. However, the pattern of behaviour among scientists shown here is not uncommon. Ideas are pursued only so far and are dropped when they fail to produce interesting results. This can mean that sometimes ideas are not pursued far enough and things of interest missed — at least for a while.

31. (May, 2001), 136–7.

32. The geometry of organisms in their environment had been discussed earlier. See (MacArthur and Pianka, 1966; Levins, 1969). The idea of heterogeneity led to the idea of modelling metapopulations — spatially separated but interacting species populations.

33. There has been much speculation on this. Are there more species in the tropics because of the heat? Or because the tropics occupy more area? Was it extinctions during the ice ages that kept biodiversity down in more temperate zones? Is seasonal change bad for biodiversity? Does environmental stability (stable temperature, stable light levels) help to increase biodiversity? For a discussion of the role of biodiversity and its value, (Wilson, 1992).

34. (May, 2001), xix.

35. If the equations governing ecological behaviour are highly sensitive to small changes it is hard to be sure in which direction one should move. It makes giving environmental advice difficult. Looking back to MacArthur and Wilson's discussions of Krakatau (see chapter five) gives some grounds for optimism. The volcanic explosion was no minor perturbation and stability was achieved soon after. It may not have been quite the same ecological world as before, but it was functional. The 1970s and 80s saw catastrophic ideas entering the discourse more generally. For example, Luis Alvarez hypothesized that the dinosaur extinction was due to a massive meteorite hitting the earth; and Stephen J. Gould (Gould, 1989) helped bring the idea of historical contingency into some of the more public ecological and evolution debates. He pondered the fact that there appear to be no descendants of the invertebrate animals found fossilized in the Burgess Shale at Field, British Columbia. He came to the debatable conclusion that these animals (they existed about 530 million years ago) were the victims of some historical accident and that natural selection had nothing to do with their demise. Survival of the lucky rather than survival of the fit was, he

believed, a possibility. The idea that humans may well have evolved by chance and that our future existence is similarly contingent is a central theme in modernist thought. They are not ideas that would have occurred to a natural philosopher of the seventeenth century.

36. John W. Terborgh spent much of his time in Peru where, from 1973, he ran the Cocha Cashu biological station in the Manu Biosphere Reserve. Today he is director of Duke University's Centre for Tropical Conservation. Peter and Rosemary Grant joined the Princeton department just before May's departure for England.

37. At that time Diamond was a professor of physiology in the medical school but he had a serious interest in ornithology, something he had been encouraged to cultivate by his Harvard classmate, John Terborgh. Diamond carried out important work in bird ecology in New Guinea. Southwood was later to collaborate with Diamond on some of his New Guinea work. (See chapter four for May's appointment at Princeton, and for more on Murdoch).

38. Encouraged by Southwood, May purchased a small house in Sunninghill in 1975. It was rented to visiting faculty at other times of the year. For correspondence on the visiting professorship see Bodleian Library, Southwood papers R181. May was to visit for about six weeks each July and August, to give a couple of seminars, and to work also with postgraduate students.

39. (Southwood, 1966). This book went through three editions, the most recent, in 2000, in collaboration with Peter Henderson, a former student of William Hamilton.

40. (May, Conway, Hassell, and Southwood, 1974; Southwood, May, Hassell and Conway, 1974; Southwood and Comins, 1976). The first two of these papers deal with questions relating to why some insect populations remain steady from year to year while others show marked oscillations. Southwood's paper with Comins described a model for the dynamical behaviour of an insect population with discrete generations in terms of population density, intrinsic population growth rate, and some predictable aspects of the environment. A three-dimensional representation of the mathematical model (made of plaster of Paris for the Royal Society Summer Exhibition of 1976) still exists in the department at Silwood. The surface shape is defined by the influence of both competition and predation. One of the burning questions of the period was pesticide resistance. Comins was of great help to Southwood and Conway, as well as to others interested in this problem (see chapter four).

41. (Southwood, 1976b). This is a review of (Cody and Diamond, 1975), a collection of papers given at a symposium in memory of Robert H. MacArthur.

42. Hutchinson's contribution gives a good account both of how earlier mathematical models led to Gause's exclusion principle, and of MacArthur's field observations and experiments, carried out to determine whether the principle was true. (Hutchinson, 1975).

43. (Krebs, 1977); review of (Cody and Diamond, 1975). To be fair, Krebs was not blindly uncritical and understood the need for experimental testing of the new ideas.

44. When I spoke with Krebs about the history of ecology at Oxford, he expressed the view that the zoology department had been misguided in hiring as Elton's successor someone interested in energetics. A more recent reflection on MacArthur's legacy by two of his close American associates sees him as a 'towering visionary ... overdue for a comprehensive review'; (Pianka and Horn, 2005). The authors mention that, around 1990, John Lawton assigned his postgraduate students a paper by MacArthur on warblers, and that the students had no idea who MacArthur was (for the original story see Lawton, 1991). Understandably, given their close association with MacArthur, Pianka and Horn would like to see him and his warbler papers ensconced among the classics, along with Darwin and other greats — but reception and historical memory are fickle.

45. (May, 1976b).

46. (May, 1976b), 4.

47. Classic work of this type had been carried out by A. J. Nicholson on the sheep-blowfly (*Lucilia euprina*); (Nicholson, 1954). For work by members of the Silwood Circle along these lines see, for example, (Hassell, Lawton and May, 1976). This paper discusses 28 populations of insects with non-overlapping generations in light of the new models for which they provided some support.

48. Southwood contributed the second chapter, following May's introduction; (Southwood, 1976). The term 'bionomic' was archaic by the 1970s. It was coined by E. Ray Lankester, Linacre professor (1891–98), and referred to the observation of, and collection of data on, organisms in their natural environments — not unlike what today would be termed 'behavioural ecological', though I am not sure whether that terminological mouthful is ever used.

49. (Beddington, Free and Lawton, 1975 and 1976; Diamond and May, 1981). Debate over the ideal size and shape of nature reserves took off after the publication of MacArthur and Wilson's book on island biogeography. It remains a topic of great importance.

50. J. L. Harper, who reviewed the book favourably, wrote of a 'Princeton-Imperial College-Harvard axis'. See (Harper, 1977).

51. (Hassel, 1976; Conway, 1976).

52. Bodleian, Southwood papers A70–77 for job offers in this period, including one of a chair at Cambridge. For lectures see papers in section E. During the 1970s Southwood gave lectures at universities in the USA, Greece, South Africa, Australia and Canada, including at the University of British Columbia where he was invited by C. S. Holling.

53. May replied that 'nothing could be as good as appointing Hassell as Director [at Silwood] ... there is no other ecologist with the stature and ability....'. Bodleian, Southwood papers, R181, May to Southwood, 17 April, 1978.

54. For Way see chapter 4. Hassell was given a chair in 1979 just before Southwood left Imperial College, and Conway was given one a year later in 1980. One of Hassell's referees was George Varley, his doctoral supervisor at Oxford. The mathematician, Harry Jones, who was then Pro-Rector at Imperial College, stated that a reference from Varley told one more about Varley than it did about the candidate. Nonetheless Varley wrote that 'Hassell ... has been my most outstanding student over the years at Oxford.' May, too, wrote a very positive letter supporting Hassell, a letter in which he also praised John Lawton. When Southwood left Imperial College the Rector, Brian Flowers, encouraged the botany and zoology departments to merge. The head of the new department of pure and applied biology was the botanist R. K. S. Wood. Bodleian Library, Southwood papers S160 (file on Hassell's promotion to professor).

55. Southwood wrote to John Sutton (Sutton succeeded Southwood as chair of ICCET) in glowing terms about Conway's work in pest management, also stating that he was 'the driving force and lynchpin' in the development of ICCET, and that Imperial could lose him since he had been courted by other universities. Bodleian; Southwood papers section R: T. R. E. Southwood to John Sutton 30 November 1979. For Hamilton see chapter four.

56. May wrote that Lawton had been 'not too happy' at Oxford and, besides, 'his wife likes York'. Southwood tried again later to lure Anderson to Oxford, but Anderson was unwilling. He wanted a chair of his own and had his eye on the parasitology chair at Imperial College shortly to be vacated on the retirement of J. D. Smyth. Bodleian, Southwood papers, R181 May to Southwood, 17 April 1978, Princeton; R5 Anderson to Southwood, 31 January, 1981.

57. Bodleian; Southwood papers R164, Lawton to Southwood, 25 August, 1977.

58. Bodleian; Southwood papers, Southwood to Lawton, 5 October, 1976. This was about Lawton's 'lively' presence with the 'Silwood mob' at a conference in Washington DC. For letters in support of Lawton see Bodleian, Southwood Papers, R164. For their joint work see chapter 8.

59. (Odum, 1953).

60. E. P. Odum in (Innis, 1975), 202. For another collection of systems-based papers, (Halfon, 1979). For some insight into the Odums' way of seeing things, (Taylor, 2005), chapter 3. Of H. T. Odum, Taylor writes, he was an 'agent working to make the overlapping realms he inhabited — the social, personal and scientific — reinforce one another, so that efforts made and directions pursued in one realm did not undermine those in others'. He was able to find in nature a special role for systems engineers and to see himself as working in the interests of society; quotation, 93.

61. This is not something that can be discussed further here. Much system work was carried out also on a relatively modest level. One important centre was the Hubbard Brook experimental forest in New Hampshire where the aim was to gain an understanding of a complete watershed system. See, for example, (Borman and Likens, 1979). Hubbard Brook is now a designated UNESCO biosphere reserve.

62. This general problem was discussed in an interesting paper — significantly without reference to the Odums; (Levins and Lewontin, 1980). The authors were not against computer modelling *per se* but against the idea that it could lead to any holistic picture. In connection with the idea of a system they raised the question of whether people thought they were studying ecosystems as aggregates of simple components or as collective processes in varying landscapes. And, if the latter, whether it made sense to speak of emergent properties or processes. The main thrust of their paper, however, was a critique of a paper by Daniel Simberloff published in the same issue of *Synthese*. Simberloff was not defending the systems approach but his critics saw him as trying to 'escape from the obscurantist holism of Clements' "superorganism" [and falling] into the pit of obscurantist stochasticity and indeterminism'. (p. 48). Theirs is a well argued paper, reductionist in spirit, but it has its own problems. The authors construe the dynamics of community in terms of a Marxist dialectic where the whole and the parts of a community act in reciprocal, but not fully deterministic, fashion. Southwood had a paper in the same issue and was more sympathetic towards Simberloff; (Southwood, 1980). Simberloff, like R. H. Peters (see chapter 5, note 98), was a Popperian falsificationist and a proponent of the use of null models.

63. (Anderson, Turner and Taylor, 1979). Among the contributors were Anderson, Hassell, Lawton, Beddington and — summing things up and speculating on the future — was a paper by Robert May, 'The structure and dynamics of ecological communities'. The Silwood Circle was well represented.

64. One of the contributors was John L. Harper (1925–2009), professor of plant ecology at the University of Wales, Bangor. Harper, the most esteemed British plant ecologist of his generation, brought evolution theory firmly into plant ecology, (Harper, 1977). Harper placed himself in the MacArthur camp and was critical of system theorists. He wrote 'one of the dangers of the systems approach to community productivity is that it may tempt the investigator to treat the behaviour that he discovers as something that can be interpreted as if community function is organized, optimized, maximized or stabilized,' quotation, (McIntosh, 1985), 241–42.

65. (Levin, 1974), introduction.

66. No wonder so many philosophers of science got into the act at around this time trying, among other things, to figure out how dependent biology should be on the traditions of the physical sciences. See, for example, (Ruse, 1973; Hull, 1974). As Robert P. McIntosh pointed out there were also religious undertones to some of the debates. Some of the holists saw reductionism as 'secular pedantry'; others saw as 'religious nonsense' the idea that communities are more than the sum of their parts. In this one can see also a political divide between the left and the right. (McIntosh, 1985), 253 and 256.

67. (May, 1981).

68. See, for example papers in (Resetarits and Bernardo, 1998).

69. This would have been the view of at least one philosopher of science of the period. See (Feyerabend, 1975). Feyerabend's anarchism reached beyond method. He argued that the state should stop funding science since, in doing so, it perverted the spirit of true enquiry. In his 1974 book, John Maynard Smith wrote, 'ecology is still a branch of science in which it is usually better to rely on the judgement of an experienced practitioner than on the prediction of a theorist'; (Maynard Smith, 1974), 6.

70. (May and Hassell, 2008), especially, 360.

71. In 1985 there was a further merger with the National Vegetable Research Station in Wellesbourne, Warwickshire, the East Malling Research Station in Kent and the Hop department of Wye College, to create the Institute of Horticultural Research. After yet further mergers the newly named Horticulture Research International moved to Wellesbourne in 1995 and activities at Littlehampton were closed down. The more general reorganization and consolidation of government research bodies in the UK during the later twentieth century is something worthy of further study. Bernard George Peters (1903–67) was an earlier professor of parasitology at Imperial College and a pioneer in agricultural nematology.

72. The UK government decided to set up this Royal Commission in 1970 in anticipation of the 1972 UN conference on the human environment held in Stockholm. They were lobbied to do so by Eric Ashby, who became its first chairman. Ashby had been a botany student at Imperial College and began his career as a lecturer there. Southwood wrote to a journalist at *The Independent* about the founding of the Commission stating that Ashby 'made a major contribution to environmental policy in the United Kingdom'. Bodleian Library, Southwood papers A71; Southwood to Jebb, 28 October, 1982.

73. (May and Hassell, 2008). May and Hassell present testimony by Thomas Radice, Secretary of the Royal Commission, who saw Southwood as 'a forthright champion of the report, and instrumental in getting its recommendations rapidly translated into legislation and government action' (361–64). See also 'Good riddance to lead', editorial column in *The Times*, 19 April, 1983. Southwood was asked by the government to look into the problem of acid rain. He sat on a joint committee of the Royal Society, the Royal Swedish Academy and the Royal Norwegian Academy (the Scandinavian countries had long complained that Britain's sulphur dioxide emissions harmed their forests) to discuss the problem. It was something also taken up by the Royal Commission in its tenth report, and led to some later improvements.

74. If, after seven years, someone proposed is not elected, the proposal is withdrawn. People can be proposed again, but only after an interval of several years. Southwood's election certificate shows that he was proposed by O. W. Richards, seconded by J. S. Kennedy; V. C. Wynne-Edwards, V. B. Wigglesworth, A. D. Lees, C. B. Williams, P. C. Garnham, C. S. Elton, and W. G. Penney (Lord Penney, Rector of Imperial College) were among the other signatories.

75. The Royal Society brings newly elected fellows on to its section selection committees almost immediately after their election. This allows them a major say in shortlisting those who have been proposed, and helps the Royal Society renew itself in directions of current interest.

76. (May, 1978). May acknowledged the help of a number of Silwood people, including Southwood, Hassell, O. W. Richards, J. K. Waage and N. Waloff. He also acknowledged Lawton, then at York. Among the others who gave papers at the symposium were Southwood, Lawton, and W. D. Hamilton.

77. James Munro Dodd FRS (1915–86) succeeded F. W. Rogers Brambell as professor of zoology at the University of Wales at Bangor. A marine zoologist, he was renowned for his work on dogfish and comparative endocrinology. He and Southwood were fellow trustees of the British Museum (Natural History). In the fifty years before 1977 the only ecologists to be elected to the Royal

Society were Sir Edward Salisbury (elected 1933, died 1978), Edmund B. Ford (elected 1948), David Lack (elected 1951, died 1973), Charles Elton (elected 1953), and V. C. Wynne-Edwards (elected 1970). John Harper was elected with Southwood in 1977. John Maynard Smith was elected in 1978.

78. Bodleian Library, Southwood papers; see, for example, letters from Hoyle and Blin Stoyle in section G87. Also in G87, Southwood to J. M. Dodd, 11 November, 1977; Wynne Edwards to Southwood, 28 June, 1977; Elton to Southwood, 27 June, 1977; Richards to Southwood, 27 June, 1977.

79. May's election certificate shows that he was proposed by Southwood, seconded by R. O. Slatyer. Others who signed included M. S. Bartlett, V. C. Wynne Edwards, J. S. Kennedy, S. F. Edwards, Freeman Dyson, and Brian Flowers. The mix of support from physicists and biologists may have helped in his quick election. He was proposed in 1977, elected in 1979.

80. Bodleian, Southwood papers R181, May to Southwood, 9 April, 1979.

81. In the early 1990s Sir Richard Sykes was chairman and chief executive officer of Glaxo and was working on the merger with Wellcome. He became deputy chairman of Glaxo-Wellcome in 1993 and chairman in 1997. A further merger saw him become chairman of GlaxoSmithKline, a position he held until 2001 when he was appointed Rector of Imperial College. Southwood was appointed a nonexecutive director of Glaxo Holdings in 1982 and remained a director after the merger with Wellcome.

chapter seven

The Growth of Careers 1970–1995

Part One

7.1 Introduction

This and the following chapter illustrate the growing importance of ecology in late twentieth-century science. They do so indirectly by outlining the progress of people's careers and briefly describing the kind of work they carried out. Five people who worked at Silwood during the 1970s will be discussed here, and four whose careers began elsewhere will be discussed in chapter eight.

At the start of the 1970s Conway and Hassell were young lecturers, Anderson and Crawley were doctoral students and Southwood was a well established professor. Conway's early experience with the Ford Foundation helped to pave the way for a career that bridged the academic and international development worlds. Hassell, Anderson and Crawley were to have successful, yet more conventional, academic careers. As was shown in earlier chapters, and as will be shown further here, all four were helped by Southwood's patronage and by their association with Robert May. This is not to claim that they, and those to be discussed in the next chapter, would not have been successful without the help they received. Rather that, in its absence, their careers would have been different and their early years more difficult.

Having influential mentors is of great advantage in the making of any career. Gaining a mentor's attention, however, requires a degree of intellectual and social ability, and the willingness to be something of a courtier. People usually mentor only those whom they believe will be of use to them in their own work, or those where the association can be imagined as bringing some future reward. Southwood was an excellent mentor to young scientists he saw as promising. As to May, the situation was a little different. Louis Pasteur famously stated that 'in the field of observation chance favours the prepared mind'. This is true not only in the field of observation. Several of the scientists to be discussed in this and the following chapter were helped by May because they had identified some interesting ecological problems, and were prepared to see utility in new analytical approaches. In this sense their minds were prepared. Chance entered their lives when May, whose first visit to Silwood so impressed Southwood, was invited back in 1973. Inspired by MacArthur's methodological approach to community ecology, May sought out people with interesting problems. Like Southwood, he promoted the interests of his collaborators, nominated them for awards and prizes, supported their election to the Royal Society and other bodies, and was later to support their appointments to influential government committees and jobs.

Although much of the discussion in this and the next chapter focusses on individual careers, something of the collective spirit and the more general interaction among the members of the Silwood Circle is also illustrated. The chapter ends with a brief account of Southwood's move to Oxford, and of some of his public service activities. He was a role model, not only as teacher and mentor but also in his more worldly success. Mimetic behaviour is to be expected within social circles. Southwood's career path, in its general features, was something to which others in the Silwood Circle could aspire.

7.2 Gordon Conway

As noted in chapter four, Southwood was keen to recruit Conway. He knew that Conway would help the department in the areas of insect ecology and pest management and that he would bring to Silwood some new theoretical ideas from Kenneth Watt's laboratory at the University of California. While earlier in the decade it was Conway who had courted Southwood, by the late 1960s the balance had shifted slightly. Southwood saw in Conway someone who would help promote his ecological agenda, someone with wider than usual experience and more than usual ambition. Further, Conway's large grant from the Ford Foundation would enable the department to establish a valuable new research unit. Conway's interests were more ecological than environmental but, given the political climate, Southwood early saw advantage in joining the two. For example, he asked Conway to collaborate on an environmental study of the Rhine Basin;[1] and encouraged him to make a TV programme for the BBC. Shown in 1972, the year of the Stockholm Human Environment Conference, it portrayed ways in which the environment impacted people's daily lives. In promoting his environmentalist agenda Southwood brought others alongside. Notable among them was the college Rector, Brian Flowers, who helped in persuading the Ministry of Environment to fund a new unit at Imperial College. As mentioned earlier, the Imperial College Centre for Environmental Technology (ICCET) opened in 1976 with Conway its first director.

The few years leading up to Conway's 1976 ICCET appointment were, perhaps, the most critical in determining the career paths of several members of the Silwood Circle. As we have seen it was a period of intense engagement during which those with empirical grounding exchanged ideas with May and with each other. They also hurried to familiarize themselves with the new mathematical modelling. Joint papers were published and social and scientific ties were made that lasted through life. Conway, for example, contributed chapters not only to the 1976 edition but also to the 1981 and 2007 editions of May's *Theoretical Ecology*. He and others within the Silwood Circle gave papers at many of the same meetings. Two such were the 1986 and 1988 Royal Society conferences on biological invasions.[2] Invasion and the biological control of pests were hot topics in the 1970s and 80s. Conway, interested in the applied aspects of this work, thought about how pests could be stabilized at low population levels, and what would be the outcomes of releasing natural enemies against target pest populations. His choice of research topics was heavily influenced by his earlier experiences in Sabah, and by the Ford Foundation having appointed him a consultant on sustainable agriculture. His quick promotion owed much to his association with Southwood and May, but once a professor he was freer to look beyond the college. His overseas commitments grew and, while old ties remained strong, his career developed more independently of Silwood than did those of the other scientists discussed in this chapter.

ICCET began as a centre in which academic resources were pooled so as to offer a range of MSc courses in environmental science and technology.[3] It was located at the South Kensington campus of Imperial College and Conway moved there from Silwood.[4] Many publications came from ICCET and, given the political interest in the environment during the 1970s, funding came from a number of sources including the European Economic Community, various branches of the UK government, charitable foundations, and the Royal Society. Especially important for Conway's future was a 1978 grant of £450,000, over five years, from the UK Department of the Environment to develop aspects of quantitative ecological system analysis. The grant enabled Conway to supervise a number of students on different projects, several of them overseas. Studies were carried out related to food production, pest control and climate change, and the consequences for health and food safety. Being in London allowed for easier contact with people in the international development community.

In 1978 Conway and his student, Ian Craig, began working with an interdisciplinary team of agronomists, biologists, sociologists and economists at Chiang Mai University in Thailand. The team was awarded a Ford Foundation grant to develop a multiple cropping project in connection with a new irrigation scheme in the Chiang Mai valley. Conway spent the year 1980–81 as a visiting professor at Chiang Mai and, with his Thai colleagues, began to develop a technique that he named agroecosystem analysis.[5] Conway defined agroecosystems as 'ecological and socio-economic systems, comprising domesticated plants and/or animals and the people who husband them, intended for the purpose of producing food, fibre, or other agricultural products'.[6] Four principles — productivity, stability, sustainability and equity — and the trade-offs between them, form the backbone of the analysis. The principles were interpreted in light of scientific ideas Conway had acquired over the previous twenty years, but Holling's ideas on resilience and the ecological focus at Silwood on competition between food plants and weeds, herbivory by various insect pests, predation, parasitism and disease — all in the context of population biology and biological control — were central. As to the trade-offs, they had become all too clear as a result of the Green Revolution.

The Green Revolution began in Mexico during the 1940s with the work of the plant pathologist, agronomist and future Nobel Laureate, Norman Borlaug. Borlaug's high-yield dwarf varieties of disease-resistant wheat suited to warm climates were highly productive. However, while disease resistant, the new varieties were not pest resistant. They required pesticides as well as fertilizers and good irrigation for the high yields to be realized. This was true also of new high-yield rice varieties, many developed in India by M. S. Swaminathan, winner of the 1987 World Food Prize. Since poor farmers had little access to the new seeds, to chemical fertilizers and pesticides, and in many cases had insufficient access to water, inequity was added to the environmental and ecological problems. Greater productivity was won at the cost of stability, sustainability and equity. Conway's aim was not to dismiss the Green Revolution since greater productivity meant that millions were being saved from starvation.[7] Rather it was to find a good balance between his four principles, and to achieve this in conjunction with input from local farmers, villagers, and agronomists.[8]

Productivity is not simply a matter of developing and choosing good plant varieties, but also of good husbandry. Achieving stability and sustainability depends on some understanding of local ecologies. With respect to equity, Conway appears to have held something close to Marx's view that the value of an individual's labour should not be taken and exploited by others, and that small farmers and farm labourers should receive a fair reward for their work, and have access to environmentally sound new technologies. The more recent Fairtrade movement reflects something of the same idea. What represents a good balance of the four principles will differ from place to place, but the idea was to increase social value overall. In the Chiang Mai valley it was noted that the farmers adopted a range of multiple-cropping practices, and that some of those worked better than others. But the better schemes were not always copied. Sociologists and economists in the research team helped to better understand people's motivations. A little later Conway worked with some Indonesian scientists analyzing home gardens as well as farms. In home gardens productivity may be lower but this is compensated by greater stability, sustainability and, possibly, equity.[9] Overall, the aim of agroecosystem analysis was to enable people involved in promoting sustainable and productive food-producing practices to quickly ascertain and describe the rural environments in which they were working. The method entailed the production of diagrammatic and other easy-to-read forms of the data gathered so as to give a general overview of how farming in an area worked. If well done such diagrams can be suggestive of ways to move forward and are readily modifiable in the light of new data. Since many people working in the development field are local government officials, or people employed by NGOs, the simplicity of Conway's approach, founded on what were deemed sound ecological and social principles, was one of its strengths.

In 1986 Conway took a sabbatical leave from Imperial College to become director of the sustainable agriculture programme at the International Institute for Environment and Development (IIED). This institute was another of the many positive outcomes of the UN Human Environment Conference held in Stockholm in 1972. The conference's secretary-general, Maurice Strong, persuaded his friend and fellow oilman, Robert O. Anderson, chairman of Atlantic Richfield (ARCO, now part of British Petroleum) to become the principal funder of a new research institute for Barbara Ward. As mentioned in chapter four, Ward was a British economist and advocate for the developing world. She was one of the stars at the Stockholm conference. At the time she held a chair at Columbia University in New York and the IIED began its life there in 1973, but it soon moved to London. Ward remained its director until 1980. Strong's original idea was to bring business people into the environment and development fold and that the Institute would be one way of doing so. The IIED continued seeking money for its projects from sympathetic business magnates as well as from the World Bank and other development funders, but its finances were never secure. For some of the IIED's younger associates who had grown up on the radical politics of the 1960s, the tie to business was a bitter pill to swallow. However, during the 1980s and 90s the cultural ethos changed. Business and entrepreneurship became more fashionable and many young people came to understand that while governmental input is essential, governments alone cannot solve the problems of the developing world.[10]

Conway's connection to the Institute began in the early 1980s. He was in Washington carrying out some consulting work for the World Bank and for the United States Agency for International Development (USAID), when he met Richard Sandbrook. As Conway wrote, they immediately took to each other:

> Both of us were passionately committed to changing the world, at least in helping to change the developing world, in a way that married development, the alleviation of poverty and environmental protection ... we were both, under our respectable exteriors, iconoclasts unwilling to accept the conventional wisdom whether from government servants, international bureaucrats or NGO activists.[11]

Richard Sandbrook (1946–2005) was an environmental activist already as a biology student at the University of East Anglia. Shortly after graduating he and some others founded Friends of the Earth UK and, in 1974, Sandbrook became the managing director of that organization. He was highly effective and led many successful protests. In 1976 he joined the IIED becoming one of 'Barbara's boys', as Ward's young male associates were known.[12] Sandbrook was one of her favourites and after ten years at the IIED he became its director.[13] Sandbrook who had spent a few years as an accountant after graduating was good at dealing with people in the business world. He became well connected internationally, channelled his activism in productive ways, and helped the Institute to become a centre of high activity. Those engaged in international development were keen to attend workshops at the institute. Many people were brought on side, not only those in business and government but big names such as the Prince of Wales, and Bob Geldoff who consulted the Institute on the Ethiopian famine-relief projects to be supported with funds raised by his 1985 Live Aid concerts. Barbara Ward died in 1981 but the legacy of her many international connections, and of her dedication to helping the developing world, lived on.

The interesting activities associated with the IIED make it understandable that Conway gravitated toward it when Sandbrook became director. By the time of his sabbatical leave in 1986, Conway had a number of overseas projects in hand. He moved to the IIED taking some members of his Imperial College research team with him. It was a busy year. At the invitation of the Ethiopian Red Cross, Conway and his IIED colleague, Jennifer McCracken, went to Wollo province where a major famine had occurred in 1984.[14] It was one of a series of famines across the Sahel belt (a biogeographic zone of desert

and savanna running from Senegal on the Atlantic to Ethiopia on the Red
Sea). During the late 1960s and early 1970s there were severe droughts at the
Western end of the belt. In the early 1980s it was the droughts in Ethiopia
and Eritrea that drew world attention. The droughts (and perhaps equally
the local politics) caused wide-scale human suffering. This famine was one of
the first to be well documented on television. The result was a blooming of
relief agencies on a scale never before witnessed.[15] The Red Cross expanded
its activities and many new NGOs entered the scene. Public awareness led
also to increased government development aid, much of it for ecological and
agricultural research.

It was in this context that Conway was asked to help in the development
of sustainable agricultural projects in Wollo.[16] It is worth noting that funds
from USAID had already gone toward similar work at the Western end of the
Sahel belt and that many university scientists who were engaged in that effort
had been influenced by the Odums and their systems approach to ecology,
and by new systems engineering ideas coming out of places such as MIT.[17]
Conway's time with Silwood ecologists meant that he was somewhat immune
to the Odums' influence, though he, too, had a systems approach and must
have paid attention to the earlier development work. By his own account he
learned much from the methodology of Robert Chambers of the Institute of
Development Studies at the University of Sussex who joined the IIED team
on the Wollo visit. Conway understood the importance of local knowledge
from his time in Thailand and Indonesia. From Chambers he learned more
about how to go about accessing it. As a result he sat down with Ethiopian
farmers to learn about their farming practices.[18] Conway told me that he was
surprised at how much they knew, not simply about farming but about the
wider ecology. For example they had extensive knowledge of the local trees
and their role in the ecology. And they knew exactly how many days of rainfall there had been over the past ten years, when the rain had arrived, and
how much had fallen. Conway was far from alone in coming late to the reali-
zation that local knowledges are often extensive and have great utility. But he
immediately saw that such knowledge could be built into the kinds of
diagrams he had developed for agroecosystem analysis.

Also that year, Conway visited the Philippines on behalf of USAID. The
agency asked for help in resolving a dispute among local farmers over a new
irrigation scheme. With much local anger over water rights, the trip was not

without its dangers. On his return, and at the suggestion of his IIED colleague Johan Holmberg, Conway initiated a series of bulletins titled *Gatekeeper*. The first, written together with McCracken, was on farming in the Philippines. In it they outlined the work of Michael Loevinsohn, one of Conway's Imperial College students who had studied pesticide use on rice crops, and had documented high death and morbidity rates among farm-workers. Conway also co-authored a book with another of his research students, Jules Pretty. It was a comprehensive treatment of the problems related to agricultural pollution, including pesticide poisoning. It also gives scientific accounts of the types of ecological problems noted earlier by Rachel Carson.[19]

Conway must have enjoyed working at the IIED since he extended his leave from Imperial College and stayed on as director of the Institute's sustainable agriculture programme for two more years.[20] In this he was funded by the Rockefeller Foundation. At the time there was much interest in the Brundtland Commission and its Report, *Our Common Future*, which people at the IIED helped to frame.[21] Conway took on new projects in Pakistan, funded by the Aga Khan Foundation, and further consulting work for funding bodies such as the World Bank. Earlier, two of his associates at the IIED, Robert Stein and Brian Johnson, had written a book, *Banking on the Biosphere* (1979). They were highly critical of the approach taken to development funding, especially by the World Bank. Noted was a lack of concern for social, ecological and environmental issues. The book was successful in getting bankers to think about these issues and by the time Conway came to work with the World Bank its internal culture was beginning to change.

By the late 1980s Conway had gathered a wealth of experience, so it is not surprising that the Ford Foundation wanted more of his time. He was asked by John Gerhart, head of the foundation's Middle East and North Africa Program, whether he was interested in taking over the foundation's India Office.[22] Conway was already leaning away from academia and, in 1989, he accepted the offer. In a letter to Southwood he wrote that the job would be a 'very considerable challenge' and a 'logical continuation of my work on sustainable development which has gone so well in the past two years'.[23] From headquarters in New Delhi he was responsible for programmes not only in India but also in Nepal and Sri Lanka. He had a staff of about twelve and a budget of about $10 million per annum to be used for sustainable agriculture projects, and for projects in areas such as forestry, maternal health, human rights, folklore theatre, archeology, and local governance. Conway told me that he enjoyed his time in India but that both he and his wife had health problems associated with their work there. They decided to return to England. Conway saw an advertisement in *The Economist* for the vice-chancellorship of the University of Sussex and applied.

Conway was an outsider but he had an interesting *curriculum vitae*. After a contested appointments process, he was offered the vice-chancellorship and held it for the period 1992–8.[24] I asked Conway how he remembered his time in office there. He said that he admired Asa Briggs' vision of an interdisciplinary university but was not entirely successful in keeping it alive.[25] He also mentioned being careful with honorary degrees, seeking recipients who reflected both contemporary affairs and the interests of the university. One representative choice was Thabo Mbeki who was awarded an honorary degree in 1995. He arrived together with some other senior South African politicians who, like he, had been educated at the university. Conway organized a seminar for them and invited Eddie George, Governor of the Bank of England, and some other experts in financial and economic affairs to attend. Bringing people together in this way is one of Conway's strengths. A further illustration was the founding of the Centre for German-Jewish Studies at the university. Helping to bring this about was Rabbi (later Baroness) Julia Neuberger and the Runnymede Trust.[26] In 1992 a Runnymede commission had examined anti-semitism in Britain. In 1996 the Trust invited Conway to chair a commission on British Muslims and Islamophobia.[27] Conway is proud of this work, and of working behind the scenes to have Richard Attenborough become university chancellor. In 1994 he helped persuade administrators at the other new universities (those founded in the 1960s) to come together in the 94 Group, further testament to his skill with people — testament also, perhaps, to some things learned from Brian Flowers and Richard Southwood.[28] But Conway did not leave his earlier interests entirely behind. Getting up early each morning he devoted a couple of hours before beginning his main work to writing his book *The Doubly Green Revolution: Food for all in the twenty-first century* (1997). This was to help move his career forward after his time at Sussex.

7.3 Michael Crawley

As mentioned in chapter three, Michael Crawley graduated with a PhD from Imperial College in 1973. His doctoral work, under Conway's supervision, entailed computer programming related to the statistical analyses of ecological data. Crawley had learned how to programme computers as an undergraduate in Edinburgh where, together with fellow student, Mark Westoby, he had carried out an honours project studying movement patterns in free-ranging sheep. Computing skills picked up while carrying out that project had taken them both to Utah and work on the desert biome project, part of the International Biology Programme.[29] After returning and gaining his doctorate at Silwood, Crawley joined the University of Bradford. As at many other universities in this period, Bradford was offering a new degree in environmental science. The programme was under the direction of Bernard Stonehouse, well known for his work in Antarctica on emperor penguins.[30] Crawley took a full-time teaching position at Bradford where he gave daily lectures and practical instruction to students in the ecology stream. This included teaching quantitative methods, statistics and computing. Crawley told me that the experience set him up for a 'lifelong interest in how to teach maths and statistics to biologists who typically lack any background (and often any enthusiasm) for these subjects'.

Crawley's work at Bradford allowed no time for research but he did write one paper which harked back to the brief time he spent with Holling at the University of British Columbia.[31] Interestingly, given his future direction, plants are absent from this paper on insect ecology. Also interesting is that when he returned to Silwood he did so as a lecturer in entomology where he took over from George Murdie who left to become a farmer. As at Bradford his Silwood work entailed teaching statistics, but now he had time for research. Crawley was very lucky to be given a position at Imperial College since he had no publications from his doctoral work and only the one paper from his six years at Bradford. Three points about his lucky break are worth noting. First, Crawley had kept up socially with those he met at Silwood, and had kept up also with their work. Second, Bradford was a convenient meeting place for others in the Circle and several gatherings took place at Indian restaurants in that city. Third, when attending a meeting of the British Ecological Society in Bristol, Crawley went for a pub lunch with a group of people that included John Maynard Smith and Michael Hassell. Hassell told him that Murdie was leaving and, quick as a flash, Crawley said 'I could do that job'.[32] Hassell reported this to Southwood. Southwood, keen to surround himself with mathematically competent ecologists, must have seen something in Crawley earlier. Believing he had ability for research, he decided to give him a new start at Silwood. It was one of the last appointments Southwood made before moving to Oxford.[33]

To catch up with others in his age cohort, Crawley chose a research niche that was relatively empty, one where he thought he could make an impression. Sensibly it was also one in which Southwood had a serious interest. He began an active research programme in herbivory.[34] This included field experiments on oak trees, acorn production and regeneration, on the impact of rabbits grazing on grassland, on the interaction of ragwort and cinnabar moths, and on the dynamics of seed-limitation and coexistence in annual plants.[35] Crawley brought considerable mathematical skills to the planning and analysis of his long-term manipulated field experiments. Indeed, his methodological innovations were an important contribution to this type of ecology. Before writing any papers Crawley decided that he would produce a serious book. His *Herbivory* (1983) was the first comprehensive account of herbivore ecology, and of the dynamics between plants and the animals that feed on them. In ending his book Crawley set out his methodological position, 'only by carefully designed, critically manipulative experiments can we hope to determine the causes of population change'. Simple observation of plant-herbivore systems, even long term studies, were not going to be enough.[36] But such experiments are difficult to design in ways that allow for an understanding of what truly happens in the natural world. If the population changes of interest are not simply those of herbivores but also those of the plants they devour, then the difficulties multiply. Working in the area of herbivory meant crossing the divide between zoology and botany. Crawley was able to do this well. His pedagogical interests in mathematical biology also allowed him to communicate mathematical ideas. Three books by Crawley on statistical methods have helped biology students worldwide.[37]

Crawley told me that, while totally unplanned, he was happy with the move to Silwood. He became part of what he saw as 'the most exciting ecology department in the world'. In finding a research niche that he could make his own, he also learned much from John Harper, professor of plant ecology at the University of Wales at Bangor with whom he became friendly. This was an important association since Harper was the most influential British plant ecologist of his generation.[38] Crawley's mathematical and field skills were appreciated also by other members of the Silwood Circle. A couple of joint papers reflect this. One with Michael Hassell and Roy Anderson dealt with the fact that individuals within a population have a dual tendency: to aggregate and to migrate. How this dual tendency plays out, and why it exists, became a fashionable topic in the later 1970s.[39] Crawley is an enthusiastic naturalist. One of the books that interested him as a boy was Edward Salisbury's *Weeds and Aliens*.[40] The book stayed with him and, on returning to Silwood, he began compiling a list of alien plants in the UK, noting their ecological impacts. As mentioned in earlier chapters, Silwood was a magnet for ecologists. One who came there on a sabbatical leave in the early 1980s was Cliff Moran, a specialist in the biocontrol of weeds.[41] He and Crawley joined in pulling together everything they knew about the success or failure of biological control methods worldwide. Crawley became knowledgeable on the ecology of invasions — the intentional (such as in classical biological control[42]) as well as the unintentional arrival of alien plants and animals. Moran and Jeff Waage dreamed up what became the Silwood Project on the Biological Control of Weeds. Crawley, Hassell, Conway, Lawton, David Greathead and several research students came to be associated with the project.[43] Crawley wrote a number of papers on biological control including one presented at the Royal Society symposium on the quantitative aspects of the ecology of biological invasions which, as mentioned above, was attended by almost all members of the Silwood Circle. Also in this period he co-authored a paper with Robert May on plant community structure. Later he was to contribute a chapter to the third edition of May's *Theoretical Ecology*.[44]

Crawley's interest in plant invasions led him in a new direction. During the early 1980s Richard Brown worked with Crawley while holding a postdoctoral fellowship at Silwood. Brown then left to work for Imperial Chemical Industries (ICI) at Jeallot's Hill, Bracknell (now run by Syngenta, this is the UK's largest commercial agricultural research site). There Brown became part of an ICI team, later its leader, on the global impact of insecticides. In the early 1990s the National Science Foundation in the United

States, along with some U.S. government departments, decided that it was time for some field research to determine what might be good procedures for risk assessment in connection with genetically modified (GM) crops. The British government decided that Britain should carry out some research of its own. Civil servants in the department of trade and industry contacted various biotechnical and agriculture-related industrial companies, including ICI, on how to go about doing so. A list of candidates for carrying out risk assessment work came forward but for some of the suggested research areas none met the approval of the industrial sponsors. Brown thought Crawley, by then highly experienced in field experiments, would know how to proceed and suggested he apply. Crawley's application was successful and he came to head one of the research teams of the Planned Release of Selected and Manipulated Organisms (PROSAMO) project.[45] His part entailed a number of field experiments, lasting ten years, to see what, if anything, followed the introduction of transgenic plants into the environment. Twelve different natural habitats were chosen for the introductions, including Cornish heathland, Scottish moors and Silwood Park. Crawley had entered an area of research with important political, economic and, possibly, ecological and health, consequences.

When the PROSAMO project began there were three principal concerns: first, that the modified crops would become invasive weeds; second, that they would pollinate wild species and transfer their new genetic material; and third, that this would prove harmful should the genetic modifications make plants toxic or allergenic to humans, livestock, or wild animals. Crawley's project was intended to address the first two of these concerns. At the start of his research there were not yet many genetically modified food plants. Crawley's team looked at four important ones: oilseed-rape, maize, potatoes, and sugarbeet. Samples of the first two had been modified to be resistant to some types of herbicide, whereas those of the second two had been given some insect-repelling properties. Plants, both modified and conventional, were grown in a number of manipulated plots. Some plots were fenced, some ploughed and cultivated, others not. The conditions were designed to range from the benign to the hostile. In all the chosen habitats none of the modified populations persisted for longer than did their conventional counterparts. All died away quickly after a year due to competition from native perennial plants.

Since some of the modifications were intended to make plants repellent to insects, the fear of the new, while understandable, is interesting in light of the environmental movement's position on the use of insecticides. In theory such modification should lead to a reduction in insecticide use overall. Whether herbicide resistant crops would encourage greater use of herbicides was then an open question.[46] The debate over GM crops is ongoing, but Crawley's work allayed many people's fears that they would run rampant, spread their genes among neighbouring plants and cause harm.[47] PROSAMO was a landmark study but, as Crawley remarked, it was only a start. Other potential GM crops could behave differently and prudence requires the careful testing of each.[48] During and after the PROSAMO study Crawley served on the government body regulating the intentional release of plants and animals into the environment. He also continued with his long-term herbivory studies at Silwood. As we will see in chapter nine, another member of the Silwood Circle, John Krebs, became centrally involved in the GM food debate when he was appointed chairman of the new Food Standards Agency in 2000. His decisions were, in part, influenced by Crawley's work.

7.4 Michael Hassell

As a postgraduate student at Oxford Hassell joined a population study of the winter moth (*Operophtera brumata*) that had been underway since 1950. He helped to round out the field work, though much analysis of the overall data remained to be done. This long-term study was directed by George Varley with the assistance of George Gradwell.[49] Like Elton and Lack, Varley was interested in population numbers, but counting insects in the field is far more difficult than counting mammals or birds, and he had to develop techniques for doing so. Varley was interested in oak tree defoliators of which the winter moth, his principal insect of study, is a notable example. Together with Gradwell he initiated a census of the moth at different stages in its life cycle, with counts being made three times a year on some designated oak trees in Wytham Woods. Little was known about the interaction of the oak tree, the winter moth, and the moth's many enemies. What, for example, kept the moth population in check and which factors had the greatest influence on its population? One type of check, not the most important in this case, is parasitism.[50]

Hassell was to figure out the mechanism by which the winter moth population was (partially) kept in check by a parasitic fly, *Cyzenis albicans,* and to relate this to the population dynamics of the two species. The scenario is interesting both in itself, and because this very simple system is suggestive of the overall, and enormous, complexity of the natural world. It illustrates the kind of difficulties faced by ecologists and for that reason is worth briefly describing. Oak leaves are attacked in late Spring by the moth caterpillars after they hatch from eggs laid during the previous November and December. The attacked leaves ooze a sugary sap along their raw edges. This attracts the fly which lays its eggs on the damaged leaves. Some of those eggs are then consumed by the caterpillars, along with their oak-leaf diet. The fly eggs hatch in the caterpillar's gut. The larvae then find their way into the caterpillar's salivary glands where they mature. When ready, the moth caterpillars lower themselves into the soil on silken threads where they pupate. The pupae of the fly parasites end up inside those of the host. The host pupae are destroyed as the flies hatch in the Spring. The surviving moth pupae become adults and mate late in the year when the male moths seek out the flightless females as they climb the tree trunks toward the higher branches where their eggs are laid.[51]

Even though Hassell was later to adopt a more experimental approach in his work, it is probably fair to say that the work carried out at Oxford on the moths and their parasites determined his future research path. Insect parasitism was a fecund field both because it raised so many problems of basic ecological understanding, and because it was useful in thinking about the biological control of pests. Hassell was to contribute much to the fundamental science. Over his career he carried out many population studies using controlled laboratory experiments. And he developed mathematical models in an attempt to capture what was observed. These models were used to predict what would happen under specified conditions and, when adjusted in light of experimental observations, they improved over time.

Like his doctoral supervisor, Hassell took a mathematical approach to his work. However, the models then available, such as those developed by Lotka, Volterra, Nicholson and Bailey (see chapter two), were old and inadequate. Holling, Varley, and others had tinkered with them and there was much debate over the resulting and differing minutiae.[52] Progress up to then had been slow and incremental. Hassell will have learned of this history while at

Oxford where he received a good theoretical grounding. Together with his field work at Wytham, it gave him the skills needed to join contemporary ecological debates. His postdoctoral work with C. B. (Ben) Huffaker at the University of California (Berkeley) taught him more about pest control, and introduced him to laboratory population studies.[53] He returned to Oxford to complete his NERC fellowship having thought seriously both about further laboratory work and about theoretical approaches to prey-predator systems, especially simple ones involving insects attacked by parasitoids — parasites of a type that lead almost inevitably to the death of their host. Hassell also settled to writing papers based on his research, and he helped Varley and Gradwell in teaching a course on insect population ecology. The course, and the years of research at Wytham, led to an influential textbook published in 1973.[54] And, as will be discussed in chapter eight, Hassell exchanged ideas with David Rogers and John Lawton who were also working in Oxford at that time.

When Hassell moved to Silwood in 1970 he will have faced the problem of how to develop his work further. He began some laboratory studies, but the arrival of Robert May with his analytical approach to ecological problems was a stroke of good luck. It had a huge impact on how Hassell was to approach his future work. May helped Hassell to move forward from what, by then, were rather stale theoretical models. Not only did Hassell need some new ideas, he was well prepared to receive them. Much of his new experimental work was with parasitoid wasps that preyed on flies and moths. It suited May's need for simple empirical material to ponder, puzzle over, and model. Others could have provided May with something similar, but it seems unlikely that they could have done so in as lively and congenial a setting as Silwood. And for that Southwood was largely responsible. Reflecting Southwood's major role in bringing people together, the first papers to come out of Silwood with May's participation were inclusive. As we have seen, May joined with Southwood, Hassell, Conway, Comins and others in a range of projects. But May and Hassell soon began publishing joint papers of their own, and continued to do so over an extended period. Hassell told me that this early period working with May was the most important of his career, and that it set the stage for all that followed.

Like Conway, Hassell contributed a chapter to May's *Theoretical Ecology* (1976). It was on arthropod predator-prey systems and focussed heavily on aspects of predator searching behaviour and how it affected prey and predator populations. He was able to refer to laboratory data on the topic, especially data on parasitoid predators. But laboratory systems are usually unstable whereas real world predator-prey systems appear to be stable. In his chapter Hassell considered some ways in which more of reality could be captured in his models. His methodological approach was fairly straightforward. It entailed taking prey-predator interactions apart, and devising experiments to look at one component at a time. Such components included predator searching behaviour, demographic parameters, susceptibility to parasitism and other features of the life-histories of insects of interest. In some experiments competing host or parasite species were introduced. Overall, by keeping the experiments fairly simple, Hassell was able to construct models that reflected the dynamical effects observed, and to do so in a stepwise and rational manner.[55]

As mentioned in chapter six, May and others had drawn attention to the spatial distribution of populations, and to so-called meta-population behaviour. Meta-populations are linked to each other in some historical way since they form as a result of population dispersal. Such dispersal is not entirely random. The patchy population patterns that result, and the complex population dynamics with which they are associated, was a topic that interested Hassell for many years. From the start he was helped by May in thinking how to model patchiness as it applies to parasitism. In the 1970s empirical questions of interest included whether predator parasites (including parasitoids) take steps to concentrate their search for prey in high prey density areas, or whether they stay put in lower density areas so long as they can find enough prey. And how do such behaviours and heterogeneity in the distribution of parasitism affect the two populations? Hassell devised an experiment to see how parasites distribute themselves among patches of host moth pupae. David Rogers came from Oxford to help him with it. Based on this and similar experiments, models were introduced that predicted host-parasitoid populations would be unstable in the absence of spatial heterogeneity, but would persist in patchy environments — with overall density remaining roughly steady — provided the levels of parasitism varied from patch to patch.[56] The total picture is complex with population regulation being dependent on factors other than simple population density. In 1978 Hassell published a book, *The dynamics of arthropod predator-prey systems* which brought together many of the ideas of that time. In the same year, he was given a chair in insect ecology. In 1987, with the help of Southwood and May, he was elected to the Royal Society.[57] Hassell became director of the Silwood campus in 1988 and the head of the biology department in 1993. He continued to collaborate with May into the 1990s. For example, together they expanded on some of their earlier work on spatial distribution and showed how ecological systems can self organize so as to allow even incompatible predators to coexist if spatially segregated.

The kind of science with which Hassell engaged demanded much experimentation as well as inventive modelling. He reported his work at many of the same symposia as did others in the Silwood Circle; interaction and the exchange of ideas among them was ongoing. Hassell was also a dedicated teacher and directed many postgraduate students: among them Tom Bellows, Matthew Cock, Michael Bonsall and Jeff Waage. Bonsall and Waage were to become part of the wider Silwood Circle and joined in group activities such as the summer walks which Hassell told me were enormously important for the mapping out of future work. Bonsall, the youngest of the four, is a reader at Oxford University and has connections to older members of the Silwood Circle working there.[58] Waage was an undergraduate at Princeton. It was his interest in entomology that led Robert May, who was one of his professors, to suggest he study for his doctorate at Silwood. He came to England on a Marshall Scholarship and gained a PhD in two years before returning briefly to the USA. In 1978 Southwood invited him back to Silwood and to a lectureship. Waage learned much from Hassell but was drawn to applied ecology, possibly influenced by Michael Way, and by Gordon Conway whom he sees as one of the great Silwood figures of the 1970s.[59] When CABI's Institute of Biological Control moved to Silwood in 1981, in part due to Waage's own efforts, he left the department to run it. Nonetheless he retained close contact with his Silwood colleagues and carried out related research for which he received a large grant from the Leverhulme Trust. In 1986 Waage was appointed chief executive of the parent organization, CABI Bioscience.[60]

Hassell enjoyed the camaraderie at Silwood. He told me that for those who worked there, their academic and social lives revolved around the campus. After his second marriage in 1982, Hassell moved into what had been the warden's house, further solidifying his connection to Silwood.[61]

7.5 Roy Anderson

Roy Anderson studied for both his BSc and PhD at Imperial College. His doctoral thesis was on the population biology of parasite-infected bream. He had wanted to study infected fish living in the Serpentine in Kensington Gardens/Hyde Park, but permission was not given. Instead, having heard of some concerns about fish health from members of the Dagenham Angling Club, he decided to study fish living in a pond in a disused gravel pit in Dagenham, Essex. Papers that he published on his doctoral work show that Anderson's conceptual approach was similar to that of others working on prey-predator behaviour. For example, it had much in common with Hassell's doctoral work despite their working on very different animals. His work is further illustration of the complexity of even simple ecological systems. One of the parasites that Anderson studied was a helminth (parasitic worm) with a complicated life cycle. Its eggs floated in water, its larvae infected an intermediate host (a tubificid annelid worm) which was food for the bream, and the adult helminth lived in the fish. Like Hassell, Anderson undertook a quantitative study, taking the observed parasitisms apart — in his case by looking separately at the three stages of the parasite's life cycle. He also considered the spatial distribution of the tubificid intermediary and constructed models for the size of the fish population and its parasite burden. Like others, his thinking at the time was in the tradition of Lotka, Volterra, Nicholson and Bailey. As with Conway, Crawley, and other students of his generation, Anderson learned how to use computers and was able to carry out multi-variate analysis of his data to help in constructing his models. He looked to both stochastic and deterministic models in describing the host-parasite system. Underlying the modelling were questions of animal behaviour and of how the system remained stable, with all three populations surviving over the longer term.[62]

Anderson gained his doctorate in 1971 at which point he decided to learn more about biological mathematics. His won a postdoctoral fellowship to be held at Oxford. It had some travel money built in which allowed him to visit William Murdoch in Santa Barbara, as well as some other American scientists, notably G. P. Patil, a mathematician working in the field of statistical ecology at Pennsylvania State University. At Oxford Anderson learned much from the professor of biomathematics, Maurice Bartlett.[63] He then moved to a lectureship at King's College where he carried out further work on fish parasites with Philip Whitfield. In this he was helped by Andrew Dobson who had just gained his BSc from Imperial College.[64] But Anderson wanted to think beyond fish parasitism. He also wanted to return to Silwood. It is clear from correspondence in Southwood's papers that Southwood, too, wanted Anderson's return. Earlier he had recognized Anderson as a brilliant student with great potential. He wrote glowing letters in support of Anderson's application for the post-doctoral fellowship, and for a number of job applications.[65] Academic openings were then scarce because of the recession. However, Conway's effectiveness in pulling in major grants allowed him and Southwood to offer Anderson a lectureship in ICCET in 1977. In the following year Bill Hamilton moved to the United States, and Anderson was able to move back to Silwood. While continuing to teach for ICCET, his research work was largely centred at Silwood where he inherited Hamilton's office, complete with cylindrical holes drilled through the walls to the outside. The holes had once been blocked by glass observation boxes used by Hamilton to study solitary wasps. For Anderson the holes, as indeed the move back to Silwood, meant a blast of fresh air.

Even after leaving Silwood for Oxford and the Linacre chair in 1979, Southwood continued to favour Anderson. As mentioned in chapter six, he had wanted Anderson to move to Oxford with him. But by then Anderson had begun a collaboration with May and it suited them both that he remain at Silwood. As was the case for Hassell, Anderson's association with May enabled him to move forward and to think about parasitism in more general terms than earlier. Southwood was impressed with the way in which Anderson and May worked together. In a 1979 letter, supporting Anderson's application for the Royal Society's Howe Senior Research Fellowship, he wrote that already at the age of 32 Anderson had become internationally recognized as a leader in parasite ecology and that his collaboration with Robert May 'on equal terms speaks for itself'. He wanted to see Anderson freed from his teaching constraints 'in the interests of British science'.[66]

Some of Anderson's work in this period was carried out with his old teacher Neil Croll who introduced him to some helminths of medical importance. Together they studied *ascaris*, an intestinal parasite common in mammals.[67] Research carried out in Iran and India made Anderson realize that were he to continue to work on medical helminth problems (and they did interest him), he would need a large team and some large grants. A grant from the Rockefeller Foundation was the first of many to come his way.[68] It supported the work of two of his students, Melissa Haswell and David Elkins, who later married and became professors in Brisbane where they worked on parasites and public health in tropical Australia. Others who were to work with Anderson in the coming years, and who later also had successful careers, include Bryan Grenfell, Sunetra Gupta, Mark Woolhouse, Graham Medley, Angela McLean and Betz Halloran.[69]

Anderson's work bridged the areas of mathematics, ecology and epidemiology. Its medical and veterinary importance was soon recognized. His work with May extended to viral and bacterial infection which they conceptualized as being yet another form of parasitism. They were exceptional synthesizers. An example of this is their work on coevolution published in the early 1980s. In this they brought together ideas on patchy population distribution with those on the coevolution of hosts and parasites. The latter had long been a vehicle for discussing the origin of genetic diversity, as well as the origin of sexual reproduction. Further, it was widely believed that parasites had evolved so as not to seriously harm their hosts. May and Anderson attempted to generalize what was then known and came to the conclusion that while coevolution was real, there were many possible evolutionary strategies, including some that were seriously harmful to hosts. This work drew much attention. Anderson told me that he received many invitations to speak after the work was published.[70]

In 1982 Southwood wrote to the head of the new pure and applied biology department at Imperial College, R. K. S. Wood, supporting Anderson's successful promotion to professor. He noted Anderson's growing international reputation and his having brought the whole range of parasitic infection of animals, from viruses and bacteria, through protozoa to worms, within one conceptual framework. And that this allowed epidemiological generalizations to be made for a range of hosts including insects, fish and mammals. Southwood clearly saw it as a remarkable feat that Anderson had seen the patterns that allowed the unification of subjects previously studied in isolation.[71] Southwood was also behind the award of the Huxley Memorial Medal to Anderson in 1981.[72]

Anderson first met May in July 1973 at the mathematical ecology workshop held at York on ecological stability. He gave a paper at the meeting which drew on some of May's published work.[73] From then on they met regularly at the summer workshops at York and Silwood and, by 1977, had become close collaborators. Like Hassell, and indeed Southwood, Anderson was somewhat in awe of May. But he recognized that he needed May's help in order to frame the type of models needed in his parasitological line of work. May, in turn must have recognized Anderson's ability, perhaps helped in this by Southwood. In any event his collaboration with Anderson was to prove especially fecund. The reasons for this are no doubt complex, but among them were that Anderson was making a special effort to come to terms with the kind of mathematics used by May, that their personal relations were very good, and that after some earlier collaborative work they found a good topic on which to focus their attention, namely the population biology of infectious diseases. In addition to his mathematical skill, Anderson knew how to use epidemiological data and to design good studies. The choice of infectious disease as a topic was motivated by the huge body of epidemiological data in this area — relating both to people and domestic animals. They began to look at viral, bacterial, protozoan and fungal infections, and at how these disease agents regulate the numerical abundance and geographical distributions of animal, including human, populations.

May came to England on a sabbatical leave in 1977 and spent much of it at the King's College Research Centre in Cambridge thinking about the principal themes that were to occupy him and Anderson in the coming years. Two papers they published in *Nature* in 1979 illustrate where they stood at the time. Some features of direct microparasitic (viruses, bacteria and protozoa) and indirect macroparasitic (helminth and arthropod) infection were outlined, and many diseases were mentioned. Also described were some features of the dynamics of infectious diseases.[74] I asked May how he had envisaged moving on from there. He replied that there were many ideas in the literature, but that they needed to be drawn together into a coherent whole. An early result was a paper published in 1981 on the direct infection of invertebrates by microparasites.[75] It anticipated the kind of work that was to follow. First, it covered the known literature on the main types of invertebrate microparasite infections (the reference list is very long). Second, it drew its terminology from both mainstream epidemiology (infectors, susceptibles, infected but not infectious, recovered, etc.). Third, while the theory was built along ecological norms, the paper showed how such theory could inform epidemiological thinking.[76] Fourth, it anticipated future application of the theory to the diseases of vertebrates, notably human diseases such as measles. Fifth, it pointed to applications of the theory as already developed for invertebrates. For example, it picked up on the fact that arthropod pests were becoming resistant to chemical pesticides, and that pathogens could be used as biological control agents.[77] The paper owed much to the Silwood discussions; its references include some of the papers written by members of the circle during the 1970s, and hark back even further to the work of Lotka, Elton and Lack.

Anderson and May soon moved to modelling a vertebrate disease, namely rabies in foxes. Rabies was then of concern because of its spread in Continental Europe. Their paper is interesting also in that it gave some clear advice. The United Kingdom was rabies free but government scientists believed that, were an outbreak to occur in the fox population, mass culling would be necessary. Anderson and May suggested that this would not be effective in preventing transmission to humans. Instead they recommended the compulsory vaccination of all domestic cats and dogs and the culling of stray and feral animals.[78] It was a paper that put May and Anderson on the 'go to' map for government officials in need of advice on such matters. Perhaps their best known advice of this type was given much later during the Foot and Mouth epidemic of 2001 (see chapter nine).

In 1985 May asked for Southwood's support in nominating Anderson for the first David Starr Jordan prize.[79] In his nomination letter, May wrote 'Anderson has re-set the agenda for research in population-level properties of host-parasite associations'. The use of the expression 're-set the agenda' was important since one of the criteria for that prize is that winners change the direction of the field in which they are working. May also made the point that Anderson was developing models for microparasites and was thinking of their application to viral and bacterial infection control among children in developing countries.[80] This was true and will be discussed further below. But the wording was designed to appeal to the humanitarian sensibilities of the selectors. Southwood, too, wrote on Anderson's behalf and, in 1986, Anderson became the first to win this prize. In the same year, and with the same backing, he was elected a Fellow of the Royal Society. Many other awards were to follow.[81]

By the 1980s Anderson had found a field of study ripe for expansion and had the skills to make the most of it. In some ways he and May appear opportunistic, but this should not be viewed negatively.[82] They saw that something could be done, something that would bring them attention, but also something that they believed would be of considerable use in the arena of public health.

However, as mentioned, to move forward a larger team was needed and for that Anderson needed major funding. Anderson told me he was encouraged to apply for a Wellcome Unit by Bridget Ogilvie, a helminthologist then working at the National Institute for Medical Research at Mill Hill. He applied, was successful, and received funding for a new centre in 1984, just before becoming head of what was soon to be renamed the 'biology' department at Imperial College.

As to the new Wellcome Trust Research Centre for Parasite Infections, it came at a time when the Wellcome Trust was reorganizing after a lengthy legal process to get out of the conditions set by the original legacy. It had recently received permission from the High Court to sell its shares in Burroughs Wellcome.[83] A lesson had been learned from the decline in the value of shares in the Morris automobile company and the associated decline in fortune of the Nuffield Trust. The court decision ensured that the Wellcome Trust, provided it handled its investments wisely, would continue to be a major research funder for medically related projects. In 1991 the director of the Trust, Peter Williams, retired. Anderson was asked by the chairman, Roger Gibbs, whether he would take over. He declined but agreed to become a trustee. He was also the recipient of a very large grant supporting his Centre. I asked Anderson whether he saw this as a conflict of interest and he gave a reasonable reply. The Wellcome Trust, he stated, recognized the need to have a number of leading medical scientists as trustees in order that sensible decisions be made. Such people were likely to be recipients of Wellcome funding. Because of this the Trust had in place mechanisms to avoid conflicts of interest.[84]

With his new Centre in place, Anderson turned his attention to infections within human, animal livestock, and natural ecosystems. He and his team applied the knowledge gained to create a number of control–intervention programmes. Much work was carried out overseas in places where helminth infections in human populations were causing serious illness. But, as mentioned, Anderson did not stop at helminths. With their now wider view of parasitism, scientists at the research centre began modelling measles, mumps, whooping cough, rubella and HIV-AIDS.[85] Already in 1988 Anderson organized a discussion at the Royal Society on the ecology of infectious diseases where he and May presented a paper on HIV-AIDS. In it they reviewed much of the available data and outlined models suited to different geographical regions, ones depending on the dominant mode of transmission in those regions. But much data, including that on the incubation period from HIV infection to AIDS symptoms, was uncertain. As a result the models, too, were uncertain, but May and Anderson had ideas on how to improve them as new data was acquired.[86]

Anderson and May's book, *Infectious Diseases of Humans: Dynamics and Control* was published in 1991. It became a standard teaching text in infectious disease epidemiology.[87] The book is far-reaching. As the authors state, by the mid 1980s their work had 'come increasingly to deal with the detailed analysis of specific problems of human health — the evaluation of immunization programmes against viral and bacterial infections, programmes of chemotherapy, and of vector control against protozoan or helminth infections.' They had published in a range of parasitology and public health journals and believed it time to draw things together in a book where their own work could be considered alongside that of others. They wanted to provide 'keener insights into population level consequences of the dynamic interplay between infections and their host populations'.[88] The book began to take form in 1986 when May and Anderson were awarded a one-month residency at the Rockefeller Foundation's centre in Bellagio. The epidemiological field was exploding so they made the sensible decision to place a limit on looking at the new. They agreed to a cut-off date of 1990. The book was published in 1991. Its primary aim was, to show how simple mathematical models of the transmission of infectious agents within human communities can help to interpret observed epidemiological trends, to guide the collection of data towards further understanding, and to design programmes for the control of infectious disease. ... Our major goal is further understanding of the interplay between the variables that control the pattern of infection within communities of people.[89]

Their contribution, therefore, was not simply at the level of the biological or medical, but more fundamentally at the ecological. Just as in May's earlier ecological modelling, they were not interested in specific diseases *per se*; rather they were seeking a way to transcend the huge medical literature and to find some basic epidemiological patterns. But, for this, data from specific disease epidemics was needed. By adding empirical data to their population models in a stepwise manner, they were able to deal with the many and various rate parameters (birth, death, recovery, transmission etc.), so as to say something both about the epidemiological progress of an individual disease, and about epidemiological processes more generally.

Their book consists of two parts, one dealing with microparasites and the other with macroparasites. Diseases transmitted by arthropod vectors are also considered. There is much material on HIV/AIDS, measles, rubella, polio, and hepatitis B. In the final chapter there is some discussion of coevolution and the genetics of host-parasite interactions, the subject of Anderson and May's joint papers of the early 1980s. Overall, the book is a model of interdisciplinarity. It showed how to build models to forecast the path of infection, and to estimate the demographic consequences. It propelled the authors into a realm where their advice was much sought by public health officials worldwide.

In 1993 Anderson succeeded Southwood in the Linacre Chair at Oxford.[90] One year later he was invited to give the Royal Society's Croonian Lecture.[91] His lecture is a good overview of the type of mathematical models then being used to plot the path of infections, their application to epidemiological problems, and their use in the design of control procedures such as targeted vaccination. As in the 1991 book, the models followed from ones introduced twenty years earlier by May. While biologically and medically more informed, they too brought out dynamical complexity — now with respect to host-pathogen interactions. Anderson explained how attention was being paid to new patterns of infection worldwide (due to increased urban density, sexual behaviours, increased drug use and needle sharing, international travel etc.). He drew attention to advances in molecular and cellular biology, and in immunological research, and how such might impact future modelling. Much of the lecture was devoted to a discussion of HIV-AIDS. Overall, Anderson claimed that the type of models he was working with were important tools in the pursuit of public health, and that for many diseases the path of infection could be plotted in quantitative terms and with reasonably good precision.

On Anderson's move to Oxford, the Wellcome Trust supported construction of a new epidemiology centre contiguous with the Oxford zoology department on South Parks Road. It was a modern facility and attracted many good students. The new laboratory and his own research occupied much of Anderson's attention. It drew him away from other departmental duties, something that caused resentment among some of his colleagues. Anderson resigned as head of the zoology department and, as will be discussed in chapter nine, was to resign also from the Linacre Chair. He moved his research back to Imperial College in 2000.

7.6 Richard Southwood: public service and his move to Oxford

In 1977 Southwood was invited to serve on the board of electors for the Linacre Chair. Also on the board was his old friend, Henry Harris, professor of pathology at the Sir William Dunn School. They corresponded about the candidates and Harris, seemingly unhappy with those on the list, wrote 'is it at all conceivable that one T. R. E. Southwood would be interested?' Southwood replied:

> Naturally I was interested that you should mention one T. R. E. Southwood: our list is beginning to look dangerously like the Cambridge electors list last year ... we must play with it much more subtlety than Cambridge managed ... It is conceivable that Oxford could tempt Southwood because Wytham could replace Silwood Park, which has anchored him against a number of other pulls in the last few years; it is known that he has a very high regard for Zoology ... in Oxford, and that he dislikes commuting to London for his undergraduate

teaching. No doubt he would need to be convinced that he could bring some benefit to his new department and that there would be adequate university support for his research and field work and that his wife would wish to be certain that they could substitute for their home and garden, conveniently and fortuitously tucked alongside Silwood Park.[92]

Southwood was offered the Linacre Chair in January 1978 and accepted after laying down some conditions, including for his field work at Wytham.[93] When he moved to Oxford in 1979 the zoology department was at a low point. Lack, Elton, Varley, Ford and Tinbergen who had given it so much lustre were by then either dead or retired. Further, some promising younger people had moved elsewhere. Given that the department needed to rebuild, Southwood was a good choice as its new head, but his connection to Silwood was still strong. After moving he had little time for his own research but, with one major exception, the research he did carry out was almost exclusively with his old colleagues at Silwood and with his old friend Cliff Moran.[94] He also co-authored a book with John Lawton.[95] The exception was the resumption of his early work on oak tree defoliators, something that fitted well with the work pioneered at Oxford by George Varley. With some of his new colleagues Southwood catalogued the many insects to be found on the oak trees in Wytham Woods — about fifteen hundred species. It was a straightforwardly entomological study requiring the assistance of many systematists.[96] After his retirement Southwood wrote some papers with May and with May's former Princeton student, George Sugihara, who spent some time with them in Oxford.[97]

At Oxford Southwood came into closer contact with two people who were associated with the Silwood Circle from the start, John Krebs and David Rogers. Five others whom Southwood brought to Oxford in the 1980s also had Silwood connections. First, Southwood invited J. S. Kennedy to whom he was much indebted for help early in his own career. Kennedy, recently retired from his chair at Imperial College, was awarded a fellowship and continued his research at Oxford. Second, Southwood redressed an old mistake (see chapter four) and invited Bill Hamilton to Oxford. Hamilton returned from the United States to a Royal Society research professorship in 1984. Third, Southwood tried to get May to move to Oxford from the moment he moved there himself.[98] He finally succeeded in 1988. May left

Princeton for Oxford and a Royal Society Research professorship held jointly at Imperial College. Southwood also brought Paul Harvey to Oxford.[99] Harvey, an evolutionary ecologist, first met May when May visited John Maynard Smith at the University of Sussex in 1975. In the following year May invited him to join the Silwood Circle on one of its hiking trips. Harvey told me that from then on he joined the hikes almost every year and that, before moving to Oxford in 1985, he spent a sabbatical year with May at Princeton.

Southwood also appointed Charles Godfray as a demonstrator. Godfray represents a younger generation of scientists helped by an association with the Silwood Circle. He, too, joined the hiking expeditions and attended other gatherings. As an undergraduate at Oxford one of his professors was John Krebs. For his PhD he moved to Silwood where he worked under Valerie Brown, graduating in 1983. He stayed at Silwood on a two-year NERC postdoctoral fellowship before returning to Oxford as a demonstrator.[100] Back at Silwood in 1987 he rose quickly through the ranks, from lecturer to professor.[101] When I spoke with him he mentioned having been especially helped by Michael Hassell and Jeff Waage and that he published papers with both. He was to succeed John Lawton as director of the Centre for Population Biology, and became head of the department of biology in 1995. In 2006 he returned again to Oxford and the Hope professorship.

Southwood held a professorial fellowship at Merton College, the college with which the Linacre chair has long been associated. He enjoyed college life. He was held in high regard, won many awards, was awarded a knighthood in 1984 and, as a senior academic with a reputation for management skills, was much in demand.[102] For example, he was invited to join the National Radiation Protection Board and became its chairman in 1985. Robert May later joined him on the Board. The Chernobyl disaster of 1986 meant much pressure was placed on board members. Not only did they have to assess the more general health risks from low-level radiation, but also the consequences for Britain of this catastrophic event.[103] The British nuclear industry was then responsible for about one third of the nation's electricity supply. More power was needed, but opposition to increasing nuclear output was considerable. Chernobyl increased people's fears, but in Southwood the Board found a chairman able to handle the pressure.

One of the few things that turned out not so well for Southwood was his chairmanship of the MAFF/HSS working party on Bovine Spongiform Encephalopathy (BSE).[104] It would be unfair to say — though it was said — that the advice Southwood gave the government was poor. But there was a communication breakdown. The working party was set up in 1988, two and a half years after the first case of BSE had been identified in British cattle, and after many infected animals had already entered the food supply. It began reporting its findings already in 1988 but they were not made public until 1989. The main findings were misused or misread. The Minister of Health, for example, publicly denied that there was any danger to human health when Southwood had actually claimed that there was some danger, but that it was 'remote'. The denial by those in government continued for eight years until, in March 1996, it was announced that ten young people had contracted the fatal new variant Creutzfeldt-Jakob Disease (v-CJD) from eating infected beef. Southwood, with the help of Anderson's team at Oxford's new Centre for the Epidemiology of Infectious Disease, published a paper in *Nature* in August 1996 outlining what had happened and what the future could bring.[105] In 1997 the European Union banned the export of British beef. By 2000 there were about eighty known cases of v-CJD and, given the slow incubation of the disease, many more cases, possibly in the thousands, were feared. It was a national disaster, but who was to blame?

It would appear that the government had been reluctant to curtail the sale of beef because of pressure from the industry, and because of the daunting prospect of having to pay a huge amount to compensate farmers if a major cattle cull were ordered.[106] Both in 1988 and in 1989 Southwood stated correctly that, as yet, there was no evidence of danger to humans. But he also warned that this did not mean that precautions were unnecessary. Indeed, he stated that if the disease were to cross over to humans the consequences would be severe. He pointed out that cows are herbivores and should not be fed animal parts unfit for human consumption, as was then the practice. Cattle, he claimed, would not have immunity to diseases transmitted via unnatural food sources. Later Southwood was blamed both for not forecasting the danger, and for not having been more forthright in his dietary advice both for cattle and for people. But in 1989 there was no evidence of transmission to humans and the committee knew that such evidence, if it came into existence, would not do so for a few years. Further scrapie, a disease of sheep with a similar etiology to BSE, appeared not to harm people who ate infected lamb or mutton.[107] Southwood early recommended that the sale of brains (where the BSE prion was likely to be concentrated) and beef offal for human consumption be outlawed. He also recommended that all BSE infected cows be slaughtered and removed from the food chain, something the government began to move on from 1989.

Perhaps Southwood, who was politically astute, should have been more insistent on more drastic action. Perhaps the committee's make-up was too conservative.[108] Southwood must have known that there was huge pressure from the beef industry to do as little as possible, and that it would be convenient for those in government to misread his report as a total dismissal of the dangers of BSE infected meat. Later, Southwood made some interesting comments on this period in his life in a letter to Lars Walløe, a professor at the University of Oslo. They had earlier worked together on the management group of the Surface Water Acidification Programme, set up to investigate the effects of acid rain reaching Scandinavia from the United Kingdom.[109] Southwood's letter to Walløe contains a long passage on the importance of recognizing uncertainty in science — something much discussed within the Silwood Circle at that time (see chapter nine). Southwood reminisced both on some events of the 1970s and on his more recent involvement with the BSE outbreak. Comments made in the letter put me in mind of Voltaire's statement that being doubtful is an uncomfortable position to be in, but being certain is an absurd one. As to the 1970s, Southwood commented on what he saw as the reactionary attitude of an older generation of left-wing British scientists. Believing that science would improve the lot of humankind, he wrote,

> they were therefore made deeply unhappy by the development of environmental concern and "limits to growth concepts" and *Blueprint for Survival*. I recollect many senior and respected scientists pointing to uncertainties in the hypothesis with great vigour. ... I too must plead guilty to failing to emphasize the uncertainty surrounding the risk that some persons might have caught CJD (BSE form) *prior* to 1989. The risk was low and there was nothing anyone could do to protect oneself. Why worry people?[110]

The 'why worry people' sentiment fitted the paternalistic ethos of the period in which Southwood grew up; perhaps he was unable to shake it off. His letter suggests that, in retrospect, he recognized that he should have exerted more pressure on the government to publicly acknowledge uncertainty in the safety of the beef supply already in 1989. He was correct in stating that he could have done nothing to protect those who had consumed infected beef in the three years prior to November 1989. (The first infected animal was identified in November 1986.) But he should have warned them of the possible, albeit unlikely, danger associated with having done so.

Southwood was lucky in that the delay in the appearance of v-CJD meant that his reputation remained unclouded until the late 1990s. He impressed many at Oxford and was appointed vice-chancellor of the university in 1989, the first non-head of a college to hold that office.[111] Merton College commissioned a portrait to honour the occasion. The artist, Mark Wickham, had to take note of the many items Southwood wanted him to portray. The result is a highly iconographic portrait.[112] As vice-chancellor, Southwood retained the Linacre professorship and continued teaching the introductory biology course.[113] His work schedule was intense and included leading a major fund-raising campaign for the university. Just a couple of his many extra-mural activities will be mentioned here. First, in 1990 George Soros asked Southwood to chair a steering committee for a proposed department of environmental studies at his Central European University.[114] One year later Southwood stepped in after a planned summer school in environmental studies was about to be closed for lack of staff. Not only did he find the staff to keep it open, he himself participated. Impressed, as were others, Soros soon appointed Southwood to head the department at the university, centred in Budapest.[115] The department grew into the United Nations Centre for Environmental Programmes for Central Europe. Its well trained students were an important resource for new governments in the post-Communist era. Second, in 1992, after the World Summit on Sustainable Development held in Rio de Janeiro, the UK government established a Round Table of about thirty people to discuss the issues raised. Officially Southwood and the Secretary of State for the Environment were co-chairs. But much of the work fell to Southwood. At the same time he helped to set up the Environmental Change Unit at Oxford (it lasted from 1991 to 2002) and gave a lecture course on environmental problems that drew on the course Gordon Conway had given at Imperial College.

The 1990s saw Southwood receive many honours. In 1994 he was awarded the Gold Medal of the International Congress of Entomology and, in 1995, he was honoured by the Royal Society when invited to deliver the Croonian Lecture.[116] In his lecture he conveyed his environmental concerns and also illustrated, not always consciously, the ways in which his approach to ecology meshed with the ideas of MacArthur and May. In their biographical memoir May and Hassell write that in his final days Southwood reminisced about those he had gathered around him during the 1970s, the Silwood 'mob' as he sometimes called it. One of the things he said was 'they all swam in my swimming pool'.[117] I do not know whether this was a simple recollection, or whether Southwood intended some double meaning. Swimming in Southwood's pool surely meant more than simply swimming. Those included in his circle, those who swam in his pool, were privileged. It helped them on their way to becoming big fish in the larger pool of British science. They followed Southwood and each other into positions of influence such as becoming trustees of the Natural History Museum, and up the awards ladder gaining national and international recognition on their way. Much recognition came in the period after 1995, but the pattern was already set. For example, Hassell and Krebs (the latter perhaps did not literally swim at Silwood) won the Science Medal of the Zoological Society of London in 1981, Anderson won it in 1982 and Harvey in 1986. Among younger associates, Charles Godfray won in 1993, Bryan Grenfell 1994, Andrew Purvis in 2005 and Timothy Coulson in 2007.[118] Anderson was to win the Frink medal in 1993, May in 1995, Krebs in 1996, Lawton in 1998, Hassell in 2002 and Godfray in 2009. Maynard Smith, May, Krebs, Lawton and Beddington were all made Honorary Fellows of the Zoological Society.[119] And, like Southwood, others in the Silwood Circle moved toward public service. These mimetic patterns will be discussed further in the following chapter and, in chapter nine, we will pick up on the later careers of the people discussed here.

Endnotes

1. (Southwood and Conway, 1975).
2. (Kornberg and Williamson, 1986; Wood and Way, 1988). Southwood, May, Anderson, Lawton, Hamilton, Conway and Crawley were all participants at

the first conference. The new focus on invasion harked back to one of Charles Elton's major interests (see chapter 5). See also (Conway, 1977 and 1984; Hassell, 1984; Anderson, 1984). R. K. Wood and M. J. Way were both professors at Silwood Park. Other Silwood contributors to the 1988 volume include J. K. Waage, D. T. Greathead, May, Hassell, Lawton and Beddington.

3. ICCET was a considerable success and the number of MSc students rose from 24 in the first year to over 100 by the 1990s. A PhD programme was added later. The centre has since been amalgamated with some other smaller units in the Centre for Environmental Policy.

4. The major part of the zoology and applied entomology department was in South Kensington and much teaching took place there. Later the department was to join with botany in a new biology department. Ecological research was (and remains) located primarily at Silwood.

5. For his analysis Conway owed something to his Imperial College colleague, Geoffrey Norton, and to another environmental economist, Edward Barbier. See, for example, (Conway and Barbier, 1990). Barbier worked at the IIED (see below). He also held a position at the Environmental Economics Centre, University College London. Ian Craig stayed in Thailand where he became a farmer while continuing to work as a consultant, recently on the introduction of agroecosystem analysis in Cambodia.

6. (Conway, 2003), 109. Conway's chapter, in which agroecosystem analysis is discussed, is a contribution to a book edited by Nigel Cross. Collectively the chapters in the book track the concept of sustainable development from its articulation at the United Nations Human Environment Conference held in Stockholm in 1972, through the follow-up UN conferences at Rio de Janeiro (1992) and Johannesburg (2002). See also how Conway arrived at his analysis based on work carried out in Thailand, (Conway, 1985).

7. The Green Revolution worked well in Mexico where the country went from importing half its wheat in the 1940s to being self-sufficient — even exporting some wheat by the 1960s. The new wheat varieties were also highly successful in India where yields almost doubled. New corn/maize varieties were also developed.

8. (Conway and Barbier, 1990).

9. Stability and sustainability in garden systems relates to their containing many plant species and being buffered against pests and disease as a result. Equity depends on many factors. For example, on whether home gardens are owned or rented, and on the relationship between those with access to garden plots and poorer people with no access. Home gardeners presumably profit more

from their own labour than do farmworkers, but I have not looked deeply into Conway's agroecosystem models and cannot comment further. For some of the Indonesian work, (Conway and McCauley, 1983; Soemarwoto and Conway, 1992). McCauley was an economist working for the Ford Foundation and Soemarwoto, who died in 2008, was professor of plant physiology at Padjadjaran University, Bandung.

10. This type of tension often manifests itself within NGOs. There are those who want to see governments doing much more than they do, and who resent the intrusion of business interests in their activities. But when government takes on too many roles, the results are no better. Indeed, the logical end-point of such intervention is a totalitarian regime. This is not to deny the important role that governments can and should play, but there has to be space for civil society, voluntarism, and entrepreneurship.

11. (Conway, 2003), 108.

12. Young women also worked at the IIED, but at that time feminism had won only a few of its many battles. The early environmental movement and its academic offshoots were still largely male dominated. Ward was exceptional in many ways, not only as a successful woman.

13. Sandbrook obituary, *The Independent*, 12 December, 2005.

14. Since secession in 1993, Wollo is now in Eritrea. As I write (in 2011) it appears that another major drought in the region is causing serious problems yet again.

15. Geldoff is said to have been moved to act by the graphic reports from Ethiopia by Brian Stewart of the Canadian Broadcasting Corporation.

16. The revolutionary movement that began in Tigray led after many years of struggle to a new government in Addis Adaba in 1991. Agricultural and food-delivery improvements were then made. The drought of 2002–2003 was worse than the one in the 1980s but there were far fewer deaths.

17. See (Taylor, 2005), chapter 5, for an interesting analysis of some USAID work in this period.

18. (Conway, 2003; Chambers and Conway, 1991). Asking local farmers and forest managers simple questions about what worked and what did not and organizing their replies in a matrix format became standard procedure. McCracken later took the idea of participatory appraisal to India.

19. (Conway, 2003). Michael Loevinsohn worked for some years in the Netherlands and is now at the Institute for Development Studies at the University of Sussex. Jules Pretty stayed on at the IIED where he later became director. A prolific author, he is now professor of environment and society at the University of Essex. (Conway and Pretty, 1991) notes problems with integrated pest man-

agement (biological, chemical, and cultural) and with integrated nutrient management (organic and chemical). For a general account of some post-war developments along these lines, (Gay, 2012).

20. Conway was clearly thinking of making some sort of move from Imperial College at this time. As correspondence in the Southwood papers makes clear, he applied for a number of positions, including one in the United States.

21. The UN World Commission on Environment and Development convened in 1983 and was chaired by Gro Harlem Brundtland. Sandbrook and others at the IIED drafted the Brundtland Report, *Our Common Future* (1987). Its principal message was that development should meet the needs of the present without compromising those of the future. Nonetheless, it received much criticism from some quarters of the Green Movement as being too sympathetic to business and to its role in developing green technologies in support of the developing world.

22. A major development specialist, John D. Gerhart (1943–2003) ended his career as president of the American University in Cairo.

23. Bodleian Library, Southwood papers, R52; Conway to Southwood, 9 July, 1988.

24. Lord Flowers was a behind-the-scenes emollient. Both Flowers and Southwood wrote reference letters for Conway. (A copy of Southwood's is in the Bodleian Library and is very supportive; Southwood papers, R52). Conway told me that he also asked Lord (Roger) Nathan, a lawyer and keen spokesperson on environmental matters in the House of Lords, for a reference. Nathan was supportive but could not oblige with a letter since he was on the selection committee. In his eulogy for Flowers, at the memorial held at Imperial College on 11 November, 2010, Conway mentioned that Flowers was a frequent guest at Sussex and gave him useful advice on a number of occasions. For Conway's Nathan memorial lecture see chapter 4.

25. For example, his attempt to unite the physics and chemistry schools was far from popular and did not succeed. Perhaps wisely, chemistry moved rather toward the life sciences.

26. The Runnymede Trust, founded in 1968, was dedicated to supporting research on multi-ethnic Britain. Neuberger was a trustee.

27. Conway told me that the Trust wanted a total outsider to chair the commission and that he learned much from the experience. His report, *Islamophobia: A Challenge for us all* was published by the Trust in 1997.

28. The universities' group formed in 1994, hence its name. Conway arranged for Southwood to be given an honorary degree by the University of Sussex in 1994.

29. Mark Westoby decided to stay at the University of Utah for his PhD. He then moved to work at Macquarie University, Australia, where he made his career.

30. Bernard Stonehouse began studying penguins when he was a navy pilot for the British Antarctic Survey. He later studied zoology at University College London and then took a PhD at Oxford's Edward Gray Institute of Field Ornithology. In 1983 he left Bradford for the Scott Polar Research Institute at Cambridge and the editorship of *Polar Record*. For more on Stonehouse, (Cruwys and Riffenburgh, 2002). Stonehouse spent a couple of years at the University of British Columbia at around the time Crawley was there, though they met only later. While in Canada Stonehouse studied some Yukon mammals. He received much criticism for his view that the function of deer antlers was thermoregulation.

31. (Crawley, 1975).

32. Crawley, personal communication. Crawley claims that he never planned on returning to Silwood, that his wife had an interesting job with the Bradford Museums, that they were happy in Bradford, and lived in a beautiful house in the Yorkshire Dales. Proximity to the Dales and good walking country helped in bringing the others to Bradford on short visits.

33. Southwood was to write a letter very supportive of Crawley's promotion to reader in 1989. Bodleian Library, Southwood papers, S46.

34. For Southwood's work in this area see chapter four. On his Imperial College London website Crawley describes his own work as seeking answers to two questions. 'How does feeding by herbivorous animals affect the distribution and abundance of plants? And how do changes in the abundance and quality of plants influence the distribution and abundance of herbivores? The aim is the development of theory to take explicit account of the features that make plant-animal interactions so distinctive.' The theory he uses is not unlike that used by insect ecologists in the Silwood Circle and described elsewhere. However, such models have to be adjusted since plants usually fare better than animals when attacked by predators. The models must also reflect the feedbacks which Crawley's two questions imply.

35. See, for example, (Islam and Crawley, 1983; Crawley, 1985 and 1990).

36. (Crawley, 1983), quotation, 345.

37. (Crawley, 1993, 2002 and 2007).

38. In the preface to his *Herbivory* (1983), Crawley thanks Harper 'who started it all by rekindling the spirit of Darwinism in plant ecology'. In Crawley ed. (1986a), he further expressed his indebtedness to Harper. Crawley contributed two of his own chapters to this volume which became a much used textbook. When Crawley gave his inaugural lecture as professor, Harper gave the vote of

thanks and made some amusing comments on Crawley's 'rather odd career trajectory', (Crawley, personal communication).

39. (Anderson, Gordon, Crawley and Hassell, 1982). This paper considers stochastic aspects of species distributions, arguing against some behaviourally deterministic ideas suggested by others. Crawley also co-edited a collection of papers following the 26th symposium of the British Ecological Society: (Gray, Crawley and Edwards, 1987). Crawley contributed a paper 'What makes a community invasible?', (424–453) in which he asks what attributes of a community make it likely to be invaded and what are the attributes of a successful invader. The volume includes also a paper by John Lawton and a joint paper by Southwood and Valerie Brown.

40. (Salisbury, 1961). This book was part of the Collins New Naturalist Series. Sir Edward Salisbury FRS (1886–1978) had grown up near the Rothamsted Experimental Station and his family had connections to people working there. Later he became a trustee of the Lawes Agricultural Trust. He was Quain professor of botany at University College London, and Director of the Royal Botanic Gardens at Kew. Crawley dedicated his book on plant ecology (1986a) to Salisbury and H. A. Gleason, two pioneers of plant ecology.

41. V. Cliff Moran was then professor of entomology at Rhodes University in Grahamstown. He later moved to the University of Cape Town. Moran had been a postdoctoral fellow at Silwood earlier. Later he spent another sabbatical year (1980–1981) with Southwood at Oxford. Southwood and his wife Alison formed a close friendship with Moran and his wife Peggy. This can be inferred not only from Moran's visits to the UK but from correspondence held in Southwood's papers in the Bodleian Library. The Southwoods made a couple of trips to visit the Morans in South Africa.

42. Classical biological control entails the introduction of an alien species in order to control a pest species. More recent biological control methods attempt to ensure that conditions for native control species are effective, and possibly enhanced. Both approaches are in use; the more recent by Conway in some of his overseas work.

43. Work along this line from the period 1984–1987 was reviewed in (Crawley, 1989). For Jeff Waage see below. David Greathead was an Imperial College graduate, and director of the Commonwealth Institute of Entomology, later restructured as part of the International Institute of Biological Control (IIBC). Greathead moved to Silwood with the IIBC in 1981. He was a specialist in biocontrol with much experience working in Africa. He was also editor of *Biocontrol News and Information*.

44. (Crawley, 1986b and 1992); the second of these, a collection of papers edited by Crawley, is interesting for the range of natural enemies with which it deals: from wolves to bacteria; from specialist predators to generalists; from gregarious predators to solitary ones. Whether there were some underlying and unifying explanations to all these behaviours was an open question, as were some of the evolutionary implications. Contributors to the volume included May, Krebs (a chapter written jointly with Crawley), Hassell and Beddington. There were also some contributions from people who can be seen as part of an outer Silwood Circle: Andrew Dobson, Charles Godfray, Jeff Waage and Paul Harvey (for more on these people see below and chapter eight). See also (Crawley and May, 1987) and (Crawley, 2007).

45. PROSAMO arose after much discussion. Already in the 1980s people in both industry and government recognized the need for regulatory regimes for GM crops and for the use of GM microbial agents such as bacteria that could prevent the freezing of strawberry crops. In addition to Crawley's research, funding was provided for microbial research carried out at the Universities of Essex and Aberdeen, and for further work on plants carried out at the University of East Anglia. For a sociologist's take on the project, (Moroso, 2008). Moroso claims that the people he interviewed who were working for industrial partners such as ICI, Unilever and Monsanto believed that GM crops were harmless, but that the research was necessary because industry needed to calm the fears of the public (ie consumers). A regulatory regime would provide both public assurance and lay down the rules for industrial competition, something also desired by industry. It was, he claims, for these reasons rather than scientific discovery, that PROSAMO was undertaken. It is not clear whether all of the academic scientists involved would agree that the scientific value was minimal, though it is true that many ecologists played down the dangers of GM crops even before much testing had taken place. For an outline of Crawley's contribution, (Crawley, Brown, *et. al.,* 2001).

46. See, for example, an interesting guest editorial by Helmut F. van Emden, a professor of horticulture and of applied ecology at Reading University. (Emden, 1998) Emden was a zoology student at Imperial College where he gained his BSc and then his PhD under Southwood. For a brief period he was also Southwood's research assistant. The situation is now much changed with greater understanding of what it will take to feed the growing world population, as well as greater acceptance of new insect resistant crops, crops suited to drought conditions, crops engineered to produce micronutrients, and so on.

47. Some gene movement is bound to occur; it has done so since life began and the genes of engineered crops will be no different. There is some risk, but Crawley's work did not find any harmful results.

48. Crawley's remarks were made after one of his papers received much discussion in the press. He was quoted in an article by William K. Stevens in the *New York Times*, 22 June, 1993. There was also concern about changes to the behaviour and populations of insect pests that feed on transgenic crops. It was a concern addressed in (Crawley, 2000).

49. George Gradwell was a forest insect ecologist who began work at Oxford in the forestry department before moving to join Varley in the Hope department.

50. The major controllers of the moth population are vertebrate and invertebrate predators that eat the pupae located in the soil — the most important of these is a beetle. Since the beetle does not live in North America, the winter moth is a more serious pest there. It is possible that I am using the terms 'control' and 'check' in ways that Varley would have disapproved. In a paper critical of a scientist whom they believed was using Varley and Gradwell's data incorrectly, Hassell and some of his Silwood colleagues cite one of Varley's papers on the control of the 'Ruritanian bean bug', in which he showed how the term 'control' was used in six different ways, leading to semantic confusion. Since Ruritania is a fictional country I assume the bug is also fictional and was invented by Varley to make a point about clarity of meaning. See (Hassell, Crawley, Godfray and Lawton, 1998). This paper summarizes well the purpose of the winter moth study and takes issue with a particular interpretation of the data. As to semantic clarity, that is bound to be more elusive than Varley as well as Hassell *et. al.* would like. It is a price we pay for the flexibility of language. In science good observation and accurate data collection is what is most important. Interpretation, even with care taken over terms, will always be contested. This is not to say that there is no difference in the quality of interpretation, a point well made in their paper.

51. Only one fly per moth pupa survives, which means that if the moth caterpillars consume more than one egg there must be some competition within the fly population for survival. Further, the timing of the opening of the oak buds will vary a little from year to year, something to which the hatching moth larvae must somehow accommodate. These added complications — let alone interactions with yet further species, and other forms of mortality — illustrate something of the complexity of population biology. The system discussed here is relatively simple in that after mating and egg-laying the moths die and the generations are distinct.

52. In one of his early papers, Hassell engaged with the kinds of problem that interested C. S. Holling, and that were described in the previous chapter.

It shows him following in Varley and Gradwell's footsteps, yet adding ideas of his own. (Hassell, 1966).

53. Southwood's papers includes some correspondence with Huffaker.

54. For thesis work see, for example, (Hassell, 1969); see also (Varley, Gradwell and Hassell, 1973). In this clearly written book the authors discuss not only population studies but also applied aspects such as insect pest control.

55. This stepwise approach is well illustrated in an early paper, (Hassell and May, 1973). Another joint paper from that period is (Hassell, Lawton and May, 1976).

56. See, for example, (Hassell and May, 1974; Comins and Hassell, 1987; Hassell, Pacala, May, and Chesson, 1991). Interest in patchiness continues and remains an important area of study for those interested in what has become known as landscape dynamics.

57. (Hassell, 1978). Hassell's certificate at the Royal Society shows that he was proposed by Southwood, seconded by May. His certificate was also signed by the Rector of Imperial College, Eric Ash, as well as by his former departmental colleague, J. S. Kennedy. The two Silwood Circle members who had been recently elected fellows, R. M. Anderson and J. R. Krebs, also signed, as did J. Maynard Smith and a number of others. Proposed in 1985, Hassell was elected in 1987.

58. Tom S. Bellows, a specialist in biological control, is now professor of entomology at the University of California at Riverside. He gained his PhD in 1979. Matthew Cock is chief scientist at CABI's European centre in Switzerland and a specialist in the Integrated Pest Management (IPM) of both weeds and insects. Since 1986 CABI has stood for Centre for Agricultural Bioscience International, but before that it stood for Commonwealth Agricultural Bureaux International. This body has a complicated history going back to the early twentieth century. One of its ancestral roots was the Imperial Bureau of Entomology, and its offshoot the Anti-Locust Research Centre, with which Imperial College entomologists had close connections in the 1930s and 40s. In 1981 the CABI Institute of Biological Control settled in a new building at Silwood. CABI also has centres around the world; its head office is in Wallingford, Oxfordshire. Bonsall and Hassell wrote a joint chapter 'Predator-Prey Interactions' in (May and McLean, 2007).

59. Waage, personal communication.

60. Waage took this position against the advice of Southwood. Bodleian Library, Southwood papers B22, letter from Waage to Southwood, 20 August, 1986. Later Waage briefly returned to Imperial College as head of its short-lived agricultural sciences centre at the Wye campus, the result of a merger with the

Wye agricultural college. He is now Director of the London International Development Centre, run jointly by six University of London colleges. Waage's earlier work at the Institute of Biological Control picked up on the research of the locust specialist, J. S. Kennedy (see chapter 4). Waage applied biological control to the desert locust. His team developed sprays for the delivery of local insecticidal fungi. By then Kennedy was long retired from Imperial College and had moved to Oxford, invited there by Southwood.

61. The position of warden had been abolished.

62. See, for example, (Anderson, 1974a).

63. Maurice S. Bartlett FRS (1910–2002) was a Cambridge wrangler and briefly a lecturer at Imperial College before moving to work with R. A. Fisher and J. B. S. Haldane at University College London. A brilliant man, a specialist in mathematical modelling, he held several positions in and out of universities before spending the last few years of his academic career at Oxford. Perhaps his best known work related to ecology is (Bartlett, 1960). Anderson's IBM fellowship was administered by Bartlett.

64. Some of the work at King's College was on a zebra fish infected by a flatworm that attached to the outside of the fish. See, for example, (Anderson, Whitfield and Mills, 1977). Andrew Dobson would soon move on from King's College to work on his DPhil at the Edward Grey Institute at Oxford. Later he held postdoctoral fellowships with Anderson at Silwood and then with May at Princeton. Today he is a professor in the department of ecology and evolutionary biology at Princeton. One of Philip Whitfield's PhD students was Don Bundy who later became Senior Scientist at the World Bank. Bundy, a friend of Anderson's, spent some time at Imperial College after Anderson became head of department. But most of his work was carried out overseas. He has commented on Anderson's early move to include epidemiological ideas in his ecological work on parasites, and states that Anderson drew on some ideas in two papers published by the veterinary parasitologist, Harry Crofton, in 1971, (Esch, 2007). Anderson told me that he was influenced by Crofton's 1971 papers and that he met Crofton before Crofton's premature death in 1972. For his own take on Crofton's influence see Esch, p. 70. Anderson's work with May led to a generalization of Crofton's population ecology work, and to the introduction of models applicable to a wide spectrum of disease agents. As will be discussed, their work led to a more targeted approach to infectious disease control than was earlier the case.

65. Bodleian Library, Southwood papers R5. See, for example, letter dated 17 December 1970. In addition to praising Anderson's scientific ability, Southwood wrote that he was active in student affairs, especially with the

rugby football club, and that he 'has a charming and open personality and has been popular with both staff and students.' Also in this file are a number of other letters written in support of Anderson's early job applications. Anderson made many requests for letters.

66. Bodleian Library; Southwood papers R5, copy of referee's form. Anderson withdrew his application after receiving some other generous research grant.

67. Croll was then working at McGill University. According to Britannica on-line, about 25% of the human population is infected by *Ascaris lumbricoides*, often as a result of eating infected pork. The infection is easily treatable when detected but about 20,000 people die of it each year. The round nematode worm is one of the largest intestinal parasites. For some work on it see, for example, (Croll, Anderson, Gyorkos and Ghadirian, 1982).

68. Kenneth S. Warren, Director of Health Sciences at the Rockefeller Foundation, recognized the public health importance of helminth studies and supported a large field study in India.

69. Bryan Grenfell FRS was a doctoral student with Anderson. He spent some time as a professor in Cambridge but is now at Princeton. For Dobson see note 64. Others who held research fellowships with Anderson include Mark Woolhouse, now professor of infectious disease epidemiology at Edinburgh University, Angela McLean FRS a professor in the zoology department at Oxford and Elizabeth (Betz) Halloran, now a professor at the University of Washington in Seattle. Sunetra Gupta was a student of May's at Princeton before moving to work for her doctorate with Anderson. She is now a professor at Oxford.

70. (May and Anderson, 1983). This paper includes data from the infection of Australian rabbits by the myxoma virus, looking back to the classic work of Frank Fenner. It also threw some doubt on W. D. Hamilton's earlier claims on the origin of sexual reproduction. To an outsider the term 'evolutionary strategy' appears anthropomorphic.

71. Bodleian Library, Southwood papers, R5; letter to R. K. S. Wood, 6 January, 1982.

72. This medal, awarded earlier to Southwood, is awarded in recognition of major achievement in the biological sciences. The recipient must have some connection to Imperial College London but need not be a current student or staff member.

73. (Anderson, 1974b).

74. (Anderson and May, 1979).

75. (Anderson and May, 1981). Anderson and May also edited a volume of papers from a workshop held in Dahlem, (Anderson and May, 1982). See also (Anderson, 1982).

76. This connection had been made much earlier. As mentioned briefly in chapter two, it was something discussed by Lotka, and even earlier by Ronald Ross and others.

77. Anderson contributed a paper to a conference on this topic hosted by the International Institute of Applied Systems Analysis; (Anderson, 1981). The Institute was founded in 1972 as a way to bridge the divide between East and West during the Cold War. Located near Vienna, it continues to support international scientific cooperation. C. S. Holling was director of the institute, 1978–1981.

78. (Anderson, Jackson, May and Smith, 1981; Anderson, 1986).

79. This prize, presented by Cornell, Indiana and Stanford Universities, was named in honour of David Starr Jordan (1851–1931), a biological scientist who was educated at Cornell and later worked as a professor, then president, at Indiana University before moving to become the first president of Stanford University. The prize is awarded every three years to a biological scientist under the age of 40.

80. Bodleian Library, Southwood papers, R5; May to Southwood, 26 Nov, 1985, Princeton. Letter and nomination form.

81. Already in this period Anderson was awarded the Scientific Medal of the Zoological Society of London (1982), the C. A. Wright Memorial Medal of the British Society for Parasitology (1986), the Chalmers Memorial Medal of the Royal Society of Tropical Medicine and Hygiene (1988) and the Weldon Memorial Prize for Biomathematics from the University of Oxford (1989). Anderson's election certificate at the Royal Society shows that he was nominated by May, seconded by Southwood. Among the others who signed were W. D. Hamilton, M. S. Bartlett, J. L. Harper and J. Krebs. Anderson was proposed in 1984 and elected in 1986.

82. May told me that he thought of himself as an academic r-strategist, that is an opportunist in the survivalist sense (see chapter five for MacArthur and Wilson's ideas on r and K strategies).

83. Sir Henry Wellcome, who died in 1936, left both shares and instructions for setting up the Wellcome Foundation and its sole shareholder, the Wellcome Trust. As the shares appreciated money was to be allocated 'to advance medical research and understanding of its history'. In 1985, the year of the court decision, the Foundation was valued at about £1 billion and the allocation to the Trust was £26.5 million. Renamed Wellcome plc, the Foundation was floated on the London Stock Exchange and the Trust sold about one quarter of its shares. In 1992 the Trust reduced its holdings to about 40%. In 1995 the remainder of its shares were sold to Glaxo which then became Glaxo Wellcome.

In 2000 Glaxo Wellcome merged with Smith Kline Beecham to become GlaxoSmithKline. The Wellcome Trust invested in the new company so as to hold about 3% of its shares. By 2007, and by diversifying its holdings, the Wellcome Trust had assets valued at about £15 billion and, in that year, gave about £65 million in research grants. See Wellcome Trust website.

84. Trustees receive a small remuneration for their services. All attempts are made to exclude those with interests in a given project from any decisions related to its funding. Anderson was far from alone among trustees who were also beneficiaries. Perhaps the most outstanding example from roughly the same period was the funding of Sir David Weatherall, and his Institute of Molecular Medicine at Oxford (now named for him the Weatherall Institute).

85. See, for example, (May and Anderson, 1987; Anderson, Blythe, Malley and Johnson, 1987; Nokes and Anderson, 1988; Grenfell and Anderson, 1989).

86. (May and Anderson, 1988).

87. (Anderson and May, 1991).

88. (Anderson and May, 1991), v.

89. (Anderson and May, 1991), 8.

90. Anderson was also offered the Rectorship of Imperial College at this time. Robert May appears to have been instrumental in getting Anderson to Oxford. Southwood returned to the department after his term as vice-chancellor but not to the Linacre chair (see below and chapter 9).

91. Anderson, (Croonian Lecture, 1994).

92. Bodleian Library, Southwood papers C1; Harris to Southwood, 5 October, 1977; Southwood to Harris, 9 October, 1977. Southwood was earlier offered the chair at Cambridge but declined. He also wrote to Harris that, 'thinking as an elector', John Gurdon was far more distinguished than himself, and that since the university would soon also have to find someone for the Hope Chair it was perhaps not a good idea to appoint another entomologist to the Linacre chair. The politics appear to have been complicated, Oxford decided to consider the appointments to the Hope and Linacre chairs together and to bring the Hope Department into the same building as the Linacre Department. (See letter from J. W. S. Pringle to Southwood, 19 October, 1977). Harris then asked Southwood which chair he would prefer and Southwood replied he would prefer the Linacre, ideally as a chairman for five years, not as a 'life sentence'; and provided budgetary and space conditions were made clear, and that he was able to make some appointments of his own in the near future. When things looked promising Southwood resigned from the Board of Electors. (Southwood to Harris, 19 November, 1977). He and his wife found a new home on Upland Park Road in Oxford.

93. A sign of inflation over the past thirty-three years: Southwood's salary in 1979 was £9294 for the Linacre Chair plus an extra £1241 for being head of department.

94. See, for example, (Southwood, Brown and Reader, 1979 and 1983; Hassell, Southwood, and Reader, 1987). Acknowledged in the 1987 paper were Crawley and May. See also (Moran and Southwood, 1982). This paper covered work carried out at Silwood and in South Africa and entailed the help of many systematists. For Moran see note (41) above.

95. (Strong, Lawton, and Southwood), 1984). Strong was then working at Florida State University. This collaboration will be discussed in chapter eight.

96. (Southwood, Wint, Kennedy and Greenwood, 2004). The study was conducted over about five years and the authors speculated on the reasons for plants being susceptible, or not, to insect pests.

97. See, for example, (Southwood, May and Sugihara, 2006). Published posthumously this was Southwood's final publication. Interestingly it harks back to some of the concerns of Robert MacArthur about the relationships between species numbers and area, body size and area, and the dimensions of niche space. George Sugihara's doctoral work at Princeton entailed a network analysis of food webs. It required complex computing, carried out with people at Bell Laboratories. Later he went to work in Tennessee and became interested in fisheries. He also began looking at chaos theory and at seeking ways to distinguish what was truly random from what was not. After a period working with Deutsches Bank on securities, he returned to academia and fisheries at the Scripps Institute where he is a professor. It was through Sugihara that May began working with bankers. See chapter 9.

98. In reply to an early invitation, May wrote that he might consider accepting in a few years time after his daughter, then in high school, was in college. Bodleian Library, Southwood papers C6, May to Southwood, 25 January, 1979.

99. Harvey, who later became head of the Oxford department, had been a biology student at the University of York where he stayed for his doctorate under Mark Williamson. After working as a lecturer at Swansea for two years, he moved to join John Maynard Smith at the University of Sussex.

100. While on his postdoctoral fellowship, Godfray worked for six months with CABI on some pest management problems in the Philippines, Malaysia and Fiji.

101. From its start in 1907 Imperial College supported its promising young lecturers and, on average, promoted people to professorships at a younger age than happened elsewhere. Typically in the first half of the twentieth century most British universities had only one or two professors in each department.

Imperial College's promotion practises became more difficult to sustain after the college became part of the University of London in 1929, but they did not totally disappear. In the 1950s, on the insistence of Patrick Blackett, all those seen as deserving in his department (physics) were given chairs. This opened the way for other departments to do the same. From the late 1950s to the 1980s, until other British universities caught up, Imperial (by then independent of London University with respect to national granting agencies) was almost alone in promoting relatively young and deserving people to professorships, allowing several in each department. This pattern helped with Anderson's meteoric rise. By the time Godfray came along early promotion to professor was becoming more common, but to some of his seniors his rise, too, will have seemed meteoric. For a festival held at Silwood in the late 1990s some students designed a t-shirt with a picture of Hassell leading a dachshund (poodle?) on a leash. The dog's head was replaced by Godfray's.

102. Southwood won the Linnean Society's gold medal for zoology in 1988 and, in the same year, was elected a foreign associate of the US National Academy of Sciences. He was awarded many honorary degrees. In 1987 he was especially pleased to be appointed a trustee of the Lawes Agricultural Trust, thus helping to guide the Rothamsted Experimental Station (now Rothamsted Research) where he began his career. He was also on the board of Oxford University Press.

103. In this connection Southwood was the chairman of an international conference organized by Friends of the Earth and Greenpeace International. Its proceedings were published; (Southwood and Russell Jones, 1987).

104. The Ministry of Agriculture, Fisheries and Food (MAFF) and the Department of Health and Social Security (HSS). For Southwood's papers on the Working Party, Bodleian Library, Southwood papers, section M. For the report, see *The BSE Inquiry*, vol. 4 'The Southwood Working Party, 1988–89' in the National Archives. Much can be read on the archive website.

105. (Anderson, Donnelly, Ferguson, Woolhouse, Southwood, *et al.*, 1996). Two of the authors not listed were veterinarians working in public laboratories. The authors, especially Anderson and Southwood, were very much in the public eye during this period, interviewed in the Press, radio and TV. Donnelly, Ferguson and Woolhouse were young members of Anderson's team, all are now professors with their own research groups. Cristl Donnelly and Neil Ferguson are at the Department of Infectious Disease Epidemiology at Imperial College London and Mark Woolhouse is at the University of Edinburgh. By August 1996 two more cases of v-CJD were confirmed. The paper, however, dealt mainly with BSE and, in light of the nature of its spread among cattle, evaluated different culling policies for the eradication of the disease.

106. I have not read the Phillips Report, only some short reports on it. I have looked at some of Southwood's papers on BSE, at an account of the events by a senior civil servant, and at one or two academic sources. Bodleian Library, Southwood papers; section M; Packer (2006). Packer was the permanent secretary at the Ministry of Agriculture, Fisheries and Food (MAFF), 1993–2000. Lord Phillips delivered his voluminous report 'Inquiry into BSE and v-CJD in the UK' to the government in 2000. See also (Jasanoff, 1997; Seguin, 2000).

107. Eating infected sheep brains was later shown to carry some danger.

108. Something along these lines appears to have been the view of Sheila Jasanoff who was critical of British advisory committees more generally for not always including the best scientific experts, but relying rather on the 'great and the good' — people seen as trustworthy, as a safe pair of hands etc. (Jasanoff, 1986 and 1997).

109. The programme was set up in 1983 by the Royal Society of London, the Royal Swedish Academy of Sciences and the Norwegian Academy of Sciences to address the problems caused by sulphur emissions from British industry affecting Scandinavian lakes and forests. In 1984, after the death of Maurice Sugden, Southwood became chairman of the group.

110. Bodleian Library, Southwood papers, A171, Southwood to Walløe, 13 June, 1997. The older scientists to whom Southwood was referring were those close to J. D. Bernal, and others who were influenced by his ideas; see (Werskey, 1978).

111. In a letter to his wife and a family friend, Wendy Laffer, Southwood wrote that while it was a great honour to be asked to be vice-chancellor and that it will 'help my pension', there was so much to give up. He was 'torn, very excited and *very* depressed' at the thought of all the entertaining, fund-raising in the US, speeches, graduations etc. Bodleian Library, Southwood papers, A119, letter to Allie and Wendy; Port Moresby, 18 March (no year). The Southwoods were staying with the Laffers in Australia; Southwood was on a side trip to New Guinea where he was working on a project together with Jared Diamond.

112. Bodleian Library, Southwood papers, C25. Southwood asked for inclusion of the following: the first edition of his book, *Ecological Methods*, a number of entomological books that were important in his youth, history books that he enjoyed as a boy, and insects that were important to him at various points in his life. These included some Hawkmoths related to his work on their interactions with host plants; a Red Admiral which he found on 3 September, 1939 while the adults in his family were listening to the declaration of war on the radio; some female stag beetles that he collected in the garden of his childhood home, Parrock Manor; and an adult ant lion found at his home in France at

L'Elzieres Mars Le Vigan (in the Languedoc region of Gard). He also wanted the portrayal of some shield bugs to represent his first scientific interest in the *Heteroptera*, including a Hawthorne shield bug that was given to him by his uncle in 1943 and that fostered his interest in the taxon. Also to be included was his hand lens, given to him as a birthday gift in 1944 and later modified so he could wear it around his neck on field trips.

113. By all accounts he was an excellent teacher, especially for young beginning students. His introductory lectures at Oxford were the basis for a textbook; (Southwood, 2003).

114. Earlier Soros had approached C.S. Holling to take on this position, but they did not see eye to eye on how to proceed. (Holling, personal communication).

115. Southwood was not expected to be resident in Budapest, but to direct operations from Oxford.

116. (Southwood, Croonian Lecture, 1995).

117. (May and Hassell, 2008).

118. For Grenfell see above. Andy Purvis was a DPhil student at Oxford with Paul Harvey who then went to Silwood on a Royal Society University Research Fellowship, was appointed lecturer and soon rose to professor. Tim Coulson was a PhD student at Silwood with Michael Crawley. After a brief period at the Zoological Institute of London and the University of Cambridge he returned to Silwood where he, too, is now a professor. A rising star in ecology he can be seen as carrying the Southwood tradition forward. Coulson and Godfray contributed a chapter 'Single species dynamics' to May and McLean (2007). Bryan Grenfell and Matthew Keeling also contributed a chapter, 'Dynamics of disease: Host-pathogen associations'.

119. For some of Southwood's input on awards by the Zoological Society of London see Bodleian Library, Southwood papers, S237–38.

chapter eight

The Growth of Careers 1970–1995

Part Two

John Lawton, John Krebs and David Rogers were at Oxford during the late 1960s and early 1970s — as was Michael Hassell. When Rogers began working on his D. Phil, Hassell was on the first year of a three-year postdoctoral fellowship. As mentioned, he spent the second year at C. B. Huffaker's laboratory in California. By the time he returned for his third and final year at Oxford, Krebs had entered the doctoral programme. During the same period Lawton held a demonstratorship at Oxford where he also completed writing his University of Durham doctoral thesis and started a new research project. This chapter gives brief overviews of the careers of Lawton, Krebs and Rogers into the mid-1990s. It also looks at the career of John Beddington, a colleague of Lawton's at the University of York and, later, also at Imperial College.

8.1 John Lawton

John Lawton was an undergraduate and then doctoral student at the University of Durham. His supervisor was John Phillipson, an entomologist who, inspired by the Odums, specialized in ecological energetics. In the late 1960s the zoology department at Oxford wanted such expertise and Phillipson was appointed to succeed Elton in the readership. With the closure of the Bureau of Animal Population, Phillipson began work in the newly formed Animal Ecology Research Group. Like Elton, he carried out research in Wytham Woods, in his case on insect carnivores, herbivores and detritivores.[1] Lawton was appointed Phillipson's demonstrator and followed him to Oxford in 1968. His thesis was on damselfly larvae, and his early publications show Phillipson's energeticist influence.[2] Some of his papers were co-authored with Stuart McNeill who was working at Silwood Park and had similar interests.[3]

Lawton was also interested in the work being carried out at Silwood on insect herbivory. This, and his meeting Hassell at Oxford, led him to a more expansive view of insect ecology. On completing his doctorate Lawton began a study of insects that feed on bracken, despite receiving criticism from both Varley and Elton for choosing a plant that they believed attracted few insect herbivores. Lawton was to find, however, that in different areas of Britain, and in various combinations, at least forty species of insects feed on bracken foliage, and that over half of them are fairly common. As Lawton later mentioned, both in the introduction to his International Ecology Institute Prize book, and in his Japan Prize lecture, there were two reasons for his choice. First, he wanted to look at a relatively simple community of insects and McNeill suggested that he look at bracken-feeding insects since little was known about them. Second was that bracken appeared to contradict the widely-held view that crop monocultures are vulnerable to both disease and insect attack. Lawton was curious as to why large areas of bracken, with all the appearance of monocultures, thrive largely unharmed.[4] He also saw that bracken patches could be viewed as islands and be studied in ways suggested by MacArthur and Wilson in their book on island biogeography. And, in light of the diversity-stability debate that was just taking off, it was interesting that the insect populations seemed fairly stable. Lawton discussed the more general problem of stability with Robert May when May first visited Oxford in 1971.[5] But he never came to fully understand why bracken insects rarely, if ever, reach population levels leading to major defoliation.

Lawton also began collecting data on insects that feed on other common plants to determine whether the number of species feeding on bracken was atypical. Some of Southwood's work was useful in this regard and helped to show that the number was not unusual. To pursue this a little further, and to plug his own data base into something larger, Lawton worked for a short period with Dieter Schröder in Switzerland on host-plant geography.[6] There he learned more about the geographical ranges of host plants, as well as about taxonomic isolation and growth forms, all factors in the number of insect species that plants attract. Donald Strong, working on similar problems in Florida, suggested to Lawton that, together with Southwood, they write a book on community patterns of insects on plants. Their book was published in 1984.[7]

By the time he began working on the book Lawton was a lecturer at the University of York. He moved there in 1971 after spending three years at Oxford. The head of the biology department at York was Mark Williamson, an Oxford-trained zoologist. Williamson wanted to build a strong ecology section and he made some good appointments. John Beddington was appointed there at the same time as Lawton. Unfortunately a fire destroyed much of the department in 1973, along with personal libraries, notebooks, and much else. One consequence was that, having lost their laboratories, Lawton and Beddington decided to work together and to carry out some experiments on a bench in one of the teaching laboratories. Laboratory experimentation was becoming fashionable among ecologists wanting to test new theoretical models. Prey-predator studies of the type that Hassell was carrying out were drawing interest and they decided to do some work in that area. With Hassell's cooperation they studied the interaction between two aquatic predators (backswimmers and damselfly larvae) and their prey (water fleas [*Daphnia*]). Beddington had command of the necessary mathematics while Lawton was a good naturalist and had studied damselflies for his PhD. In designing their experiments, and in thinking about ecological dynamics, they were beginning to be influenced by May's ideas.[8] Lawton also worked with Stuart Pimm.[9] Collectively the work had implications for the bracken work since it alerted Lawton to different ways in which the populations of insects feeding on bracken could be controlled, including by parasites.[10]

Lawton's field work on bracken and its insects was carried out on Skipworth Common, situated about 15 km south of York. He conducted a study there from 1972–1990, began a national insects-on-bracken survey, and later included some international work. Lawton's problem was to come up with a model that, at least roughly, predicted what was observed — species load, for example. It appeared that MacArthur and Wilson's claim about island size applied to patches of bracken. The larger they were, the more species and the more individual insects were found. But area was not the sole determinant. As on other islands, Lawton found that insects are drawn from local, regional and more distant pools, so locality has some impact on community make up. Further, island communities, as MacArthur and Wilson had pointed out, are dependent on which species are the first to invade new territory and on whether or not they survive there — often a matter of chance. While Lawton's work supported many of MacArthur and Wilson's ideas, there was much still to be understood. Like Southwood and Crawley, Lawton looked also at succession in plant and insect communities.[11] A further connection to the ecologists at Silwood arose when Lawton considered whether bracken could be controlled by using biological agents.[12] He also noticed that some parts of the bracken plant were exploited in some localities, but not in others. This suggested that there could be unfilled niches which, as discussed, was a possibility pondered by G. E. Hutchinson.

It appears that the 'unpromising' bracken plant (*Pteridium aquilinum*) was a good choice for study. It allowed Lawton to arrive at a larger understanding of phytophagous insects and of the genesis and maintenance of biodiversity. The search for patterns led to claims linking, among other things, population abundance, local and regional species richness, and the size and life-histories of insects. For example, with respect to abundance it appears that rank order (from most to least populous species) changes over time, sometimes only slowly, suggesting differing responses to larger environmental factors. Clearly macro-ecological work is complicated and leaves room for much speculation. It is important, however, that there be ecologists working at different levels and with different approaches. For Lawton the turn to macroecology was productive. It helped that he was also a serious naturalist.

As noted in chapter five, G. E. Hutchinson drew attention to the architecture of living space in some of his papers and argued that the greater the structural complexity of a plant, the more species it could accommodate. Lawton, too, recognized the importance of plant form to insect communities. He even considered the use of fractals to mathematically describe plant architecture.[13] Southwood was clearly impressed with Lawton's work, and with his breadth of understanding. He helped him to win grants, and to progress in his career. In a letter supporting Lawton's promotion to a readership in 1982, he wrote that this 'was long overdue', that there were only four outstanding British ecologists in their late thirties or early forties and 'of these Lawton has the broadest grasp of ecology and has made a contribution over the widest front'. Three years later Lawton was promoted to professor.[14] He was not to stay at York for much longer.

As mentioned in chapter six, Southwood had thought of inviting Lawton to Oxford when he moved there in 1979. Close to ten years later, the recession was over and more money was flowing to university research. By the mid-1980s some large projects were being funded in physics, molecular biology, marine biology, and the earth sciences (the latter two in connection with North Sea oil and gas exploration which was helping to fuel the economic recovery). Southwood was at the centre of an attempt to do something big for ecology and was successful in helping to bring the Natural Environment Research Council (NERC) on side.[15] The issue was raised at the British Ecological Society where there was much discussion over what type of projects would be worth doing. It appears that Southwood was torn. On the one hand he properly put his support behind a University of Oxford bid to do something interdisciplinary involving agriculture and the environment. On the other, he wanted to see something big for Silwood. Roy Anderson was the newly appointed head of the biology department at Imperial College and Hassell had recently succeeded Michael Way as director of the Silwood Park campus.[16] They decided to compete for a new centre for population biology and, if successful, to bring Lawton in to direct it. NERC was persuaded to fund such a centre, but both Imperial College and the University of York were in the running.[17] From Southwood's correspondence one can infer that, even before the decision was taken to support the Silwood bid, he lobbied the Imperial College administration to appoint Lawton to a professorship, perhaps with an eye to tipping the NERC decision in Silwood's favour.[18] Whatever the case, Lawton was offered a chair at Imperial College before NERC announced that it would fund the centre at Silwood. Lawton accepted and, shortly after, became director of the new NERC Centre for Population Biology which opened in 1989.[19] John Beddington had moved to Imperial College five years earlier. No doubt those in the York department, and Mark Williamson who had done so much to build it up, were upset at the poaching of two of its leading lights.[20]

In 1987 Lawton was awarded the gold medal of the British Ecological Society and, in 1989, he was elected to the Royal Society.[21] At Silwood he built up the Centre for Population Biology which, at its peak in the late 1990s, had about 100 people working in it.[22] Its centrepiece was the ecotron, a major installation for carrying out controlled ecological experiments. The idea for a facility of this sort had been dreamed up earlier by Lawton, Anderson and Hassell after attending a meeting of the British Ecological Society. Over a few beers in a pub they discussed the possibility of Lawton moving to Silwood and what he might do when there.[23] The ecotron must still have seemed a good idea in the morning — and so, by and large, it turned out to be. In 1993 its purpose was described as follows:

> The Ecotron serves as an experimental means for analysing population and community dynamics, and ecosystem processes under controlled physical conditions. Within the chambers, terrestrial experimental communities are assembled into food webs of desired complexity from a pool of species selected for their preadaptations to the physical conditions of the Ecotron.[24]

The ecotron was a conceptual achievement as well as one of engineering design. Its sixteen chambers contained simple terrestrial ecosystems that could be studied over varying time periods. They were computer controlled for diurnal light-dark cycles, temperature, humidity, rainfall, and so on. Each ecosystem was established in 0.3 cubic meters of soil laid over a gravel base and held subsets of about thirty plant and animal species, all of which could be found in weedy, disturbed, ground in southern England. The organisms were selected not to simulate any actual ecosystem, but rather to collectively simulate a 'set of universal features of terrestrial ecosystems'.[25] Among the plants in each chamber was at least one nitrogen fixer and some fungi. Also present were some decomposers such as microbes and earthworms, some insect herbivores and some parasitoids. Because the various physical environments were under computer control, the experiments were relatively easy to replicate.

By the standards of the time the ecotron was not exactly a scientific mega project, but it was large for ecology. It took over three years to build and cost over £1 million. Its running costs in the early 1990s were about £70,000 a year, not including the salaries of the people who worked there.[26] Needless to say there was criticism of the project and of the expenditure. The critics claimed that artificial environments would reveal little about the real world, that the species studied were not ones that would arise together naturally, and that the systems were closed to emigration and immigration. But Lawton and others at Silwood held the view that one has to start with simple ecosystems before one can begin to understand the more complex ones of the natural world.[27] Many young scientists began their careers working at the ecotron and Lawton mentored people who went on to have successful careers.[28] I cannot judge the overall success of the many experiments carried out, but the ecotron has provided data for modellers and for debates on climate change, biodiversity, population dynamics and community ecology.[29]

The Centre for Population Biology was not, however, just about the ecotron.[30] Lawton continued with his field work and others, too, engaged in a wide range of projects. Many papers were published.[31] Lawton also began writing a regular column for *Oikos* titled, *View from the Park*. In these short journalistic pieces he discussed a range of issues that came to the fore in the 1990s. Looked at retrospectively it is clear that the issue with which he was most engaged was biodiversity and its loss. Perhaps this was in part because of his love of birds and the sharp decline in the populations of some British species, and in part because of his engagement with another research programme, the European Union's BIODEPTH Project on Biodiversity and Ecosystems Processes.[32] Lawton was the lead scientist on that project, the purpose of which was to study the impact of the loss of plant diversity on primary productivity. Teams of scientists studied grasslands in eight different countries and their field experiments showed that a reduction in biodiversity came with an overall loss in biomass.[33] It was a topic of interest also to Robert May and led to some collaborative work. For example, in 1995 May and Lawton organized a discussion on biodiversity at the Royal Society.[34] Lawton's concern with the loss of biodiversity can be seen in many papers that he contributed to this and to other symposia during the 1990s.[35]

Lawton is a longtime member of council of the Royal Society for the Protection of Birds (he joined its junior arm, the junior bird recorder's club, in his mid-teens). This voluntary society has a membership of over one million. Its principal concern is the conservation of bird habitats, something with which Lawton has been involved for much of his life. He was appointed chairman of the council for a five-year term in 1993 and became vice-president in 1999. In these capacities he was to have a say in Britain's environmental policy and, indirectly, also in the agricultural and conservation policies of the EU. Like May he pressed for nature reserves to be larger than was previously thought necessary. In 1990 he joined the Royal Commission on Environmental Pollution which allowed for further input on similar issues. Lawton was also a member (later vice-president) of the British Trust for Ornithology, and so was involved with the long-term bird census in Britain. In the later 1990s he moved yet further into the public sphere as a member of various boards and committees. He left Imperial College in 1999 when appointed chief executive officer at NERC. Lawton told me that the eleven years he spent at Silwood were the happiest of his scientific life. His later work will be discussed in chapter nine.

8.2 John Krebs

Like Lawton, Krebs was a keen birdwatcher already as a child. He collected eggs and he hand-reared birds, presaging his later interest in bird behaviour. In a memoir he wrote that mathematics and the sciences were not his strong suit at school. Nonetheless his father pushed him toward the sciences and he reluctantly agreed to specialize in them for his A-Level GCE examinations.[36] To encourage the turn to science, and to build on his interest in birds, Krebs' father arranged for him to spend a short time working at Konrad Lorenz's Institute for Behavioural Physiology at Seeweisen in Bavaria. This teenage experience appears to have been persuasive, and the father's wish to see his son become a scientist was fulfilled.[37]

Krebs grew up in Sheffield and Oxford. He attended Oxford University as both an undergraduate and doctoral student in zoology. Given his experience in Bavaria it is perhaps not surprising that he decided to carry out his DPhil research in Niko Tinbergen's unit. He did so under the supervision of J. M. (Mike) Cullen. By several accounts, including Krebs' own, Cullen was a good choice. According to Richard Dawkins, Cullen was a 'shining intelligence', the unit's 'intellectual powerhouse'. He has also been described as eccentric but, more importantly, as a kind and good teacher.[38] Judging from the acknowledgments in a couple of Krebs' early papers, Cullen appears to have given his student some mathematical and computing assistance. Given the period, the latter is an interesting reversal of what was then the norm.

Krebs was fortunate to enter the field of animal behaviour when he did. It was undergoing a major conceptual shift, one with which he was able to engage. As mentioned in earlier chapters, new ideas from evolution theory and population biology were entering ecology. The new synthesis was also to unite ecology with ethology in ways that would change the way people thought about animal behaviour. The foundations for a new behavioural ecology were laid in the decade following the publication of W. D. Hamilton's theory of kin selection in 1964 (the theory so named a little later by John Maynard Smith). Subsequent to his 1964 papers Hamilton examined the conditions favouring selfish behaviours. Maynard Smith and others introduced game theoretic models to account for altruism and to explain why competitors moderate their selfish or aggressive behaviours. Robert Trivers explored reciprocal altruism, ideas on parental investment, the causes of intra-sexual competition, and the conflicts that exist between parents and offspring. Together, Hamilton, Maynard Smith and Trivers were to influence an entire generation of behavioural scientists.[39] Hamilton stated that he was inspired by R. A. Fisher's *Genetical Theory of Natural Selection* and by Fisher's insistence on individual selection. He and the others emphasized that Darwin, himself, believed in individual selection.[40] The presumed Darwinian association helped to give their ideas greater legitimacy. As we have seen, Hamilton's ideas were accepted only slowly. They reached a wider public with the publication of E. O. Wilson's *Sociobiology* (1975) and Richard Dawkins' *The Selfish Gene* (1976).[41] Although controversial, the ideas could no longer be ignored by behavioural scientists with career ambitions. Behavioural science was later bolstered by connections to both genetics and neurobiology, but in the 1970s genetic details were not yet much discussed. Behavioural solutions to the problems of survival and reproduction in the natural environment were, however, viewed from a genetically functional perspective — namely that of gene survival and propagation.

In the late 1960s, when Krebs joined the study of great tits, the new ideas were not yet widely disseminated. They play no role in a paper that he published on his doctoral work.[42] As mentioned in chapter five, it was David Lack who began the study of great tits in Wytham Woods. Lack observed that though these birds were non-migratory there were periodic drops in their population, especially during the winter months. Lack thought these drops, which varied from year to year, could be due to any combination of predation, disease, starvation and territorial behaviour. Krebs' doctoral work was to test the relative importance of territorial behaviour and available food in controlling population. This related to the dispute between Lack and Wynne-Edwards (see chapter five) as to whether bird populations were regulated by density dependent mortality due to food shortages and other deprivations, or whether they were able to self-regulate their populations. Krebs' results were not clear cut but it appeared that territorial behaviour had a greater role to play than many at Oxford then believed. For example, when breeding pairs were removed from the woods, their place was soon taken by immigrants from less desirable hedgerow habitats, birds that previously were prevented from entering the woodland area by the resident population.[43] In all likelihood the reason for territorial behaviour was to protect access to prime nesting sites and to food sources, but it was difficult to know exactly what to conclude. The outsider is tempted to say that explanations of behaviour based on territoriality and access to food are rather obvious. This reaction is not unlike common responses to the 'obvious' theories arrived at by students of human behaviour. But the seemingly obvious can turn out to be false (the earth really does move, as Galileo stated) and our intuitive beliefs need to be examined.[44]

During Krebs' final year as a doctoral student he worked as a demonstrator in ornithology at the Edward Grey Institute. While there he heard a talk given by Dennis Chitty, a former member of Elton's research group who was visiting from the University of British Columbia (UBC) where he was a professor of zoology. Chitty, like Hassell, was an advocate of the experimental approach to population ecology, something that appealed also to Krebs. Chitty told Krebs of some job openings at UBC, associated with C. S. Holling's Institute of Resource Ecology. Krebs was to move there as an assistant professor in 1971. Another North American to influence Krebs at the time was Eric (Ric) Charnov. In 1970 Charnov was visiting Oxford from the University of Washington in Seattle where he was a doctoral student under Gordon Orians. By the time Krebs was in Vancouver, Charnov by then back in Seattle was a frequent visitor to UBC. Holling, an expert on predatory behaviour, was a member of his thesis advisory committee. Charnov carefully studied Holling's earlier work, especially that on the praying mantis.[45] He was to come up with a mathematical model to describe the optimal movements of a predator whose various prey were distributed through the environment in a patchy formation. Once again the situational logic shines through. It was surely no accident that so many young and ambitious biologists focussed their attention on patchy population distribution at this time. For Charnov the idea was to capture the breadth of a predator's diet and the role of the patch. He told me that especially important was that Orians introduced him to the work of Hamilton and George Williams.[46]

Charnov's 1976 paper on what came to be known as optimal foraging theory was declared a citation classic in 1989.[47] In his introduction to the 1989 reprint Charnov gave a brief account of his developing interest in optimal foraging behaviour. He noted that, at the time, 'Krebs criticized the new ideas from the viewpoint of an experimental animal behaviourist,' claiming that they were not easily testable. This prompted Charnov to find ways of opening up the theory to testing.[48] Producing testable theories in this area was a breakthrough and, in this, Charnov and Krebs were among the pioneers. Their collaboration resulted in four papers that preceded Charnov's classic paper of 1976.[49] One published in 1975 dealt with the supposed altruism of birds who signal danger to others by calling out, and thus possibly attracting the attention of predators to themselves. Charnov and Krebs theorized why such behaviour could be of advantage to the birds who display it.[50] According to Charnov they were jointly responsible for the conceptual development of this work, that he did the mathematical modelling and that Krebs, with a different skill set, designed and carried out the experiments together with his students.[51] This illustrates, yet again, the pairing of mathematical and experimental talent that has been so important in the development of late twentieth-century ecology. A notable feature of their 1975 paper is its reference to Hamilton's work and its acknowledgment of helpful comments received from John Maynard Smith and Robert Trivers. While at UBC Krebs also carried out a short study of a heronry with the help of Rudolf Drent, the eminent goose ornithologist.[52] And, with Bruce Falls, he began work on bird song, work that he was to carry back to Britain. Falls later came to Oxford on a sabbatical leave from the University of Toronto.[53] It seems clear that Krebs early showed an ability to collaborate with other knowledgeable people, something that helped to forward his career.

By the autumn of 1973 Krebs was back in Britain as a lecturer at the University College of North Wales in Bangor. The quick return was perhaps the result of family pressure.[54] Bangor was a good place to be. The department had built up considerable strength in ecology, a theory-based approach was in place and Krebs believed he could learn from it. But he did not stay long. His homing instinct was strong and in 1975 he accepted a job as research officer with the animal behaviour research group at Oxford.[55] A year later he was given a position as a college lecturer and, in the following year, was appointed to a university lectureship at the Edward Grey Institute of Field Ornithology. He held that position until 1988 when he was awarded a Royal Society research professorship.

On his return to Oxford, Krebs continued to experiment and to engage with new ecological and sociobiological ideas, including the further testing of optimal foraging models.[56] One interesting paper on great tit foraging was co-authored with the mathematician Peter Taylor and with Krebs' doctoral student, Alejandro Kucelnik. Kucelnik is now a professorial colleague in the Oxford department.[57] Krebs also reconnected with others in the Silwood Circle, joining them at dinners, walks and meetings.[58] He also continued his work on bird song and signalling. A 1975 paper with Falls related the volume, and variety of song repertoire to territorial behaviour; other papers were to follow.[59]

What brought Krebs to yet wider attention was the decision to bring many of the current ideas in his field together. He did so in two books published together with Nicholas Davies in 1978.[60] The books were among the many manifestations of a new behavioural ecological synthesis. One was a collection of review articles.[61] While the contributors were largely from the United Kingdom, they brought together various strands of thought, including some North American ideas, linking animal behaviour to new ideas in evolution theory, population biology, and community ecology. MacArthur and Pianka's ideas on foraging and patchy population distribution, and MacArthur's ideas on r and K selection, were taken up by several of the authors. The book played a role not unlike Robert May's *Theoretical Ecology* published two years earlier. It was a good introduction to some new and important developments in the biological sciences. As with May's book, it went through several editions with the content changing over time. By the third edition, published in 1991, sexual selection had moved to the fore and, with it, greater emphasis on the mechanisms by which animals perceive what others are up to. Two associates of the Silwood Circle, Charles Godfray and Paul Harvey, contributed to a section on sexual selection and reproductive strategies. Further, new molecular techniques were coming into play. For example, some of Davies' DNA studies of birds indicated that polygynous behaviour was far more common than had been inferred from simple observation. It was explained in terms of territoriality among male birds and the control over good nesting sites and food sources. This was suggestive of further study to determine how male birds allocate their time in parental care. Krebs and Davies' other book was a good introductory text, suited to undergraduate teaching. It, too, has gone through several editions.[62]

One of the articles in the 1978 book, 'Animal signals: information or manipulation?' was co-authored by Krebs and Richard Dawkins.[63] It brought information theory to the debate over the interpretation of animal signalling. While animal signals were traditionally seen as ways of conveying information, new theory suggested that the release of information could be counterproductive and give advantage to competitors. In light of this Krebs and Dawkins saw signalling as a form of manipulation. Animals, they claimed, exploit each other, including members of their own species, just as they exploit other resources in the environment. Signalling, they argued, could have evolved much as an arms race escalates — the result of individuals continually trying both to keep pace with, and outwit, their rivals. The survivors pass on their behavioural legacy to future generations. It is ironical that in this context Dawkins later wrote, 'I have always enjoyed collaborating, and wish I had done more of it. My three joint papers with John Krebs showed me how immensely valuable joint thinking can be — like a kind of mutual tutorial.' Perhaps it would have been more consistent to call theirs a case of mutual exploitation, albeit seemingly enjoyable. Dawkins claims that the articles he and Krebs wrote for the 1978 and 1984 editions of Krebs and Davies (eds.) were 'influential in redirecting the attention of students of animal communication in two ways. Instead of treating communication as a mainly cooperative enterprise, we stressed deception (mainly under John's influence) and manipulation (this was my contribution)'.[64] It is not really for the historian to say, but perhaps the Kropotkin-Darwin pendulum (see chapter five) had swung a little too far in Darwin's direction.

When I spoke with Krebs I asked about his interactions with others in the Silwood Circle since his interests, while related, were different. Using a metaphor from ecology I asked whether members of the circle could be seen as a small community with individuals occupying slightly different niches, so reducing competition between them. He thought this not a bad way of seeing things but added that there 'was a measure of trust between us which made for good and productive scientific exchange'. In 1979, two years after Krebs' appointment in the department, Richard Southwood was appointed to the Linacre chair. As has been discussed, Southwood was keen to see ecology front and centre on the science map. As at Silwood so at Oxford, Southwood encouraged ecologists and promoted their interests. Krebs, competent and ambitious, was attracting students to the new field of behavioural ecology, a field he had helped to establish. He was a good experimentalist, a keen naturalist, and a good courtier. It was inevitable that he would catch Southwood's attention.

Luck plays a major role in people's lives. However, only those with the ability to exploit their good fortune can enjoy the kind of success Krebs has achieved. Krebs was lucky that his father recognized his ornithological potential and had connections to Lorenz. But it was the younger Krebs who had the good sense to recognize that ecology was an expanding science, one demanding the attention of behaviourists. He was lucky to have entered the behavioural field just as it was about to explode, something he could not have foreseen. Had the ideas of Lorenz and Tinbergen remained dominant for even another decade, Krebs's career would have been very different. He was also a beneficiary of the timing of the environmental movement, that his working life occurred during a period of growing public and political concern in this area and that, as a result, funding for ecological research was forthcoming. Being around at the right time does not, however, ensure success. Krebs understood the need to enter a steep learning curve, and to keep abreast of many new ideas. He also sought and found good collaborators. As a result, his experiments were well conceptualized and well designed. He was fortunate also in his strong attachment to Oxford. It drew him back to the department at a time when it was restructuring, and at a time when he could make a difference. And it drew him close to Southwood which helped his election to the Royal Society in 1984 — at the age of 39, young for an ecologist.[65] As discussed earlier, Southwood saw it important to increase the presence of ecologists, especially younger ones, in the Royal Society. By the 1980s he was well on his way to seeing this brought about. Krebs told me that being fellows of the Royal Society was the *sine qua non* for the later success of members of the Silwood Circle. While perhaps true in his own case, more generally it is only partially so. Gordon Conway, for example, achieved success in a range of important administrative roles well before his election to the fellowship.

In the early 1980s Krebs became interested in food storage and in how the hoarding of food by birds for their own use could be seen as an evolutionary stable strategy. With the help of David Sherry, a postdoctoral fellow from the University of Toronto, Krebs's research moved in a new direction. In his memoir Krebs noted that in 1988, 'Sherry turned up in Oxford ... in great excitement to show me sections of the brains of non-storing (great tit) and storing (chickadee) parids, that he had prepared in Toronto with Anthony Vaccarino'. This led to their seeking a connection between hoarding behaviour, ways of memorizing, and neural anatomy. The aim was to find some distinguishing features both of memory and of brain anatomy in birds that stored food.[66] In 1990, together with Gabriel Horn, professor of zoology at Cambridge and an expert in neurophysiology, Krebs organized a discussion on the brain and cognition at the Royal Society. At the meeting Krebs summarized some research findings, including his own, on food storage and memory. It appeared that at least some food-hoarding birds have an enlarged hippocampus region relative to their brain and body size. According to Krebs it was the 'first documented example of an evolutionary specialization of the brain associated with memory'.[67]

As mentioned in chapter seven, Krebs was awarded the Scientific Medal of the Zoological Society of London in 1981; other awards were to follow. He was invited to give lectures and to become a visiting professor at several universities in North America. In 1988 he was elected president of the International Society of Behavioural Ecology. Several job offers came his way in the late 1970s and 80s but he remained in Oxford.[68] In 1989 he became director of the NERC unit of animal behaviour and the AFRC unit of ecology and behaviour.[69] But by then he was beginning to move away from scientific work and toward science administration. He took on much committee work and joined a number of boards. In 1994 he relinquished his Royal Society research professorship and was appointed chief executive of NERC. It was the start of a more public career to be discussed in chapter nine.

8.3 David Rogers

David Rogers, too, was an undergraduate and then doctoral student in zoology at Oxford. He told me that as an undergraduate he was among the last to be taught by Charles Elton. However, the person who inspired him to carry on to a doctorate was John Whittaker, later appointed to a chair in biology at the University of Lancaster. Rogers began his doctoral research on insect ecology under George Varley in 1967. He said that as a younger man Varley had been 'dashing, tall and athletic' and that he was still charismatic in the late 1960s. Rogers also said that, as a research student, he was much helped and influenced by Hassell. When Hassell returned from California he brought back many good ideas on how to combine laboratory system work with theory. He chose to carry out a population study of flour moths whose larvae were preyed on by a parasitoid wasp. Rogers was working with the same moths and helped Hassell to culture the specimens.[70] Around 1970 Hassell had yet to be influenced by Robert May. He and Rogers approached prey-predator ideas from within the Lotka-Volterra, Nicholson-Bailey, statistical mechanical tradition. In addition to Varley they were influenced by others, including C. S. Holling and Kenneth Watt. Holling had written on foraging but the full-blown optimal foraging approach of the behaviourists, based in part on Charnov's theoretical work and on Krebs's bird foraging experiments mentioned above, was still a few years off. When that work arrived patchiness was given a new twist, with a new emphasis on the reproductive and survivalist behaviours of individual predators and individual prey.

Perhaps even more important to Rogers' future work was the presence in Oxford of John Ford. Ford had spent much of his working life in Africa studying the tsetse fly and was back in Oxford working on an important book on the subject.[71] In his book Ford criticized the work of earlier colonial entomologists and their failure to understand the ecological aspects of the fly and its trypanosome protozoan parasites.[72] A consequence of that failure, he believed, was that little had been done to control sleeping sickness and nagana (a similar disease affecting cattle), both of which carry an enormous human and economic cost. Ford had been a student of Elton in the 1930s. He was charming and a good raconteur. Rogers told me that Ford's stories of the British colonial experience in Africa, and of his own work were 'totally fascinating'. Rogers helped Ford with preparing maps for his book, and in analyzing some recent data on the tsetse fly. Ford, in turn, encouraged Rogers to study the tsetse fly and helped him to gain a Wellcome Trust grant for work in Uganda. In 1970 Rogers went to Uganda for several months where he began a study of the population ecology of the tsetse fly using some of Varley and Gradwell's theoretical ideas. He returned in the following year for what was planned as a longer stay. But Ugandan politics intervened and safety concerns made field work there increasingly difficult. Luckily when a lectureship opened up in Oxford, Hassell notified him of it. Rogers applied and was appointed. He returned to the UK in 1972, one year ahead of schedule. Nonetheless, his future work was determined by this early overseas experience and he was later to carry out research on insect disease vectors in several African countries.

As mentioned in chapter four, Robert May visited Oxford in 1971 during his sabbatical leave from the University of Sydney. May had already met Hassell at Silwood. While in Oxford he met Lawton but Rogers was away in Africa. Nonetheless, after Rogers returned in 1972, his connection to Hassell and to Lawton drew him into the Silwood Circle. He continued to help Hassell with some of his experiments. He attended the mathematical ecology workshops at York and at Silwood, as well as other meetings attended by Silwood Circle members. He joined some of the collegial dinners, though not the hiking trips — something that may relate to family commitments. Rogers married Sarah Randolph in 1974. Like her husband, Randolph was beginning a career in the zoology department at Oxford. They were to have three children. Gordon Conway was another who did not join the hiking trips, for what may have been similar reasons. Susan Conway was explicit on this issue. She told me that if her husband was going to go on summer hikes, he was going to go on them with his family, not with his scientific colleagues.

Rogers was alone among the early Silwood Circle in being married to a fellow ecologist. He told me that when Southwood arrived at Oxford in 1979 he expressed his opinion on academic marriages: 'I've seen a lot of academic marriages fail when husbands and wives work together'.[73] This may well have been true in Southwood's limited experience, but it was far from being a universal truth. Southwood's attitude is reflective of the period in which he matured as a scientist. As was noted in chapter four, his own research depended heavily on the help of women co-workers but he did little to promote their careers. I speculate that he saw women as capable of good supporting roles in science, but not as leaders.[74] 1979 was also the year that Rogers and Randolph had their first child. It was a time when some joint career encouragement would have been welcome. Many years later Randolph took part in a discussion on family and careers with other women scientists.

The discussion was reported in the *Times Higher Education Supplement*. She spoke of the career difficulties that she and other women of her generation faced.[75]

Like her husband, Randolph was a zoology undergraduate at Oxford. After doctoral work at King's College London, she began working on ticks as vectors of encephalitis.[76] She, too, became an expert on the insect vectors of human disease. One way of coping with family and career pressures is for husbands and wives to work together, something that, *contra* Southwood, can be very productive.[77] From the late 1970s Rogers and Randolph carried out research together in Africa, first on the mosquito trypanosome vector, work begun earlier by Rogers. Later they were to work on ticks.[78] These African interests were to draw Rogers away from the others in the Silwood Circle. He was, however, to become a professor at Oxford and was appointed curator of the entomological collections.

In 1993 Rogers became the leader of the Trypanosomiasis and Land-use in Africa (TALA) research group, set up with funding from the Department of International Development. The group carries out basic research, but is active also in applying ecological knowledge to the prevention and spread of disease. One tool used in that work has been remote sensing from satellites. This allows for the collection of environmental data used to explain and predict the distribution and abundance of the trypanosome vectors. The work has since expanded to include also malaria mosquitos.[79] Square grids of territory, each with sides about 7 km in length are surveyed. By studying the topography, climate and vegetation within a grid, the environmental conditions both conducive and hostile to the spread of disease vectors can be determined. It is another area of ecology for which mathematics is essential. Brian Ripley, professor of applied statistics at Oxford, has helped Rogers with pattern recognition and spatial statistics. TALA's advice has been sought on conservation and land use in relation to disease containment by a number of international organizations, including the World Health Organization (WHO). Rogers has trained a number of doctoral students in this type of work. Two, Chris Dye and Diarmid Campbell-Lendrum, both work for the WHO. Simon Hay, who has mapped and studied malaria mosquitos in a number of African countries, is a colleague in the department at Oxford.

8.4 John Beddington

John Beddington's attendance at the mathematical ecology workshops and his working with Lawton, Hassell and May, led to his inclusion in the Silwood Circle. Beddington joined the others at dinners as well as on many of the hiking expeditions, and has remained in contact with them throughout his working life. As mentioned, after the fire at York, Beddington worked with Lawton and Hassell on insect population dynamics. For him this was a departure. His earlier doctoral work was on Red Deer populations, and on age-specific harvesting models.[80]

Beddington was, however, to make his name in the area of fisheries ecology. This came about, in part, because York University held courses for scientists working at the fisheries institute in Lowestoft.[81] Beddington instructed them in mathematics and statistics and came to learn of some interesting problems associated with their work. He thought he could make a contribution. At about the same time Beddington met May at the mathematical biology workshops held at York and was able to interest him in the fisheries work. They were to publish a number of papers together.[82] This new interest led Beddington to attend a UN Food and Agriculture Organization (FAO) fisheries conference in Bergen in 1974. One of the people whom he met there was Colin Clark. Clark, an applied mathematician, was to become an expert on the Canadian Pacific fishery. Clark, an emeritus professor at UBC, told me that, already at the time he met Beddington in Bergen, his own research interests combined the fields of resource management economics and ecology.[83] Like so many others, he was inspired by Kenneth Watt's book on ecology and resource management, a book that must surely count as a major factor in the intellectual development of a generation of ecologists and resource management specialists. In particular, Clark was challenged by Watt's claim that what was needed was a good optimal harvesting model, and that existing ones were inadequate. By the time Clark met Beddington in Bergen he had been working on such a model for a few years. Further, as Beddington then discovered, Clark had refereed his paper on age-specific harvesting that was about to be published in *Econometrica*. Clark, who had learned much from his association with C. S. Holling's group at UBC, encouraged Beddington to read more of the current literature in resource ecology and to visit the university. Beddington paid a visit in 1975 and, while there, met Holling and other members of his group. He also exchanged ideas with Carl Walters, the professor of fisheries management who was modelling the dynamics of fish populations. In Bergen Beddington also met the conference organizer, Sidney Holt, and fellow attendee John Gulland. They, too, encouraged him to continue with fisheries research and to think also about whaling, one of Holt's major concerns.[84]

By the end of the 1970s Beddington was firmly focussed on problems related to fisheries and whaling. He joined Holt as a United Kingdom representative on the International Whaling Commission's (IWC) scientific advisory committee. This committee, with about two hundred scientists from countries around the world, meets annually and issues reports on a range of whaling-related topics. Beddington's contribution was in assessing the populations of some key species, namely sei whales, fin whales and, later, sperm whales, work that resulted in the closure of hunting for those species.[85] His work should be considered also in the politico-scientific context in which the IWC moratorium on whaling came about in 1986, and the declaration of a Southern Oceans whaling sanctuary in 1994.

For Beddington, the conference at Bergen was clearly formative. So, too, was his early work with May. Beddington was critical of existing ideas on maximum sustainable yields and, together with May, thought about how population stability changes over harvesting periods, both with a constant rate of fishing and with quotas (they believed the latter to be more dangerous from the sustainability viewpoint). They also saw the need to focus on the problems of multi-species fisheries. Beddington organized a meeting in Rome with May, Clark, Holt and some others to discuss this problem. One outcome was a joint paper drawing attention to fishing practices in the southern oceans. The paper puts forward a rough model for an Antarctic marine ecosystem. At the time Baleen whales were being over hunted leaving what some people saw as a 'surplus' of krill. The authors argued against plans to harvest the krill, noting that other species, too, were dependent on krill for food. They stressed the importance of taking into account the interactions among species when making management decisions.[86]

This work took Beddington and May in a new direction. In chapter six May's attempt to distinguish environmental effects from those having to do with ecosystem dynamics, was briefly mentioned. It was a problem of increasing significance. As May retrospectively pointed out, traditional maximum sustainable yield theory ignored 'all manner of environmental fluctuations and vagaries'. It was important, he stated, to include probability relations when theorizing.[87] In other words, those managing whale populations (and fisheries) needed a better understanding of the reasons for fluctuations in the resource so as to devise improved management strategies. May and Beddington were to work on these problems together with May's former Princeton student George Sugihara, a professor at the Scripps Institute of Oceanography. They thought about fish stock variability and on how one could distinguish the effects of overfishing from those due to environmental change. Fisheries were traditionally managed by allowing only larger (older) fish to be caught. But it appears that age-truncated (juvenescent) populations may be less able to buffer environmental events than those with undisturbed age distributions.[88] The older theory had limitations which Beddington and his colleagues attempted to overcome. Their models took into account a range of new studies, including those that examined fished and unfished populations living in the same general area.[89]

While I have read only a few of the many papers written by members of the Silwood Circle, it is apparent that the same basic theoretical ideas recur, even when the particulars are very different. For example, removing whales from the ocean is not unlike removing blue tits from Wytham Woods. Up to a point 'surplus' animals can move in to take their place. However, removal beyond a certain point leads to instability in the population dynamics. Also, patchy distribution and predator behaviour play a role in whaling theory. For example, the aggregation of predatory whaling ships in particular ocean areas is a pointer to instability. Population densities are not simply linearly related to catch per unit of harvesting effort, but are affected by how the harvesting ships are distributed. The aggregation of ships can lead to non-linearities, possibly to sharp and irreversible declines in whale populations. Such ideas come from predator-prey theory but May, Sugihara and Beddington took the work to a new level. Their work allowed for a better differentiation of environmental contributions from those resulting from chaotic population dynamics. But much remains to be understood.

Beddington told me that when a student at the London School of Economics he was influenced by the ideas of Karl Popper. He came to understand that theory falsification is important to the progress of science, and that new theories require serious testing. Because of the complexity of ocean ecology, theories about the management of fisheries, whaling, and sealing are based on very limited evidence. They are bound to be falsified as new evidence arises, something that should lead to a gradual improvement in theoretical understanding. Beddington is presumably of the view that a working theory, however weak, is better than none at all. In this he is probably correct; we have to make decisions based on the best supported theories available. His older colleague, Sidney Holt, was more cautious. 'We humans', he wrote in 2008, 'are not ready to manage marine systems'.[90]

Given his work on whale and krill harvesting, Beddington was asked to head the UK delegation to the scientific committee of the Commission for the Conservation of Antarctic Marine Living Resources and attended their annual meetings. After the Falklands War a new fishery began to open in the waters surrounding the islands, but it was uncontrolled. The Foreign and Commonwealth Office asked Beddington to look into the problem, principally one of overfishing for squid.[91] He wrote a report on the problem and developed a licensing and management regime for the fishery which remains largely in place today.[92]

Beddington moved from York to a lectureship at Imperial College in 1984.[93] He joined Conway and the Centre for Environmental Technology (ICCET), where he led the renewable resources assessment group. In 1987, and with Anderson's support, he was promoted to a readership and, in 1991, to a chair in applied population biology.[94] In 1994 he became director of ICCET. But he also had other interests. Earlier, during a sabbatical leave from York, and while working on whale conservation, Beddington met Jeremy Cherfas, a science journalist whose wife worked with a large law firm in the City. Realizing that Beddington had some skills that could be of use to her colleagues, she introduced him to some of the senior partners. He began working with them both on marine insurance and on the detection of fraud. This included work for British Airways on the class action suit that followed the bankruptcy of Laker Airways.[95] These new interests led to Beddington's temporary withdrawal from Silwood Circle activities, but he was to join with them again in the mid-1990s by which time he was a well established professor at Imperial College.

Beddington sees Robert May as a having been an excellent mentor throughout his career. He also credits May with helping his election to the

Royal Society in 2001.[96] Beddington, mathematically competent, was able to work well with May. His early drawing of May into fisheries ecology paid off for both of them. Their partnership illustrates a slightly different pattern from that exhibited by the other collaborations I have described. It was one between two theorists. Like the others, Beddington found May's mathematical ideas suggestive. May, in turn, was able to turn his attention to yet another set of problems in population dynamics. When May's gifted doctoral student, George Sugihara, was drawn into the fisheries work, their collective modelling became yet more sophisticated. The models had application also in the financial and business sectors to which Beddington, May and Sugihara were all drawn. Unlike Lawton and Krebs, Beddington did not have the patronage of Southwood. Nonetheless his close connection to May and to others in the Silwood Circle affected the path of his career.

8.5 Coda

Krebs is unusual among the Silwood Circle in that he co-authored few papers with others in the group. This can be explained by his behaviourist focus. Nonetheless he stayed in close contact with the others, joined them at social events, attended many of the same meetings and joined in some of the hikes. Importantly he drew the attention of Southwood and received his patronage. Rogers, who fell under the spell of John Ford, found interesting work in Africa; but, over the longer term, by distancing himself from the others he may have paid a price in more worldly success.[97] Lawton's closeness to Southwood paid off, as did Beddington's to May. Overall, as the footnote references in this and the previous chapter have shown, there was much scientific collaboration among members of the Silwood Circle. There was also a move, not always conscious, to promote a certain way of doing ecology. Fleck's idea of the solidarity of the 'thought collective' in the service of an idea, appears to have some merit. But solidarity affects not only cognition, it affects also recognition by the wider community, something to be discussed further in the next chapter. Even when not collaborating on research projects, members of the Silwood Circle exchanged ideas and met socially. And they met also at many ecology meetings during the period 1970–95. Late in the twentieth century, however, some had reached a point in their careers when they wanted something more than active research, when they envisioned a change of scene. In achieving such, a common outlook, a round-table ethos, and mutual trust built up over a quarter century were important. Mutual trust counted as much as scientific expertise in Silwood voices being heard nationally, and in helping the circle gain interesting and influential positions within the scientific establishment. This is not to downplay the importance of expertise simply to claim that, while necessary, it is far from being sufficient for gaining influence. However, this chapter and the previous one were designed to show that those belonging to the Silwood Circle were excellent scientists, and to illustrate the nature of their expertise. As was the case with Southwood in his role as advisor to various bodies, the opinions of others in the circle were sought on matters far beyond those strictly related to their research fields. This will be illustrated in the next chapter where we will look at some of the more public activities with which members of the Silwood Circle engaged.

Endnotes

1. For his work on the trophic energetics of ecosystem components, Phillipson used a bomb calorimeter to determine the calorific content of the foods consumed by various insects.

2. (Lawton, 1969a, 1969b, 1970a and 1970b).

3. See, for example, (McNeill and Lawton, 1970). This paper points to a common interest in the trophic dynamics of complex communities. In some ways it augurs Lawton's later involvement with the BIODEPTH project (see below). See also, (Lawton and McNeill, 1979).

4. (Lawton, 2000 and 2004). The 2000 book is an accessible account of community ecology.

5. (Lawton, 2000), 23. Lawton writes that he met May in 1970 but other records indicate that the meeting took place in 1971.

6. (Lawton and Schröder, 1977). This was a statistical study based on a survey of some existing literature on British plants other than trees. Schröder was working at the Swiss centre of the International Institute of Biological Control.

7. (Strong, Lawton and Southwood, 1984).

8. (Beddington, Free and Lawton, 1975; Lawton, Hassell and Beddington, 1975; Beddington, Hassell and Lawton, 1976; Hassell, Lawton and Beddington, 1977). The last of these papers takes issue with Holling's characterization of the functional response (see chapter six).

9. See (Pimm and Lawton, 1977). In this paper they argue that the number of trophic levels is constrained by population dynamics and not by energy constraints. For a response to some criticism of this argument, (Lawton and Pimm, 1978). Lawton and Pimm were joining in a larger debate on trophic levels, food webs and overall ecosystem stability, raised in part by May's work. They suggested some qualifications to May's general claim that complexity leads to greater instability. Collaboration with Pimm began at a Silwood workshop, probably in 1976. Pimm, an undergraduate student of Lawton's at Oxford, was then a graduate student in the United States where he stayed to make his career. He now holds a chair in conservation ecology at Duke University. He and Lawton have published a few papers together over the years, most recently on issues of biodiversity.

10. Lawton was not entirely confident about the bracken food supply and considered frond size, density, and tannin content for example. Some insects fared better in shaded patches, others in sunny ones.

11. (Lawton, 1987) was a contribution to a volume containing papers given at the 26th Symposium of the British Ecological Society held jointly with the Linnean Society of London. The volume includes also papers by Crawley and by Valerie Brown and Southwood.

12. Bracken is a serious and invasive weed. Carcinogenic, and toxic to sheep and cattle, it has ruined much pasture land. In (Heads and Lawton, 1986), Lawton and his co-author considered the possibility of introducing the bracken-eating caterpillars of some South African moths in an attempt at biological control. (Lawton, 1988) was among the papers presented at a Royal Society discussion on 'Biological Control of Pests, Pathogens and Weeds: developments and prospects' attended, as mentioned in chapter 7, by many people working at Silwood. In this paper Lawton discussed the many problems associated with bracken and possible means of its control. He acknowledged the help of people at Silwood, notably Cliff Moran who suggested using the South African moths.

13. For a review of the literature on this subject, (Lawton, 1983). See also (Morse, Lawton, Dodson and Williamson, 1985). This paper was written with colleagues at York. (Dodson is a mathematician).

14. Bodleian Library, Southwood papers R186; Southwood to S. Berrick Saul, vice-chancellor of the University of York, 30 March, 1982 and 31 January, 1 February, 1985. The 1985 letters concern Lawton's promotion to professor; in the latter Southwood noted that Lawton's reputation 'continues to grow'. (Southwood's papers include other letters of support for Lawton with regard to research grants, jobs etc.)

15. In 1986 Southwood and Anderson wrote a joint letter to John Bowman, Secretary of NERC, stressing the importance of ecological work and that more funding should come its way. Southwood sent copies of the letter to some eminent international scientists and asked them also to write to Bowman. Among those who did were E. O. Wilson who wrote that research in ecology is 'vital to the economic — and even social — welfare of the world'. Robert May, writing from Princeton, stated among other things that the Silwood group of ecologists was 'arguably the most influential single group anywhere'. Paul Ehrlich and C. B (Ben) Huffaker also wrote letters. Bodleian Library; Southwood papers, G56. This correspondence must be seen in the context of a power struggle within NERC between earth and biological scientists over the allocation of funds. Ecologists believed that earth science had had more than its fair share and that it was now their turn.

16. Zoology and botany were merged in a new department in 1981; Anderson was appointed head of the department in 1984 (renamed department of biology from 1989).

17. Oxford's bid was not totally neglected by NERC. A new unit in behavioural ecology was funded with a modest grant. Bodleian Library, Southwood papers; see papers in section G57. See also section below on John Krebs who was to head the new Oxford unit.

18. Bodleian Library; Southwood papers, R186. There are several letters related to this episode. See, for example, Southwood to J. L. Smith (College Secretary, Imperial College), 5 October, 1988. Southwood was supporting what Anderson and Hassell had already decided on.

19. Anderson, May and Hassell were appointed assistant directors of the new Centre; Lawton wanted Southwood to join the management team. He declined but agreed to chair the Centre's advisory board. Bodleian Library, Southwood papers, R186; Lawton to Southwood, 14 February, 1989. Lawton invited Philip Heads to become the business manager of the new Centre. Heads, one of his PhD students at York, had left academic work on graduating and taken business management training.

20. Lawton told me that moving was a difficult decision. His wife was also working at the university and was reluctant to leave York. They kept their home there, purchased a small terraced cottage in Ascot, and commuted back and forth. During Lawton's subsequent tenure as chief executive at NERC (see chapter 9), they lived in Swindon, moving back to York in 2005.

21. Lawton's election certificate shows that he was proposed by May, seconded by Hassell. Among the others who signed were Southwood, Anderson, Krebs and Hamilton. Lawton was proposed in 1987, elected in 1989.

22. There was a permanent staff of about 25. In addition there were research students, postdoctoral fellows, and many visitors who came for short periods to conduct their experiments.

23. Lawton, personal communication.

24. (Lawton *et al.*, 1993). Two of Lawton's co-authors came from other institutions in the UK and one came from France.

25. (Lawton *et al.*, 1993), 183.

26. By 2009 the running costs were about £1.3 m per year, now including technician and secretarial salaries, (uncatalogued information on the Ecotron in the ICA).

27. For some response to criticism, (Lawton, 1998).

28. For example, Shaheed Naeem, now professor of ecology at Columbia University, Kevin Gaston, professor of biodiversity and conservation at the University of Sheffield, and Susan Harrison, professor of ecology at the University of California (Davis).

29. For experiments related to global warming and varying CO_2 levels see, for example, (Bezemer, Jones and Knight, 1998). As to biodiversity, a set of experiments in which species richness at four trophic levels was studied stands out, (Naeem, Thompson, Lawler, *et al.*, 1994a, 1994b and 1995).

30. The Centre closed down in 2010 after its twenty-year funding came to an end. The ecotron is still functioning but will soon also close. However, as of writing, there are plans to build a new one.

31. By the mid 1990s, aside from visitors' publications, about 75 papers appeared annually in refereed journals. People also contributed chapters to books and material to other publications.

32. The BIODEPTH project was a finalist for the Descartes Prize in 2001. This prize, founded in 2000, rewards outstanding collaborative scientific work within the European Union. For the work of Lawton's team, (Hector *et al.*, 1999).

33. Lotka had predicted this; see chapter 2.

34. (Lawton and May, 1995). This book, based on the Royal Society discussion, contains papers that are a little different from those published in *Phil. Trans. Roy. Soc. Lond.* after the meeting.

35. For example, (Lawton, Letcher and Nee, 1994). Letcher and Nee were then at Oxford. The paper was given at the 55th Symposium of the British Ecological Society held together with the Society of Conservation Biology at the University of Southampton in 1993. The authors argued for the need for better data on range distribution. They discussed the ranges of birds and mammals from different parts of globe and the implications for conservation.

36. I am grateful to Krebs for sending me a copy of his memoir 'Luck, chance and choice'. An abbreviated version can be found in (Drickamer and Dewsbury, 2010), chapter 14. Krebs' father, Hans A. Krebs, was the 1953 Nobel Laureate (with F. P. Lipmann) in physiology/medicine. (A-Level GCE; Advanced-Level General Certificate of Education).

37. From Krebs' memoir one can infer that his father would have preferred him to have become a medical scientist. Krebs spent two summers working in Bavaria and, while there, stayed with Jürgen Aschoff who headed a department at the Institute. Later Krebs wrote a biographical memoir of Lorenz. It includes, as does his own memoir, a description of life at Seeweisen; (Krebs and Sjölander, 1992).

38. John Michael (Mike) Cullen (1927–2001). Obituary, *The Independent*, 18 April, 2001, and Richard Dawkins' eulogy, given at a memorial service in Oxford, 13 November, 2001 (Richard Dawkins' website). Cullen left Oxford for a chair in Australia.

39. Some of this history was discussed in chapters 4 and 5. I do not wish to imply that Hamilton, Maynard Smith and Trivers were alone in bringing about this change in thinking, but they were among its leaders. In one of several papers he published in the early 1970s, Trivers used iterated versions of game theoretical models to indicate how the gradual evolution of reciprocal altruistic behaviour was possible since, within small groups such as our ancestors may have lived in, there could be some future reward. Such models, even if not perfect, can be good heuristic devices. For example they prompted May to think also about non-optimal behaviour. See, for example (May, 1987).

40. This was a justifiable claim, but Darwin's views were not entirely clear cut.

41. In his memoir Krebs (see note 36) wrote that Eric (Ric) Charnov (see below) sent him a pre-publication copy of (Trivers and Hare, 1976) on the evolution of social insects. Krebs showed it to Dawkins 'with great excitement'. The paper helped to bolster the central argument in Dawkins' book. The first edition of that book includes a foreword by Trivers, mysteriously dropped in later editions.

42. (Krebs, 1971). This paper became a Citation Classic.

43. Krebs' thesis included a population model for the Great Tit, constructed with the help of Michael Hassell who had learned key factor analysis from George Varley. They used a Facit calculator for this work.

44. I was amused to see in Krebs' memoir (note 36) that *Private Eye* had much the same reaction. Krebs gamely quotes a passage from the magazine, 'under the influence of brilliant theoretical biologist Eric Charnov, [Krebs] found a new subject for research, optimal foraging, the question of whether birds tend to feed in the areas where — er — feeding is best'.

45. Holling gave Charnov all of his data on the praying mantis. Holling had studied the behaviour of this predatory insect under a wide range of laboratory conditions.

46. Gordon Orians was himself an important theorist. In 1970 he wrote a review paper that helped to inaugurate a new journal and which became a Citation Classic; (Brown and Orians, 1970). It shows something of the complex history of behavioural/evolutionary ecology. For some early work on bird behavioural ecology, (Orians, 1966).

47. The first papers to be published on optimal foraging were in the November–December 1966 issue of *The American Naturalist*. One was by John M. Emlen and the other by Robert MacArthur and Eric Pianka. They were to influence much work including Charnov's; (Charnov, 1976 and 1989).

48. This was in 1972 when Charnov took a one-year postdoctoral fellowship at UBC. Charnov and Krebs spent much time discussing foraging ideas. Charnov told me that another person with whom he recalls having 'zillions of talks' was Stephen Stearns, a graduate student with Holling at the time. Stearns, a specialist in life-history evolution, is now the Edward P. Bass professor of ecology and evolutionary biology at Yale University.

49. See, for example, (Krebs, Ryan and Charnov, 1974). Ryan was a graduate student at UBC who carried out the experiments; see also (Charnov and Krebs, 1974; Krebs, 1973). This latter paper on foraging behaviour showed that chickadees could learn how to extend their range and access to food by observing birds in related species.

50. (Charnov and Krebs, 1975). Krebs also wrote a short piece with a another UBC colleague on the calling patterns of male tree frogs in their attempts to attract females. The purpose was to see whether the signalling could be associated with fitness and, if so, how; (Krebs and Whitney, 1975).

51. Charnov, personal communication. Krebs was later to publish a major collation of foraging ideas with one of his doctoral students for the Princeton Monographs in Behaviour and Ecology. It served as a specialist textbook and became a much used resource; (Stephens and Krebs, 1986).

52. Rudolf Drent (1937–2008). Krebs also joined Drent on a bird survey in the Northwest Territories, in connection with environmental concerns over oil exploration in the area. A paper that Krebs wrote on the great blue heron proposes an interesting hypothesis on head tilting to avoid sun glare; (Krebs and Partridge, 1973).

53. James Bruce Falls (1927–) is an emeritus professor of zoology at the University of Toronto. An early behavioural ecologist, he specialized in bird behaviour and population dynamics. Krebs was to spend some time working with him in Toronto.

54. In his memoir (see note 36) Krebs wrote that his mother sent him the job advertisement, and that his wife, who came from Wales, wanted to return to Britain. He also wrote that while intending to return at some point, had they stayed in Vancouver a few years longer they would probably have remained there.

55. In his memoir, Krebs writes that this job offer was the result of a chance meeting with Tinbergen's replacement, the new reader in animal behaviour, David McFarland.

56. Much of this work entailed manipulative experiments by Krebs and his students. For example, foraging behaviour was studied by observing birds in pens. Given patchy distribution of their prey (for example, leatherjackets) it was possible to see whether birds foraged for them in ways the models deemed optimal. Krebs was still collaborating with Charnov. See, for example (Krebs, Erichsen, Webber and Charnov, 1977). The other two authors were doctoral students at the animal behaviour research unit at Oxford. By this time Charnov was an assistant professor at the University of Utah. He was later to hold a chair at the University of New Mexico.

57. (Krebs, Kacelnik and Taylor, 1978). The paper showed that tits sampled sites before opting to forage at the best one. They appeared to behave as predicted by a model for the optimal solution for maximizing food intake over a given foraging period.

58. Krebs did not co-publish much with the others. One co-authored paper is (Crawley and Krebs, 1992). As noted elsewhere, other contributors to the same volume included May, Hassell, Beddington, Charles Godfray, Paul Harvey and Jeff Waage. The collection appears to have been directed at advanced ecology students.

59. (Falls and Krebs, 1975; Krebs, 1976); the second paper was published in a new journal, *Behavioural Ecology and Sociobiology*. E. O. Wilson published an article in the same issue. Work on the great tits showed that there are many elements to their songs, and that birds recognized the singing of their neighbours, reacting less strongly to them than to the songs of strangers.

60. Nicholas Davies had carried out doctoral research at Wytham on territoriality in speckled wood butterflies before moving to a demonstratorship in the department of zoology at Cambridge. Now a professor at Cambridge, Davies has been a major contributor to the behavioural ecology of birds, as field biologist, experimentalist and theorist. He has used game theoretic ideas and DNA fingerprinting in analysing mating systems.

61. (Krebs and Davies, 1978). A fourth edition was published in 1997. The later editions have more international content.

62. (Krebs and Davies, 1981). For this textbook the two authors pooled their teaching materials.

63. (Dawkins and Krebs, 1978). An updated version of this article, 'Animal signals: mind reading and manipulation' appeared in the 1984 edition of the book.

64. Richard Dawkins website, autobiography. Given the earlier paper (Charnov and Krebs, 1975), I doubt that Dawkins' claim about who contributed what — always a difficult assessment — is totally correct. See also (Dawkins and Krebs, 1979), one of a collection of papers given at a discussion meeting held at the Royal Society in December 1978 and organized by J. Maynard Smith and R. Holliday. For other contributors see (Maynard Smith, 1979; Harvey and Clutton Brock, 1979) and, for two dissenting voices, (Gould and Lewontin, 1979). The latter argues, among other things, that behaviour cannot simply be viewed through an adaptationist lens.

65. Krebs' election certificate shows that he was proposed by Southwood, seconded by May. He was proposed in 1983, elected in 1984.

66. This work continued over several years together with David Sherry and Sara Shettleworth working in Toronto. See, for example, (Sherry, Krebs and Cowie, 1981; Shettleworth and Krebs, 1982; Krebs, Sherry, Healy et al., 1989).

67. (Krebs and Horn, 1990; Krebs, 1990), quotation, 159. See also (Harvey and Krebs, 1990). In this paper the authors speculate as to whether food-storing birds pay a price for their enlarged hippocampus. Given overall size restrictions, perhaps other areas of the brain are diminished.

68. In his memoir Krebs writes that the two most tempting offers came from Harvard and from the Max Planck Institute in Seeweisen.

69. See the above discussion of Lawton and the creation of the Centre for Population Biology at Silwood. The units at Oxford had less secure funding. Functionally a single unit, the separate names were a bureaucratic consequence of being funded by two research councils. The Agriculture and Food Research Council (AFRC) was later disbanded. Its work came under the Biotechnical and Biological Research Council.

70. (Hassell and Rogers, 1972; Rogers and Hassell, 1974).

71. (Ford, 1971).

72. The trypanosome parasites were being studied also at Imperial College from the cell biological perspective, including by Robert Sinden, who joined the college in 1971, and who later turned to work on the malaria parasite. More recent work on malaria from a population genetics and ecological perspective has been carried out at the college by Austin Burt. Burt co-published with Robert Trivers, and with Charles Godfray, a younger associate of the Silwood Circle who, too, is interested in insect disease vectors. Much recent work in this area has been supported by the Bill and Melinda Gates Foundation. For Godfray see chapter 7.

73. I am quoting Rogers who was paraphrasing. Southwood's career began at a time when many institutions did not allow married couples to hold faculty positions in the same department. Imperial College was an exception. When Southwood began teaching at Silwood in the 1950s, the head of department was O. W. Richards. Richards' wife, Maud Norris who had a PhD in applied entomology, worked together with him (see chapter 3). It appears to have been a good working relationship, something Southwood must have known. Sadly Norris died prematurely. She did not advance beyond a lectureship.

74. This is a topic addressed in my history of Imperial College London. Many scientists of Southwood's generation had good women scientists assisting them in their research. Some were better than he was in promoting their interests. A prevailing view of the 1950s and 60s was that women, while capable of being good scientists, were not suited to the world of affairs, and were unsuited to playing leadership roles, to running research groups, becoming professors or heads of department, or sitting on major committees. In my view this prejudice, rather than one to do with scientific ability *per se*, had to be overcome for women to advance.

75. (Saunders, 2001) and (Pritchard, 2006), 299. Despite a good research and publication record, Randolph, now a professor at Oxford, has had a precarious career, dependent largely on soft money. One problem alluded to in the two sources is that with joint work outsiders often pass judgement on who did what. Something I cannot pursue further here is the suspicion that some people, including Southwood, held the view that Randolph was the junior partner in the work she carried out with her husband. This could well have been wrong and may have slowed her academic advancement.

76. The head of the zoology department at King's College, Don Arthur, was an internationally recognized expert on ticks and disease.

77. I give a few examples of married couples working together in my history of Imperial College London. In addition to Richards and Norris mentioned in note 73, a notable case was that of the geologists John Sutton and Janet Watson, both FRSs, both well known to Southwood. Theirs was a difficult marriage though not, in my view, because of their joint academic work. For some good work on scientific couples, (Pycior, Slack and Abir Am, 1995).

78. It is worth noting that the pioneers in work on nagana and sleeping sickness, joint discoverers in the 1890s of the trypanosome parasite in cattle, were another married couple, David and Mary Bruce. Rogers and Randolph published many joint papers from 1978 onwards. For example, (Randolph and Rogers, 1978 and 1997; Rogers and Randolph, 1985 and 1986).

79. See, for example, (Rogers, Hay and Packer, 1996; Rogers, Randolph and Hay, 2002).

80. (Beddington and Taylor, 1973). Taylor worked at the computing centre at the University of Edinburgh. Theirs was a theoretical paper on how to determine the maximum, yet sustainable, yield in a population of constant size, by means of altering the age distribution of the cull. It drew on some data from Beddington's doctoral work on red deer populations, and on demographic theory (including the Leslie matrix, mentioned in chapter four) learned at the London School of Economics. A further paper, (Beddington, Watts and Wright, 1975) led to some new associations with fisheries scientists (see below). Beddington also applied the matrix in assessing the stability of interacting species; see (Beddington and Free, 1976).

81. This institute known earlier as the Directorate of Fisheries Research is today the Centre for Environment, Fisheries and Aquaculture Science. It is an arm of the Department of Environment, Food and Rural Affairs.

82. The first, and ground breaking paper, was (Beddington and May, 1977). It drew both on Beddington's earlier papers and on May's ideas on how to treat environmental variables. Later work provided corroboration for this early paper; see refs. in note 89.

83. Clark is the author of an influential text of the period; (Clark, 1976).

84. John Alan Gulland FRS (1926–1990), was a Cambridge trained mathematician who became chief of marine resources for the UN FAO, working at its headquarters in Rome. In 1984 he was invited to Imperial College by Beddington, then running the renewable resources assessment group in the Centre for Environmental Technology. Gulland, a major figure in the marine resource area, left many boxes of papers, now in the Imperial College archives. They are of much interest in the areas of whaling, sealing and fisheries conservation. Sidney Holt (1926–) was co-author of a major work in fisheries science; (Holt and Beverton, 1957). The authors were perhaps the first to take an age-structure approach to fisheries management. Holt worked at the fisheries research laboratory in Lowestoft, and later in Rome for the FAO as director of fisheries resources and operations. He served on the scientific committee of the International Whaling Commission from 1960–85. Now retired, Holt devotes much time to whale conservation. Raymond J. H. Beverton FRS (1922–1995) was deputy director of the fisheries research laboratory at Lowestoft, later the first Secretary of NERC, and then head of the department of applied biology at the University of Wales.

85. The IWC rules were such that if a stock could be shown to be depleted below a certain level, then harvesting had to stop. Beddington and his group showed

that this was the case for the whale species mentioned. Beddington used some of the earlier predator-prey modelling in this work. It is clear from even a brief look at IWC documents that Beddington's scientific work came under criticism from Japanese scientists keen to maintain their country's whaling quotas, and to keep open access to whales, especially those in the Southern oceans.

86. (May, Beddington, Clark, Holt and Laws, 1979). Laws worked with the British Antarctic Survey in Cambridge. See also (Beddington, 1980). For a historical overview of whale hunting in the Southern oceans and on the fate of baleen whale populations from the 1920s when Norwegian factory fleets began operations there, (Clark and Lamberson, 1982). Another person with whom Beddington and May engaged when considering multi-species fisheries was Daniel Pauly, now a professor of zoology at the UBC fisheries centre. During the 1970s and early 1980s the balance between minke and blue whale populations in the Southern oceans was a serious concern. Blue whales were in serious decline while minke numbers were increasing, but no one fully understood why.

87. (May, 2000).

88. For a start, mature fish are the ones that lay the eggs that keep the population going; (Beddington, 1974). The age distribution theme is discussed also for whaling in (Lankester and Beddington, 1986).

89. These issues are discussed in two fairly recent papers: (Hsieh, Reiss, Hunter, Beddington, May and Sugihara, 2006; Anderson, Hsieh, Sandin, Hewitt, Hollowed, Beddington, May and Sugihara, 2008). (The titles of these papers — see bibliography — put one in mind of Vito Volterra). The kind of mathematical modelling used is not unlike that used in very different areas, such as in theorizing the financial markets with the aim of demarcating random environmental causes from systemic ones. For Sugihara see chapter 7.

90. (Payne, Cotter and Potter, 2008). The book has a foreword by Sidney Holt; quotation, xviii. Holt's sentiment may look a little hypocritical coming from someone who spent his career in fisheries management. Perhaps he was writing with the wisdom of old age. The book was written to celebrate the fiftieth anniversary of (Holt and Beverton, 1957). This book was reprinted in 2004 with a new foreword by Holt.

91. The continuing dispute between Argentina and the United Kingdom over sovereignty extended to the South Atlantic waters and the fishery. The squid fishery attracted vessels from many countries, especially from Japan where there is a huge domestic market. It appears, however, that Polish ships were the first to seriously exploit the resource. The Russians were also fishing for squid even though their domestic market was small. They bartered the squid for fish that they could sell domestically, such as herring from the North Sea.

92. Beddington told me that the UK government wanted a controlled and profitable fishery to provide a more sustainable economy for the islands which were then overly dependent on sheep farming. See (Cherfas, 1987); also (Rosenberg, Kirkwood, Crombie and Beddington, 1990; Beddington, Rosenberg, Crombie and Kirkwood, 1990) Their models took into consideration stock sizes, number and size of fleets, exploitation rates and so on.

93. Given the histories of Lawton and Beddington at York, it seems the biology department there was slow to promote its local talent.

94. Interestingly Southwood, who knew little about Beddington or his work, was asked for his opinion on the promotion by Eric Ash, the Imperial College Rector. Southwood wrote an ambivalent letter. Anderson and May were supportive. Anderson wrote '[Beddington] is regarded as one of the world's leading authorities in the quantitative analysis of fisheries and sea mammal management problems. He is often asked to advise the British government and international agencies... on questions and problems in the harvesting of fish and mammal stocks. His research is widely quoted in the ecological and fisheries literature'. Bodleian Library, Southwood papers S311; Southwood to Ash, 4 December 1986, Anderson to Ash, 19 December, 1986.

95. Many gifted mathematicians moved to work in the City during the 1980s. Beddington's City connections led him to become something of an entrepreneur. He and some others set up a company to buy vacation properties for rental purposes. More recently, in his role as Government Chief Scientific Advisor, he has set up a project using the Foresight team from the Government Office for Science. The project, sponsored by the Treasury, is to examine high frequency computer trading.

96. Beddington's election certificate shows that he was proposed by May, seconded by Lawton. Others who signed include Southwood, Krebs, Anderson, Hassell, and Imperial College Rector, E. R. Oxburgh. Beddington was proposed in 1996, elected in 2001.

97. This is not a value judgement, simply a speculation.

chapter nine

Voices in the Larger World

Responsibilities, Awards and Rewards

With due modesty we have all done rather well (Southwood, 2003).[1]

In 1993 Richard Southwood's term as Vice-Chancellor of the University of Oxford came to an end. He resigned from the Linacre Chair and returned to his department as professor of zoology. His formal retirement from the university came five years later in 1998, but he remained active until his final illness, resumed his scientific interests, published papers into the twenty-first century, and served on a number of committees.[2] Many voluntary hours were spent in the service of various bodies but, as he himself recognized, he had joined too many organizations, taken on too many responsibilities, and had to send too many notes of apology for not attending meetings. Toward the end of his life two bodies appear to have been especially important to him. One was the Rhodes Trust where he was a trustee from 1986, and chairman from 1999 to 2002; the other was the Lawes Trust where he was chairman from 1991 — fitting repayment to an institution that had helped set him on his way almost forty years earlier. Shortly before his death, he asked Gordon Conway to join him as a fellow trustee. As the previous two chapters have shown, by the 1990s the working lives of members of the Silwood Circle were largely independent of Southwood. All had found their way, and some were moving into positions of authority in the larger scientific world. However, Southwood remained the wise counsellor, joined with members of the Circle in a few scientific projects, and continued to help them in their later careers. After his death in 2005 many condolence letters were sent to his widow; they tell something of the respect in which he was held and of people's indebtedness to him.[3]

It is perhaps no surprise that Southwood's successor in the Linacre chair was one of his favourites, Roy Anderson. Southwood had admired Anderson since spotting him as a gifted student at Silwood thirty years earlier. Both he and May will have been pleased to have Anderson join them in Oxford. However, while for Southwood the move from Silwood to the Linacre chair and head of the zoology department worked out well, for Anderson it did not. Anderson resigned the chair in 2000, after just six years and returned to Imperial College with roughly thirty-five members of his research team.[4] Among many of those who worked with him, Anderson inspired loyalty. Further, as one of them put it, 'there's always excitement around Roy'.[5] His scientific reputation remained high and, at the St. Mary's Hospital campus, he set up a new department in infectious disease epidemiology. This was made possible by the Imperial College Rector, Ronald Oxburgh. Oxburgh recognized that Anderson had built a research team of international importance, that it needed a new home, and that a number of academic positions, including a few senior ones, should be given to it. The timing was good in that Oxburgh was keen to see the new Imperial College Medical School become a major research centre.[6]

After Anderson returned to London, and his team was settling in at the St. Mary's campus, two new infectious disease crises emerged — foot and mouth disease in 2001 and an outbreak of severe acute respiratory syndrome (SARS) in the following year. The former is discussed below. As to SARS, most of the cases were in China but there were some significant outbreaks elsewhere, including one in Toronto. Anderson's team geared up quickly during these epidemics, plotting the progress of the diseases and trying to predict their future paths. The team had to deal also with the media and an anxious public. After the various excitements perhaps Anderson felt the need of a short break from academia and the front lines of infectious disease research. In this his timing was again good. Kevin Tebbit, the permanent secretary at the Ministry of Defence, was looking for a new scientific advisor, someone who could advise on biosecurity. It was the time of the Iraq war and the government was worried that Saddam Hussein had a hidden cache of biological weapons. There were also more general concerns about bioterrorism after 9/11.[7] Anderson had the kind of scientific expertise that Tebbit wanted; he also had an interest in biosecurity. Among those in a position to advise Tebbit on the appointment were Oxburgh, himself a former chief scientific advisor at the Ministry of Defence, and May, former chief scientific advisor to the government and President of the Royal Society. Anderson was seconded to the Ministry of Defence as chief scientific advisor for a period of three years beginning in 2004. He was given a knighthood in 2006, the typical reward, indeed inducement, for high-level work in government. Later Anderson became the Rector of Imperial College, but he resigned after one year in the position. Back at the St. Mary's campus, and together with members of his research group, he has built one of the world's leading disease epidemiology departments, and has engaged in much international collaborative research.[8]

Before moving to England in 1988, Robert May held a top administrative post at Princeton University as chairman of the research board.[9] He was also active on conservation issues and served on the science advisory council of the World Wildlife Fund WWF (USA) and on the U.S. Marine Animals Commission. When May settled in Oxford as a Royal Society research professor, Southwood encouraged him to engage in science-related public affairs.[10] May joined Southwood as a trustee of the Natural History Museum in 1989 and was appointed chairman of the board for a five-year term from 1994. He also became a trustee of both the WWF (UK) and the Royal Botanic Gardens, Kew and Wakehurst Place. And, following the break-up of the centralized Nature Conservancy, and the formation of new regional bodies, he became involved with the Joint Conservancy Council in its early period.[11] Through this type of activity he met many people. When talking with me he mentioned especially two, Crispin Tickell and William (Bill) Stewart, both of whom were well connected in government circles, and involved also with many science-related bodies.[12] They encouraged May to engage yet further, and Stewart involved him in a number of government related projects. May is unusual in that while becoming heavily involved in science politics and public affairs, he remained intellectually engaged within his own discipline. Indeed, he has continued to publish papers into his seventies, and in recent years has applied ecological modelling to banking and financial systems, and to how they respond to disturbance.[13]

In a short piece in *Nature*, Martin Nowak, who worked with May in the zoology department at Oxford during the early 1990s, wrote that May had given him some advice, 'you never lose by being too generous'. Nowak went on to say,

I was impressed because Bob is a winner. To him winning a game is everything. He has thought more deeply about winning and losing than anyone else I know. As his wife once said, 'when he plays with the dog, he plays to win'.

According to Nowak, 'a mathematical analysis of human nature suggests that Bob was right', that one wins by being cooperative.[14] As a young man in Australia May was competitive at the national level in both chess and contract bridge. He was also a serious tennis player. Those early experiences may have put his mind to ways of winning, and to the rarified puzzle solving that he so much enjoys. But, as previous chapters have shown, May is also cooperative. He has worked with many people, not only with those in the Silwood Circle. In doing so he has brought his analytical skills to bear on their many and varied ecological problems. The results could be described as 'wins' for all concerned; possibly also for science — though to make that claim a longer perspective is needed. Cooperation aside, May's old Silwood croquet playing companions would surely testify to Nowak's description of him as a serious competitor, even in that supposedly leisurely game.[15]

That May is a winner can be seen by even a cursory glance at his many achievements and awards. Many of the former have been discussed elsewhere, but testament to his gaining the respect and trust of others are his election as president of the British Ecological Society in 1991 and, especially, as president of the Royal Society for the period 2000–2005. Among his many awards have been the Marsh Award for Conservation Biology (1992) and the Frink Medal (1996) both given by the Zoological Society of London; the Crafoord Prize (1996) awarded by the Royal Swedish Academy and intended to complement the Nobel prizes in areas they do not cover;[16] the Balzan Prize (1998) awarded by an Italian foundation that invites nominations from the world's most prestigious academies and institutions, and the Blue Planet Prize (2001) awarded by the Japanese Asahi Glass Foundation for important work in environmental conservation. May has also received many honorary degrees, honorary memberships of foreign scientific academies, and the Copley Medal, the Royal Society's oldest (1731) and most prestigious award. In 1995 he was appointed chief scientific advisor to the UK government and head of the Office of Science and Technology. He was knighted in 1996 and made a Companion of the Order of Australia in 1998. In 2001 he became one of the first fifteen Life Peers selected by the House of Lords Appointments Commission, established to make non-party-political appointments. In 2002 the Queen appointed May to the Order of Merit. By the early twenty-first century he had won just about all the glittering prizes open to him.[17]

May followed William Stewart as chief scientific advisor in 1995.[18] Back when Solly Zuckerman was advisor to Harold Wilson's government, the position came with no staff, and Zuckerman's advice was called on only occasionally. Earlier, as chief scientist at the Ministry of Defence, Zuckerman did have a staff and also some authority. He pushed to have the same in his new role. By the time Stewart was appointed chief scientific advisor, the position had begun to take its modern form. The Office of Science and Technology was established during Stewart's term and, among other things, was given the role of distributing government funding between the various research councils.[19] Stewart also saw to it that advice given to and by government departments was independently refereed. When May took over he paid further attention to how scientific advice should be garnered, and how delivered. The first cases of v-CJD had recently been confirmed and those within government, worried and mindful of the failure in communication over BSE, were mainly supportive of his initiatives.[20]

May spoke with the Prime Minister, John Major, about the problem of miscommunication and agreed to produce a set of 'Protocols for the Science Advisor in Policy Making'.[21] These were first issued in 1996. May especially advocated the active soliciting of dissenting views and the full acknowledgement of uncertainty in science. In a subsequent meeting of the House of Lords Select Committee on Science and Technology May again emphasized that scientific advice can only be 'robust when subject to national and international peer review'. Much of his advice was coloured by the BSE fiasco, something alluded to in his memorandum to the Select Committee:

> Bland statements of zero risk or 100% safety are much more likely to undermine the scientific advisory system than to promote confidence in it, particularly where the issue of concern is at the frontier of knowledge. [And, when

expressing uncertainty, there should be no] muffling the nature and strength of alternative views.

May also recognized that understanding 'the public (and, indeed, many different "publics") and their attitudes towards science is a critical factor in enabling better public engagement.'[22] More openness and less defensiveness was needed, he stated. 'Transparency' became a buzz word and was seen as essential in re/gaining public confidence in the scientific underpinnings of government policy. It was during this period of facing up to failure over BSE that both the Food Standards Agency and the Health Protection Agency were created.[23] Food security and infectious disease control, along with climate change, were among the most pressing scientific issues for government in the first decade of the twenty-first century.

May worked also on rationalizing the British science budget, beginning with a look at citation patterns in science publications in relation to GDP, something continued by his successor, David King.[24] The old question of what percentage of the GDP should be devoted to science returned to the fore. In the early 1970s Shirley Williams, noting that increases in science funding were outstripping the rate of increase in the GDP, wrote that 'for the scientists the party is over'.[25] It was the prelude to cuts in research funding by both the Callaghan and Thatcher governments. In his own study May looked at the period 1981–94, and at funding practices in twelve countries. He wanted to measure scientific productivity, as compared to the different levels of funding in research and development. He claimed that though Britain was relatively strong scientifically, with respect to research it was spending below its weight. In a 1998 paper he came to a further more familiar, and debatable, conclusion:

> The strong United Kingdom science base does more than its share in helping create wealth around the world ... but this science-base strength is not consistently translated into strong industrial performance within the United Kingdom itself.[26]

May has had perhaps the most glittering of the Silwood Circle careers. However, by any normal standards all in the Circle have been very successful. In the final chapter I will speculate on some reasons for that success, but here will continue with a more descriptive narrative.

When May went to Oxford, two others of the original Silwood Circle, David Rogers and John Krebs, in addition to Southwood, were already there. As outlined in the previous chapter, Rogers distanced himself from the Circle, in part because of his work in Africa. A good scientist, Rogers is not especially careerist. He has had an adventurous but otherwise conventional academic career. Krebs, who accomplished much as a young scientist, turned toward science administration in his later forties. In his memoir he states that he was surprised to receive a call in 1993 asking whether he was interested in becoming the chief executive of the Natural Environment Research Council (NERC). He was appointed after what he described as a 'daunting' job interview by a panel that included the government's chief scientific advisor, Sir William Stewart, and Stewart's soon-to-be successor Robert May.[27] In 1993 Stewart and his associates at the Office of Science and Technology published a White Paper, 'Realizing our Potential: A strategy for science, engineering and technology'. It led to a new approach to government funding, to the creation of the Council on Science and Technology, and to the UK Technology Foresight Programme. At NERC Krebs had to follow the funding guide lines set out in the White paper, even though they were not entirely popular with academic scientists because of their emphasis on utility, and on taxpayers getting their money's worth. It was not Krebs' role to micromanage grants, but he had to make decisions on the general direction of research funding. A major decision in this regard had already been made by his predecessor, John Knill, a former head of the geology department at Imperial College, and John Woods, director of marine and atmospheric sciences at NERC. (Woods, too, had Imperial College connections and would soon return there to become professor of oceanography and head of the department of earth resources engineering. Earlier he held a similar chair at the University of Southampton.) The decision that he and Knill made was for the construction of a new national oceanographic centre at the University of Southampton, to be located on the old Empress Dock. The price tag for the building was £50m and construction had begun by the time Krebs arrived at NERC. Not surprisingly, it was soon apparent that the cost estimate was too low. One of Krebs's first challenges, under strict orders from the House of Commons Public Accounts Committee, was to keep the cost overrun as low as possible. Perhaps it was experience with this project that prompted Krebs to seek more democratic procedures for arriving at funding decisions, and to avoid having forceful directors make decisions largely on their own.[28]

Despite the many budgetary and political concerns, Krebs wrote that he enjoyed his new job. Running NERC meant being responsible for several centres and laboratories as well as for the British Geological Survey and the British Antarctic Survey. It allowed him to learn what was happening in a range of sciences in addition to his own field of behavioural ecology. One of his tasks was to oversee the procurement of some oceanographic vessels for work on the British Antarctic Survey. He also visited the Survey's remote research station. More relevant to the work of those in the Silwood Circle, however, was that David Shannon, chief scientist at the Ministry of Agriculture, asked Krebs to set up, and chair, a working group to review government policy on bovine tuberculosis. The disease was on the rise and the group's remit was to determine whether this was linked to an increase in the badger population and, if so, to recommend action. This was not the first time that a government committee had looked into the problem.[29] Like the earlier committees, Krebs' group saw badgers as part of the problem but was uncertain as to whether a major cull would help to control the disease. Just after the 1997 election Krebs advised the new Labour minister of agriculture, Jack Cunningham, that a large scale experiment be carried out. The working group recommended that three sizeable areas of the countryside be chosen: that in one there should be no culling, in the second culling should be carried out only if and when a major outbreak of bovine TB occurred within it and, in the third, there should be proactive culling of badgers. Cunningham agreed to this and the trial began. But it was ended when foot and mouth disease broke out in 2001. However, by then enough data had been collected for epidemiologists to make some predictions. One who worked on this problem was Cristl Donnelly, a member of Roy Anderson's team who was working also on the BSE outbreak. Models indicated that reactive culling would not work and that proactive culling could perhaps work were the badgers to be killed in areas large enough to surround areas in which TB outbreaks were likely to occur.[30] But this would have entailed killing badgers on a scale unacceptable to the public. Some of the better advice given by Krebs's committee seems rather commonsensical, namely that farmers take measures to keep badgers out of feed and water troughs. Clearly making policy on this matter is difficult.[31]

Krebs served a four-year term at NERC. He was offered a renewal but decided instead to return to Oxford. His work at NERC was rewarded with a knighthood in 1999. Settling back as a research professor was a possibility, but evidence suggests that Krebs was actively looking around to see where he could move to next, and that he favoured some further work in the public sector, or becoming the head of an Oxford College. Correspondence in the Southwood papers shows that Krebs asked for his support in seeking a number of positions.[32] The one that came his way, and for which he had both May and Southwood's support, was the chairmanship of the new Food Standards Agency (FSA).[33] This was an important job. In 2002 Krebs became the head of a new non-ministerial department of government that was created from scratch. About 600 civil servants were transferred to the FSA and Krebs took over responsibility for matters relating to food safety and nutrition from the health and agriculture ministers. The agency was also responsible for introducing and enforcing regulations within the food industry.[34] The Secretary of State for Health, Frank Dobson, described Krebs' job as 'a defensive shield for ministers'. Working with Krebs were Geoffrey Podger a career civil servant and chief executive of the agency, and Suzi Leather, a 'fierce campaigner for the consumer movement', who was deputy chair. According to Krebs, the three worked together in a state of 'collective tension'.[35]

As mentioned, the creation of the agency was prompted by fallout from the BSE crisis. Just as May advised the government to be more open with the public so, too, Krebs wanted more openness with respect to advice on food-related matters. Both men wanted government spokespersons to cite evidence in support of any advice given, and to explain the evidence in ways that the public would understand. The problem, as Krebs put it in his memoir, was to develop an institutional culture new to the civil servants he inherited. Civil service briefing notes tended to hedge all bets. People were often reluctant to state their views, to give advice, or to recommend any definitive action. The easy path was often taken; namely setting up committees, delaying decisions, and diluting responsibility. There was also the habit of secrecy. Krebs moved to change these institutional behaviours — just as others of his generation attempted the same in other government departments. It is difficult to assess but, overall, government today allows more access to information and appears to be more open than it was fifty years ago.

The three executives were put to the test when the issues of GM and organic foods came to the fore. Krebs was much criticized for being in favour of the former and only lukewarm, if that, on the latter.[36] While this drew much public attention, the biggest food safety challenge remained that of BSE. In 2000 it was still feared that possibly thousands of people were harbouring v-CJD.[37] And there was concern that BSE could have jumped to sheep. Under laboratory conditions some sheep had become infected. Outside the laboratory, on farms around the country, many sheep had been fed the same infected feed as had been given to cattle. Once again there was secrecy at MAFF, this time out of fear of upsetting the sheep industry. Krebs decided that the secrecy had to stop and he held a press conference to explain the situation. He stated that, in all likelihood, eating lamb and mutton was safe; but he also made clear that there was some uncertainty.[38] Krebs later summarized his views on food and risk in his Croonian Lecture of 2004. He ended the lecture by noting that most decisions are made in a climate of uncertainty since consensus even among scientists is never total. While stating that a better educated public would be a good thing, he noted 'we should recognize that a scientifically literate public may well be a more sceptical one'.[39]

When Krebs was one year into his new job, Foot and Mouth disease made a comeback. Abigail Woods has written an interesting history of this disease in Britain, claiming that it made its first known appearance around 1840.[40] For many years it was largely ignored by the farming community. Most infected animals recovered and, if the sick were isolated, the disease did not spread far. But that was before the large-scale movement of animals — nationally and internationally — became commonplace in the second half of the twentieth century. Woods' book tells how the disease came to be viewed more seriously by the mid-twentieth century, who came to define it, and how mass slaughter, rather than isolation or vaccination, came to be seen as the proper means of control. At first it was pedigree livestock breeders, and later the exporters of more general livestock, who saw culling to be in their interest. This was because importers were reluctant to buy breeding stock that could be infected and foreign governments increasingly banned the import of meat and livestock from countries where the infection had broken out. Pedigree livestock breeders, especially, wanted the quick stamping out of infection and thought culling was the way to do that. They began to view the disease in draconian terms as a serious plague, began to label it as such, and soon came to believe their own propaganda. It was a means of protecting their own businesses, even at the expense of farmers who had no wish to cull their herds or flocks. Woods claims that since such breeders belonged to the rural elite they had access to the levers of power within government. Over time, she argues, civil servants in the various agriculture ministries came to accept their view as to the seriousness of the disease. By the mid-twentieth century few questioned it, or the need for the occasional mass slaughter. Vaccination was a possibility from the 1950s on, but could not be guaranteed to provide life-time immunity. Given that the disease was intermittent and with fairly long intervals between outbreaks, no one in Westminster took seriously the manufacture, stocking and distribution of a vaccine when an occasional cull seemed to be effective. Further, there was little opposition to culling among the public. In 2001, however, the massive scale of the cull was very disturbing. Further, the previous forty years had seen the rise of an animal rights movement and an increased consciousness of what was widely seen as an inhumane livestock industry. The result was much public revulsion over what happened.

As Krebs notes in his memoir the first cases of the 2001 outbreak were detected in February among some pigs. Like others he was worried and believed that the outbreak was being downplayed by MAFF. He states that when he and Bob May were on one of their regular Sunday morning runs, and thinking about the heavy movement of sheep around the country, they wondered whether the disease could have spread to sheep. This thought had already occurred to people working at MAFF but they were not letting anyone else know.[41] The pigs that were the first to be diagnosed were on a farm in Essex. However, they had recently been transported there from a farm in Northumberland. Sheep from the same Northumberland farm had been distributed to other farms in the North of England. Knowing this, and knowing that the disease was highly infectious, people at MAFF supported by the National Farmer's Union whose leaders wanted exports to resume as soon as possible, were gearing up for another mass cull (the last one had been in 1967–8). At the same time they were trying to assess the seriousness of the outbreak, trying to deal with a flood of reports on tissue samples, pondering whether a firewall cull might work, and thinking about how and when to inform the public of the situation.[42]

As more farms became infected, Krebs and May decided to act independently. Krebs brought together a group of scientists who, like May, were experts in infectious disease transmission. Present at the first meeting on 6 March 2001 were Roy Anderson and Neil Ferguson from Imperial College London, and three of Anderson's former students and colleagues, Bryan Grenfell from Cambridge, Mark Woolhouse from Edinburgh and Graham Medley from Warwick.[43] It was a closely connected group of people. MAFF was informed that the meeting would take place and was asked to send a representative but declined. Those present came to the conclusion that, were they to gain access to all the MAFF data on the various outbreaks, they could then use their own models to estimate the rate of spread of the disease. Such information, they believed, would help inform the decision makers in Cabinet. New data was being collected at MAFF but, according to Woods' more general assessment, the Ministry's overall database 'which supposedly held details of all British farms' was 'both inaccurate and out of date, and official ignorance of the state of sheep movements enhanced the inefficiency of disease tracing and control'.[44] People at MAFF were reluctant to release their new data but soon changed their minds. Krebs and the modellers were given access to the data on 13 March and were then given two weeks to come up with something before any major policy decision was taken. According to Krebs, when the modellers presented their results on 21 March the atmosphere was 'electric'. Many representatives from MAFF were present. Also present was May's successor as chief scientific advisor, David King. Ferguson presented the Imperial College results and claimed that the doubling time for new farm infections was, by then, about nine days, and that the only way to stop further transmission was by mass vaccination or mass culling.[45] All but one of those present agreed with him on the doubling time. Questions were raised as to whether there was time for vaccination. King clearly accepted the conclusions of Anderson and Ferguson. He passed the information to the Prime Minister, Tony Blair. Blair then ordered the culling. Anderson appeared on the BBC TV programme, Newsnight, on the evening after the meeting and stated that the outbreak was out of control, not something those at MAFF wanted to be said publicly.[46] Later Anderson, convinced like May and Krebs of the virtue of being open with the public, said that he would say the same again, even knowing of the negative comments that followed. He believed it to be his duty to speak out.[47] The Foot and Mouth outbreak came to an end in December. It had lasted ten months and caused much anguish.[48]

The culling led to a further concern, namely that the dioxins released from the many burning pyres of animal corpses would get into dairy products at nearby farms. MAFF wanted secrecy on this too, but others in the government had learned something from the BSE crisis. Krebs and the FSA were asked to cooperate in framing a statement for the public on what was happening. This they did and fears were allayed. According to Krebs the dioxin levels soon dropped to a level 'too low to worry about'.[49]

The Foot and Mouth episode is interesting from a behavioural perspective. On the one hand it looks as though Krebs, May and Anderson stepped in and, with the help of David King, moved to advise the Prime Minister over the heads of people at MAFF. There is little doubt they did so because they were concerned and believed, probably correctly, that they and their associates had the best models for tracking the progress of the disease. On the other hand, given MAFF's history, and that of its predecessor ministries, its advice to the Cabinet was probably little different than that of the external modellers. It is not as though people at MAFF ignored epidemiological thinking. As Woods suggests, such expertise had long been relied on, possibly overly so. Epidemiological models enable effective culling, but are useful only when there is full openness among farmers, immediate reporting of disease, and if there is good and open record keeping. Complicating matters was that within MAFF there was dissent, something very difficult for decision makers to deal with in a crisis. Veterinarians in the Ministry were divided with some wanting an aggressive vaccination programme and far less culling. This was also the desire of many farmers and, in an interesting turn around, also the outspoken wish of the owners of herds or flocks of rare breeds. But the National Farmer's Union was against vaccination. Among the recommendations of the Royal Society's Follett Inquiry was the development of better vaccines as a preventative measure, and that the farming industry be vaccination-ready for any future outbreaks, However, Follett concluded his report by stressing the need for better data bases and by stating that quantitative modelling is an 'essential tool both for developing strategies in preparation for an outbreak and for predicting and evaluating the effectiveness of control policies during an outbreak'.[50]

This episode illustrates something about trust. Krebs clearly trusted May and both, in turn, trusted the epidemiological work of Anderson and his various associates. As members of the Silwood Circle they had known each other for many years and respected each other as scientists. No doubt there were similar circles of trust within MAFF. In a crisis, however, there is little time to canvas many opinions and the government needs to act quickly. On this occasion the government went with the advice of two of its chief scientific advisors, and they, in turn, took advice from those whom they trusted most. People will debate whether it was the best advice but that misses the point of how decisions are made when time is short.

Before leaving the FSA Krebs was interviewed with a view to his becoming, like May, a cross-bench peer. Three years later, in 2007, he joined May in the House of Lords. In the same year Gordon Brown succeeded Blair as Prime Minister and asked Krebs whether he would join the Labour Benches and become the science minister.[51] Clearly Krebs had won the trust also of those highly placed in government. He declined the invitation though in his memoir he wrote, 'I will always have some pangs of regret'.[52] By then Krebs was in a new job. Before leaving the FSA he had thought ahead about what to do next. He pursued his earlier interest in becoming the head of an Oxford College and was appointed Principal of Jesus College in 2005. While tempted by Brown's offer, he decided to stay put.[53]

Others in the Silwood Circle remained more strictly in the academic sphere during their working lives. Michael Hassell, for example, was head of the biological sciences department at Imperial College until appointed Principal of the Faculty of Life Sciences in 2001. He also continued with his research.[54] Much of it in this period was in collaboration with his doctoral student, Michael Bonsall, who was to become a reader in mathematical biology at Oxford. Hassell was elected president of the British Ecological Society in 1999 and, for services to population ecology, was made a CBE in 2002. After his retirement he took on the editorship of the biological section of the *Proceedings of the Royal Society*. In 2012 he succeeded Sir Neil Chalmers as chairman of the board of trustees of the National Biodiversity Network. The Network stores, and makes available on its website, wildlife records for the United Kingdom (70 million and growing fast). Mick Crawley, too, stayed on at Silwood carrying out further work in the area of herbivory. He collaborated with May into the early twenty-first century, and remained in the spotlight when GM crops were under discussion.[55] Crawley was elected to the Royal Society in 2002.[56] His continuing interest in field ecology and natural history led to the publication of an extensive flora of Berkshire.[57]

John Lawton was still working at Imperial College in the mid-1990s.[58] His work gained international attention, especially because of its relevance to conservation and to issues related to biodiversity.[59] As a long-term member, and then chairman, of the council of the Royal Society for the Protection of Birds, he was an active participant in conservation matters. He was also vice-president of the British Trust for Ornithology, and a trustee of WWF (UK). The combination of excellence in science and activism in ecological matters won Lawton many awards. Among them were the 1996 prize of the International Ecology Institute for which he wrote his *Community Ecology in a Changing World*, mentioned in chapter eight; also in 1996, the British Ecological Society's Marsh Award; in 1998 the Swedish Kempe Foundation award for distinguished ecologists; in 2004 the Japan Prize for his contribution to the conservation of biodiversity and, in 2006, the Government of Catalonia's Ramon Margalef Prize in Ecology and Environmental Science.[60] In 1998 the Zoological Society of London awarded him the Frink Medal and, in 2007, made him an honorary fellow. Like others in the Silwood Circle Lawton received a number of honorary degrees, honorary memberships of foreign academies, and many invitations to give lectures. He was made a CBE in 1997.

Lawton was asked whether he was interested in taking on the position of chief executive at NERC in 1993 — he was not. As we have seen the position went to Krebs. But Lawton must have changed his mind and, when Krebs stepped down in 1997, he took over. He told me it was 'a 24/7 job' looking after not only about two thousand employees, but also the many graduate student and postdoctoral grant holders. He said that when he arrived he found an organization 'plagued by an "us and them" mentality'. The 'us' contingent resided at various centres on the periphery, and the 'them' contingent at the central office in Swindon. Lawton tried to overcome the resentment and discontent by bringing the directors of the main centres and Surveys onto the Management Board. When he took over, the Southampton Oceanography Centre was ready to open. At the same time Lawton closed down some smaller marine laboratories in order to consolidate resources and to save money. Such action is never easy and in this case caused yet more resentment. Overall, however, it appears that oceanography did rather well during Krebs and Lawton's tenure at NERC. Further, Lawton and Margaret Leinen, of the US National Science Foundation, persuaded the boards of their respective bodies that there should be collaboration on a major ocean monitoring programme to track the Atlantic thermohaline circulation. In this connection Lawton played a role in persuading the government to build yet another research vessel after what he described as a 'tense bid across the research councils for capital investments'.[61]

When his term at NERC ended Lawton, like Krebs earlier, was given a knighthood. Earlier Lawton had been a member of the Royal Commission on Environmental Pollution (RCEP). In 2006 he rejoined the commission as its chairman — the last of its chairmen since the government decided to close down or merge many of its scientific advisory boards and regulatory bodies.[62] Lawton oversaw a number of reports including one, released in 2008, titled 'Novel materials in the environment: The case of nanotechnology'.[63] The Commission's final report, its twenty-ninth, 'Demographic change and environment', published in February 2011, looks forward to the kinds of problems Britain could face up to the year 2050. The RCEP closed on 1 April, 2011. Founded in 1970, it had a long record of producing fine reports and providing excellent advice to government. May told me that 'the Silwood Circle would be unanimous in regarding its closure as folly'.

Until 2008 John Beddington was professor of applied population biology at Imperial College. He was also the head of the department of environmental science and technology until it closed due to college restructuring in 2007. By then he was a long-time advisor on the development and management of fisheries to the British government, the European Commission and the UNFAO.[64] In 1997 he won the Heidelberg Award for Environmental Excellence, in 2000 he was elected to the Royal Society, and in 2003 was made a CMG by the Queen for his services to fisheries science and management.[65] In 2005 Beddington was appointed chairman of the science advisory council at the Department of Environment, Food and Rural Affairs (DEFRA), the successor ministry to MAFF; and was on the executive board of NERC. In 2008 he was appointed to succeed David King as chief scientific advisor to the government and head of the Government Office of Science. He received a knighthood in 2010.

In his role as chief scientific advisor Beddington placed much emphasis on the proper use of science. His public utterances on this subject show the influence of Karl Popper with whom he had studied at the LSE as well as the later influence of Robert May. Popper, a critic of positivism, understood that there was uncertainty in all scientific claims. He believed that a critical stance toward such claims was essential and that progress is achieved through a process of elimination — by the falsification of ideas. As chief scientific advisor Beddington has spoken publicly on a number of issues but here I will illustrate the Popperian point by looking briefly at only a couple of his statements, on climate change and the world's food supply.

In January 2010, after the Intergovernmental Panel on Climate Change (IPCC) admitted that it had vastly overestimated the rate at which Himalayan glaciers were retreating, Beddington stated that climate scientists should be less hostile to sceptics, that the IPCC's report exposed a failure in the way evidence was presented to the public, and that it was important when making predictions to acknowledge uncertainty:

> I don't think it's healthy to dismiss proper scepticism. Science grows and improves in the light of criticism. There is a fundamental uncertainty about climate change prediction that can't be changed.

Beddington stated that it was wrong that some climate scientists refused to disclose their data to their critics, stressing as had May, Krebs and Anderson, the need for openness. Beddington also clearly stated that the evidence for global warming was good, while acknowledging that what the exact consequences will be is open to debate.[66] He also stated that uncertainty is no justification for not taking action to reduce greenhouse gas emissions.

Related to climate change are issues of food and water shortages. In 2009 Beddington gave a widely reported speech to a government conference, 'Sustainable Development UK'. He stated that because of growing populations, and success in alleviating poverty in developing countries, there will be a surge in demand for food, water and energy in the coming decades. We are, he said, heading into 'a perfect storm'.[67] Beddington was especially worried about food prices since, with global reserves at only 14% of annual consumption, any major drought or flood could see prices escalate rapidly. As part of his advice he suggested that government listen also to the views of independent scientists and that more of them be taken into the government fold, even if only temporarily.[68] In this connection Beddington himself brought together a group of scientists to work on a government Foresight project. Their report 'The future of food and farming: Challenges and choices for global sustainability' was published in 2011.[69] It drew on both existing and commissioned studies from around the world and, as the earlier chief scientific advisors, Stewart and May, had recommended, all were peer reviewed. While stressing future uncertainties, the report provides much advice. For example, Beddington and his co-authors favour more research and development in the area of GM crops claiming that these will be needed especially in areas where water is in short supply. Beddington is also on record as being against the use of agricultural land (and the destruction of forested land) for growing biofuels. While outlining the many types of action needed, the report shows concern for the maintenance of biodiversity and natural ecosystems around the world.[70] In this, Beddington shows solidarity with other ecologists and shares in ways of thinking long shown by others in the Silwood Circle, notably by May, Lawton and Conway. When I spoke with him, Beddington acknowledged his indebtedness to the others, especially to May for his help both intellectually and in career terms.

Among those in the Silwood Circle, Beddington's good friend Gordon Conway has thought the most about global food security and sustainable agriculture.[71] These were the concerns that Conway took to the Rockefeller Foundation when he was appointed president in 1998 — the first non-American citizen to hold the position. As mentioned in chapter seven, Conway's book, *The Doubly Green Revolution: Food for all in the twenty-first century*, published while he was vice-chancellor of the University of Sussex, drew much attention. It helped him win this new appointment.[72] J. D. Rockefeller, who set up the foundation in 1913 wanted it to promote 'the wellbeing of mankind throughout the world'. By the time Conway became president the value of the foundation was about $3 billion and its annual budget around $160 million — not huge by today's standards. One of the first things Conway did was give new focus to Rockefeller's initial mission statement. He decided to fund, and to raise further money in support of, projects that would promote the wellbeing of the 'poor and excluded' among humankind — with stress still on global inclusion. During his tenure projects were funded in crop biotechnology, sustainable agriculture, school reform, democracy and diversity, helping people to move from welfare to work, urban policy, public-private partnerships on research into HIV vaccines, the distribution of anti-retroviral drugs to HIV positive mothers in Africa, research on TB drugs and malaria medicines, intellectual property regimes, arts education, religion and civil society, training in philanthropy, and higher education in Africa.[73] Conway also chaired a number of important committees.[74] In addition the Rockefeller Foundation promoted some commercial development and the construction of low-cost housing in the poorest areas of some American cities. Conway was also approached by Buzz Holling who had helped him at the start of his career. Holling wanted the Foundation to support some work on resilience that was starting up in Sweden.[75]

In earlier times the Rockefeller funds went further than they do today. For example, they largely funded the research and development that enabled the Green Revolution to take off. Today a project of that magnitude would require massive support from both the public and private sectors. Perhaps the thing that drew Conway the most public attention was a speech he gave to the board of directors of Monsanto in 1999. It was on plant biotechnology and global food security, and was widely discussed in the media. Conway spoke in broad terms of the plant biotechnology projects that the Rockefeller Foundation was funding and how, in doing so, it was helping to train scientists in Africa, Latin America and Asia. The research, while not excluding work on transgenic plants, was largely traditional and involved tissue culture, breeding and selection technology. There had been some positive results from this, such as the crossing of some Chinese with West African rices leading to new higher-yield varieties. As to transgenic work, some funding went to research on the addition of new genetic material to rice so as to give it some insect and disease resistance, and to increase its vitamin A content. While the latter is of little interest in wealthy countries where vitamin A deficiency is rare, it is important to those parts of Asia where children are weaned on rice gruel and not much else; and where the illness and death of children due to such deficiency is not uncommon.

Like others in the Silwood Circle, Conway is not against GM foods — on the contrary. But, in the interests of global food security he wanted a balanced set of ground rules for the new biotechnology which everyone, including charitable foundations and big businesses like Monsanto, would be required to follow.[76] Conway advised a slow and careful approach, especially when it came to the introduction of genes from viral pathogens.[77] He encouraged Monsanto to invest more in research into the safety aspects of new biotechnology. But what especially drew people's attention was Conway's criticism of Monsanto's planned use of so-called 'terminator technology' that would cause seed sterility in food crops, and would ensure high annual demand for its own seeds. Conway said that his concerns were not an expression of anti-corporate sentiment but ones of justice. Like any other plant breeder, Monsanto depended on the use of existing plant stock, stock that has been developed over thousands of years by breeders around the world. The new rush to patenting types of rice and other cereals unjustly takes control of this historical legacy and, in so doing, generates fear and animosity. Further, adding 'terminator technology' to new high-yield varieties would seriously affect poor farmers who save their own seeds. Conway advised that companies like Monsanto comply with plant variety protection schemes already in place in developing countries. Poor farmers should not be under threat from intellectual property rights on new forms of traditional staples such as Indian Basmati rice or Thai Jasmine rice, nor should they be placed in situations where they have to buy seed each year.[78] Monsanto appears to have been at least partially persuaded.[79]

Conway was elected a Fellow of the Royal Society in 2004.[80] He returned to Britain in 2005 after seven years with the Rockefeller Foundation and took up the position of chief scientific advisor at the Department for International Development (DFID).[81] This was an unexpected appointment. It occurred after Conway and the science minister, Lord (David) Sainsbury, went to see Hilary Benn, the minister at DFID, to try and persuade him that the department needed a chief scientific advisor. This they managed to do, persuading also a reluctant permanent secretary, Suma Chakrabarti. Conway was asked to take on the job, even though he was past retirement age. He spent much of his time at the department trying to get the new position recognized by the other civil servants. He had to persuade them of the importance of science to their mission and that they should hand some authority to the new advisor. In this Conway was successful and, to match the new responsibilities, the budget for the office of chief scientific advisor went from £35m to £220m by the time Conway retired five years later.[82] He was given a knighthood in 2005.[83]

Conway also held a part-time professorship in international development at Imperial College London. He carried out some teaching but, since 2009, has led a small team funded by the Bill and Melinda Gates Foundation to provide advocacy, and to lobby European governments to do more, for African agriculture. In this connection Conway chaired the Montpellier Panel which produced its report, 'Africa and Europe: Partnerships for agricultural development' in 2010.[84] He also chaired a UK collaborative group seeking an increased role for science in development areas more generally. Its board consisted of the chief executives of most of the research councils, chief scientific advisors in other ministries, and the director of the Wellcome Trust. The idea was to promote joint development activity among the various funding bodies. Conway, who has a long standing interest in promoting geography instruction in schools, was also elected president of the Royal Geographic Society, 2006–2009.

As President of the Royal Society, May spoke out on many topics. Those closest to his heart were those closest also to the hearts of others in the Silwood Circle: climate change, biodiversity and ecological conservation, food and water security, and health. His wide-ranging 2005 anniversary address to the Royal Society, his valedictory, included all these themes. He began with some thoughts on cooperative behaviour as understood by evolution theorists, and stated that while kin selection can perhaps work to produce cooperative behaviour among small isolated groups of closely related individuals, it cannot be relied on for the smooth working of large modern societies. As responsible human beings we need to look elsewhere. Globally there are many different and long-standing cultural traditions from which people draw ideas on how to make societies work. May, socialized in the Western tradition, and educated at a certain time and place, drew on the liberal ideas of the Enlightenment, ideas central also to the founding and history of the Royal Society. All in the Silwood Circle could be described as having imbibed a Western liberal ideology, with its respect for individual freedom, democratic politics, and the acceptance of scientific, rather than religious or other, approaches to understanding the natural world. (It should be added, however, that modern science and Enlightenment values were the offshoots of a particular religious tradition.) Science entails an active search for knowledge through observation and experiment, and a respect for what those observations and experiments turn up. A further goal of some

Enlightenment thinkers, one not yet fully achieved, was to have new knowledge serve the larger interests of humanity, and not simply the vested interests of powerful elites. Perhaps a less rhetorical way of putting this is to say that history evolves new knowledge by deferring conflicts and final reckonings.

Respect for observational data is clear in all May's utterances, and is clear also in his anniversary speech. But he went further. He asked his audience, and scientists more generally, to face up to the various problems that cumulative scientific observation points to. One of the most pressing is climate change. May has stood up to those denying climate change and to those denying that human activity is in any way responsible for it. In his speech he pointed to what the consequences of global warming are likely to be, and what possible actions could be taken to alleviate the situation or to adjust to it. May is concerned with what he sees as a possibly even greater threat than climate change *per se*, namely the loss of biodiversity, and the collapse of ecosystems. An ecologist, he is well aware that as human animals we are totally dependent on the stability of ecosystems of which we are part.

But the main message in May's valedictory address was that scientists need to wake up. They need to fight for better science instruction in schools and play a more active role in informing the public of what they do. They should be encouraging not only democracy but also the use of reason in the making of public policy. May sees the public as largely misinformed, too easily swayed by commercial interests, too easily persuaded by religious fundamentalists — even by some post-modernist academics — that scientific claims are simply points of view. I agree with him on all of this. However, in my view, what is also needed is better recognition by scientists that they are living in an age inflected by postmodernism. Science depends on observation of the world or, as Plato more conservatively understood it, on the appearances of the world. As Plato pointed out, appearances can deceive and truth is elusive. Further, although the history of science is a history of knowledge accumulation, it is also a history of discarded ideas. Perhaps we need to set aside the word 'truth' from modern debates. Although we should not deny the reality of the world, nor that we can acquire knowledge of it, scientists would perhaps do better to focus simply on its appearances. They should convey to the public how paying keen attention to the appearances of the world has served us well; indeed, that such attention is essential to how we live our lives today. They should communicate better what is meant by the claim that science is evidence-based, and how this differs from mere opinion or sincerely held belief. And they should point to both the wonder and the great utility of science, rather than to the possibly overly ambitious Enlightenment ideal of truth seeking.[85] This is not to say that Plato was right, simply to acknowledge that his view of things better fits the cultural climate in which we now live.

Endnotes

1. In a letter to Conway, Southwood wrote, 'the present summer has certainly reminded me of 1973 — halcyon days, and with due modesty we have all done rather well'. Bodleian Library, Southwood papers R443, Southwood to Conway, 1 September, 2003. Both summers were hot and dry. For the summer of 1973 see chapter 4.

2. He was chairman of the Inter-Agency Committee on Global Environmental Change from 1997–2000, a committee set up by the Advisory Board of the Research Councils (ADRC) in 1990. He also chaired the 2003 committee to choose a new Linacre professor.

3. Included was a thoughtful letter from Gordon Conway. Several eminent people wrote letters, among them Nelson Mandela. Bodleian Library, Southwood papers, A355.

4. Included in the team returning to London were academic researchers, technical and secretarial staff. There are a number of reasons for Anderson's departure, some, but not all, of his own making. The episode puts me in something of a dilemma since most of the witnesses are still living, and such pieces of evidence I have collected conflict on many points. It is probably true, and not only in this case, that no story is told in the same way twice. As Charles Tilly would perhaps have put it, events were recast in a range of standard story forms. For the stories people tell, and the problems in interpreting them, (Tilly, 2002 and 2010). Charles Tilly (1929–2008) was both historian and sociologist. I have learned much from his work.

5. (Birmingham, 2003), quotation, 492. The person quoted was implying that Anderson turned his attention to the latest and most interesting epidemics, and that he had the tools to do so.

6. Lord Oxburgh had the difficult task of uniting the various medical schools to form the Imperial College Medical School which was founded in 1997 (renamed Faculty of Medicine in 2001). In the 1990s research at the medical schools of both the Westminster Hospital and Charing Cross Hospital was almost non-existent and the standard of research (as measured by the RAE) at

St. Mary's, already merged with Imperial College, was lower than ideal. Only at the National Heart and Lung Institute and the Postgraduate Medical School at Hammersmith Hospital was top-level research being carried out. Oxburgh needed to assure people at those institutions that any merger would not pull their standards down. In his negotiations he took measures to ensure that the new medical school would become a major research centre. This entailed terminating many people's jobs and building new research teams. Retrospectively we can see that Oxburgh created a very successful unit. I am indebted to him for talking with me about the creation of the new medical school and about his term as Rector of Imperial College.

7. Sir William Stewart, the former chief scientific advisor to the British Government, gave speeches on the potential dangers of bioterrorism to both the British Science Association and the Royal Society of Edinburgh in 2001. The Foot and Mouth epidemic of that year heightened fears of a biological attack. For more on Stewart see note 12 below.

8. The department became a UNAIDS collaborating centre for the epidemiology of HIV and AIDS, and a WHO collaborating centre for intestinal infections. Anderson served as chairman of the WHO's Science and Technology Advisory Board on Neglected Tropical Diseases. The department also houses the Partnership for Child Development, funded by UNICEF and the World Bank. Anderson is on the Advisory Board of the Bill and Melinda Gates Foundation and chairman of its Schistosomiasis Control Initiative. The department received a £30m grant from the Foundation for work on the control of schistosomiasis in sub-Saharan Africa. The departmental staff also carry out work on influenza viruses, pneumococcal and meningococcal bacteria, BSE and vCJD. Large data bases are kept for many diseases and much modelling work is done with the aid of state of the art computing facilities.

9. Princeton had an admirably lean administration. Being chairman of the research board was akin to being vice-president for research, as such positions are known in many North American universities. However, unlike in many of today's universities, meetings at Princeton were not endless. The university was comparatively small and bureaucratic tasks were kept to a minimum. May's scientific output was high during his eleven-year term.

10. The professorship was held jointly at Imperial College, but May settled in Oxford.

11. New councils were formed in Scotland, Wales, Northern Ireland and England. But because of EU obligations a joint council was still needed.

12. Sir Crispin Tickell was educated at Oxford. After taking a degree in history he joined the diplomatic service where he had a major career before his appoint-

ment as Warden of Green College, Oxford in 1990. Tickell has a life-long interest in matters related to the environment and nature conservation, (Tickell, 1986). May came to know him when they were fellow trustees of the Natural History Museum. Sir William (Bill) Stewart, a microbiologist, was the founding professor of biology at the University of Dundee before moving to London in 1987 to become Secretary of the Agricultural and Food Research Council. (The Council was later disbanded and subsumed under the Biotechnical and Biological Research Council). Stewart preceded May as government chief scientific advisor and, later, was the first chairman of the Health Protection Agency, set up in 2003 to protect the public from the threat of infectious diseases and environmental hazards.

13. See, for example, (Sugihara and May, 1990); also a short note, (May, Levin and Sugihara, 2008). Two interesting articles related to the financial collapse of 2008, and to the problem of risk management, are (Haldane and May, 2011a and 2011b). As the authors note, the financial sector could learn something from ecological theory, especially about complexity and instability: 'scaling up risks may cause them to cascade rather than cancel out. The bigger and more complex the structure, the greater the risk' (2011b). Andrew Haldane is the Bank of England's executive director for financial stability. Computers can now be programmed to simulate the way complex systems (ecological, financial, or other) change over time. People can run experiments and watch the outcomes on their monitors.

14. (Nowak, 2008). Nowak's views on cooperation and its evolutionary significance hark back to Kropotkin (see chapter five), but he also sees Darwin as a progenitor. For more on this and on May being both competitive/collaborative, and Nowak's mentor, see (Nowak with Highfield, 2011), 38–41 and *passim*. With respect to the mathematization of cooperation, Nowak has made inroads but his claim that 'the way we human beings collaborate is as clearly described by mathematics as the descent of the apple that once fell in Newton's garden.' (xvii) is overreaching. At Oxford, Nowak was appointed head of mathematical biology and briefly held a professorship before moving to the United States. Since 2003 he has headed Harvard University's program in evolutionary dynamics. Highly creative, one of Nowak's strengths appears to be the dreaming up of games which could have evolutionary significance — iterative games along the lines invented by Axelrod for the prisoner's dilemma and tit for tat (Axelrod, 1984). May has discussed the importance of noise, and of stochastic events more generally, when dealing with the real world and its games; (May, 1987). Among the results of Nowak's cooperation with May are (Nowak and May, 1992 and 2008). See also (Nowak, Anderson, Maclean, Wolfs, Goudsmit and May, 1991; Nowak and Sigmund, 2007).

15. I was told that in the 1970s some among the Silwood Circle enjoyed playing darts in the pub since it was a game that May was not so good at and at which others had a chance of winning. Another game played at Silwood was slosh. This game has many variants, usually played with a drink in hand. The version played at Silwood was introduced by May who had played the game in Australia. It requires the use of a billiards table but not necessarily any skill at billiards or snooker since the ball is propelled by hand. Players form teams and, by a system of competitive elimination, the team with the last player left standing wins. Since May had the advantage of having played the game often, and of being a non-drinker, he was usually that person.

16. C. S. (Buzz) Holling was a runner-up for the Crafoord prize in the same year and, in that capacity, was also invited to Stockholm to give some talks. Today Holling has a following among younger Swedish scientists; (see note 75 below).

17. May also has a prize named for him. The Robert M. May prize is awarded annually by the British Ecological Society to a young ecologist for the best article published in the journal, *Methods in Ecology and Evolution*.

18. Aside from Stewart, May had the support of Ronald Oxburgh, a former chief scientist at the Ministry of Defence and Rector of Imperial College who was a member of the interview panel.

19. Jeremy Bray MP (1930–2002), Labour Party spokesman on science and technology (1983–1992), suggested to Neil Kinnock that, if elected, a Labour government should introduce an office of science and technology. The idea was taken forward by the Conservative MP, William Waldegrave, who had held various science-related government portfolios. He persuaded John Major to introduce such an office. Bray had a double first in mathematics from Cambridge and was probably the only MP to have read Robert May's scientific work. For Bray, see obituary, *The Guardian*, 5 June, 2002.

20. May told me he got on well with the Cabinet Secretary, Robin Butler, who was very helpful but he found the civil servants at MAFF defensive. For Southwood and BSE see chapter 7.

21. May, personal communication.

22. UK Parliament website: Select Committee on Science and Technology, 'Memorandum submitted by the Office of Science and Technology', July 1998, sections 4.7, 4.8, 4.11. The allusion in the May quotation is to the Health Minister, John Gummer, who to 'demonstrate' the 'absolute safety' of beef was photographed in 1990 eating a hamburger together with his four-year old daughter.

23. The idea for these agencies was entertained during Stewart's period as chief scientific advisor and was developed further during May's tenure. John Krebs became director of the FSA and Stewart became director of the HPA. Proven and trusted managers, both were probably seen as safe pairs of hands. For more on Krebs and the FSA see below.

24. (May, 1997 and 1998).

25. (Williams, 1971). Shirley Williams was later appointed Secretary of State for Science (1976–1979).

26. (May, 1998; King, 2004). The idea was to correlate citation intensity with wealth creation and some other variables. May looked at the G7 countries and five others and normalized his data against GDP and population size. Interestingly his analysis suggested that countries where more research is carried out in universities, rather than in dedicated research institutes, appear to fare better in citation intensity. He speculated that this was because universities are lively places, buzzing with eager and critical postgraduate students and thus ideal places for knowledge creation. I agree that this is the case but, as to citations, suspect something rather different often goes on. In order to get on in their careers people are likely to cite the papers of those whom they see as important in that connection; or the papers they think journal editors and referees will deem important — even if the papers in question have only marginal relevance to their own work. A not always conscious self-policing goes on when it comes to citation, as Merton (see chapter 1) understood. King showed that Switzerland, Sweden, Denmark, Finland and the Netherlands with a combined population of roughly the same as that of the United Kingdom, but with a collective GDP about 6% lower, performed equally well in terms of citations. May and King also showed that emerging countries had greater rates of increase in citations than the US and UK.

27. I am grateful to Krebs for sending me a copy of his memoir 'Luck, chance and choice'. An abbreviated version can be found in (Drickamer and Dewsbury, 2010), chapter 14.

28. (Krebs, 1998).

29. Earlier reviews by Solly Zuckerman in the 1970s and by George Dunnet in the 1980s had concluded only that badgers were part of the problem. Their reviews did not include the kind of modelling used later.

30. This work has continued. More recently reactive culling has been related to an *increased* risk of bovine TB. Bovine TB is a serious health problem among British herds, affecting about one in ten. Donnelly was interviewed on the trials by *The Guardian*, 13 July, 2011 (cached on-line).

31. (Krebs, 1997; Krebs, Anderson, Clutton-Brock, Donnelly *et. al.*, 1998).

32. In his memoir (note 27), Krebs wrote that he discussed with Bob May the prospect of returning to academic work, something he had misgivings about. Both concluded 'modestly', as Krebs put it, that they were exceptional people who could return to successful research after time away. But Krebs never put that to the test. Among the letters Southwood wrote on Krebs' behalf was a very positive one to the Welsh Office with respect to an appointment with the Higher Education Funding Council (HEFC) of Wales, and another in support of his application to join the British ministerial group at the Round Table on Sustainable Development. He also wrote when Krebs was being considered for the presidency of St. John's College, Oxford and for the principalship of Brasenose College. It is not clear whether Krebs was offered any of these positions. Bodleian Library, Southwood papers S317, Southwood to J. Young (Welsh Office), 27 January, 1997; Krebs to Southwood and May, 20 January, 1998; Southwood to C. J. K. Batty (vice president, St. John's College), 21 December 1999; and Southwood to secretary of Brasenose College, 29 October, 2001. In his letter to Batty, Southwood with his usual strategic sense wrote, 'from our many conversations I know he is a strong supporter of the Oxford collegiate system and when, through the generosity of a benefactor, I gained the endowment for him to hold a stipendiary Fellowship at a college, he was most anxious that this should be at an undergraduate college where he could participate in tutorial work. He became fully involved in the various activities of the college and has maintained a high level of contact even when spending much of the time away from Oxford as CE of NERC. John and his wife (who holds a DPhil) are very sociable and hospitable persons.'

33. Bodleian Library; Southwood papers, S317, Southwood to Emma Robinson at Saxton Bampfylde Hever plc (the head hunters who had also managed the NERC appointment), 12 August, 1999. Southwood wrote 'he is particularly qualified to take on such a major public responsibility'. In his letter Southwood noted that soon after he joined the Oxford department he was struck by Krebs' sharp mind, his organizing ability, and his success in chairing committees. It was no surprise, he wrote, when Krebs was recruited by NERC. While May wanted him for the FSA job, Krebs was perhaps being coy when, in his memoir, he wrote that he was 'persuaded' to take up this new job. (It was a four-day-a-week position.)

34. One success was getting manufacturers to reduce the sodium content in many prepared foods.

35. Quotations from Krebs' memoir (see note 27).

36. According to his memoir, Krebs sees the organic food industry as having 'brilliant marketing' and having 'conned' the public that the 50% premium on organic

foods is worth paying. But perhaps more controversial at the time was his claim that the GM foods approved for human consumption by the FSA were safe. As noted in chapter 7, the public mood has since shifted and people are now more accepting of GM food research, seeing it as necessary if global hunger is to be avoided. See Krebs' chairman's report following the OECD Edinburgh conference on the scientific and health aspects of genetically modified foods; (Krebs, 2000a); also (Krebs 2000b) for comments on Gordon Conway's views on GM crops produced in the years between 1996–1999. 1996 was the year that the first genetically modified food showed up in Britain in the form of tomato paste made from GM tomatoes. In his article Krebs cited Crawley's work on GM crops.

37. The number of confirmed cases per year began falling rapidly from a peak of 28 in 2001 and fears then began to relax. By the time Krebs left the agency the total number of deaths was around 155.

38. Since then a biochemical test has been developed for the BSE prion and many thousands of sheep have been tested. There have been no positive results and since animal feed is now more strictly controlled the danger is probably over. However, in January 2002 the question remained open as can be seen from a letter to *Nature*; (Krebs, May and Stumpf, 2002). The theories referred to in this letter included those of Anderson's associate Neil Ferguson and of Anderson's former student Graham Medley. Michael Stumpf, a mathematical biologist who gained his DPhil. at Oxford, and then held a fellowship there, is a younger associate of the Silwood Circle. He has joined others in the Circle on some hiking expeditions and is now professor of theoretical systems biology at Imperial College London.

39. Croonian Lecture of 2004; (Krebs, 2005).

40. (Woods, 2004)

41. This information came out after the epidemic was over and the government ordered three separate inquiries. A further inquiry was put in place by the European Union. See, for example, the report of the inquiry by the Royal Society under Sir Brian Follett; (Follett, 2002). (The government offered to pay for the Royal Society inquiry, something unprecedented; May solicited the funding as PRS.) The inquiry focussed on the scientific aspects of the disease. Follett, vice chancellor of the University of Warwick, was a former professor of zoology at Bristol. Two members of the Royal Society committee of inquiry had close connections to May and Anderson, namely Simon Levin of Princeton University and Angela McLean at Oxford.

42. See Follett (2002). Woods makes the point that the NFU and MAFF traditionally tried to enforce discipline within the farming community by stressing 'loyalty' and 'working together' for the common good. One problem that she

points to is that historical memory at MAFF was poor. This is often the case within large organizations. Even when extensive records are kept, few people seem to know how to access them well, or even to know of their existence. Perhaps people at DEFRA (MAFF's successor ministry) will have learned some lessons from this episode. Woods also makes the point that during the 1967 outbreak the authorities did consider an emergency ring vaccination, but the disease soon died away and plans were shelved. In the 2001 case, by March the outbreak showed few signs of dying away. This could be because, as Follett later discovered, the movement of livestock before restrictions were brought in had resulted in at least 57 seedings of the disease.

43. Cristl Donnelly had recently carried out an analysis on the 1967–8 outbreak of foot and mouth disease. Her conclusion that, had a cull been carried out sooner, the scale of the outbreak would have been considerably smaller, was to influence discussion of the 2001 outbreak. She joined in the modelling of the 2001 epidemic. Donnelly is now professor of statistical epidemiology at Imperial College London.

44. (Woods, 2004), 138.

45. (Ferguson, Donnelly and Anderson, 2001a and 2001b).

46. People at MAFF believed that outbreaks were close to uncontrollable when the number of cases reaches about 150. 411 cases were known when Anderson went on TV. Over 2000 infected animals were identified late in the epidemic, but the number killed was in the millions.

47. Anderson, in testimony given to the Foot and Mouth Disease Lessons Learned Inquiry, 11 June, 2002. Iain Anderson was chairman of that Inquiry. Interestingly Iain Anderson stated that before beginning his Inquiry he was unaware that Krebs, Anderson and others had intervened, and that it had been their advice that the government had taken.

48. The epidemic could probably have ended earlier by adding vaccination to the culling but this was opposed by the NFU which wanted a quick resumption of farm exports and did not wish to see the required post-vaccination delays that would have been imposed by countries within the EU and by some countries elsewhere. See (Keeling, Woolhouse, May, Davies and Grenfell, 2003). This paper, written after the events of 2001, describes both a prophylactic and a reactive model for vaccination. The models are complicated by the range of species involved. But, roughly, the conclusions were that major epidemics could well be preventable if about 30% of animals were vaccinated. As to reactive vaccination, the model predicts it to be effective only when begun within a week of any outbreak and when there is a rapid and high up-take.

49. Quotation from Krebs' memoir (note 27).

50. (Follett, 2002).

51. This was not a cabinet position and was situated within the Department of Innovation, Universities and Skills, a department since merged with others.

52. Krebs, memoir, (see note 27).

53. Krebs had again written to Southwood asking for his help. In his letter, as later in his memoir, Krebs noted that he wanted to find an exit from the FSA and that his wife wanted to see him settled back in Oxford. Bodleian Library; Southwood papers, S317: Krebs to Southwood, 5 June, 2004. Southwood had already received a letter from P. J. Clarke, vice-principal of Jesus College, asking for his views. (2 June, 2004). Southwood responded positively, much as he had on earlier occasions when asked about Krebs' suitability as a college head.

54. (Hassell, 2000) is an excellent review of the field of host-parasitoid interactions as it stood toward the end of the century.

55. (Sugihara, Crawley and May, 2007; Crawley, 2007b). For GM plants see, for example, (Crawley, 1999).

56. His election certificate shows that he was proposed in 1999 by Hassell, seconded by May. Other signatories included T. R. E. Southwood, J. L. Harper, Roy Anderson, John Krebs and Paul Harvey.

57. (Crawley, 2005). Crawley includes a history of Silwood Park. There is also a book website where material is updated.

58. He was still collaborating with Hassell (and Comins); see (Halley, Comins, Lawton and Hassell, 1994).

59. As mentioned in the previous chapter he was coeditor with May of *Extinction Rates* (1995).

60. For his Japan Prize Lecture, (Lawton, 2004). The Japan Prize is one of four major international prizes offered by bodies in Japan. It is given for work in areas of 'pure' science. The other three are the Kyoto, Cosmos and Blue Planet prizes; the latter was won by May.

61. Lawton, personal communication. The thermohaline circulation, known as the Atlantic meridional overturning circulation (or Atlantic conveyor belt) appeared to have weakened by about 30% over the previous decade. See, for example, (Bryden, Longworth and Cunningham, 2005). The authors worked at the National Oceanographic Centre in Southampton. There was much public concern over the Gulf Stream, whether it could become drastically altered as a result of climate change, and what might be the possible consequences.

62. See 'Quango bonfire kindles advice fears', *Nature* 467 (2010) (Naturenews online).

63. Nanomaterials have long existed in the environment and some, such as fine asbestos fibres, are known to be dangerous. We now face questions about the safety of new nanoproducts.

64. He also continued to publish in this area, and to work with Colin Clark. See, for example, (Beddington, Agnew and Clark, 2007).

65. Beddington's election certificate at the Royal Society shows he was proposed by May, seconded by Lawton. Others who signed included Southwood, Anderson, Krebs, Hassell and Oxburgh. CMG stands for Commander of the Order of St. Michael and St. George.

66. Beddington, interviewed and quoted in *The Times*, 27 January, 2010. See also (Oreskes and Conway, 2010). Theirs is a good and timely book but I would be happier if the rhetorically strong word 'truth' had not appeared in its title. The evidence for global warming is excellent — just as it was in connecting smoking and secondary smoke to cancer. But scientists need to be careful in claiming to be in possession of the truth, even when opposed by those only too ready to exploit any uncertainty.

67. As reported in *The Guardian*, 18 March, 2009. The reporter must have had prior access to the speech which was delivered on 19 March.

68. As Beddington claimed, this was being done by the Obama government in the USA.

69. The Report was issued by the Government Office for Science.

70. In my view there is too little emphasis on current practice in the conservation, storage and transportation of food. While the world will need to grow more food, and do so more efficiently, we also need to face the many ways in which food is wasted.

71. Beddington first met Conway when he was a research student at Edinburgh. Later Conway brought Beddington to Imperial College. Conway told me that, together with their wives, they still get together, including spending time watching rugby matches at Twickenham.

72. Conway told me that the Rockefeller Foundation was having some difficulty finding a new president. They wanted someone with knowledge of agriculture and health issues, able to work in the developing world, and able also to encourage local arts and cultural groups. Conway clearly fitted the bill.

73. I am indebted to Conway for this information.

74. For example, Conway was co-chair of Living Cities: The National Community Development Initiative, a large consortium that provides roughly $45 million a year for development in US cities. He also chaired a pilot initiative focussing on inner-city development in Miami, Chicago, Baltimore and Minneapolis/St. Paul. He was chair of the Oversight Committee for an alliance of the International Rice Research Institute in the Philippines and the International Wheat and Maize Institute in Mexico, bodies that were exploring a possible

merger. And he was chair of an MTCT Plus (mother-to-child transmission) committee on AIDS care for Women and Infants in Africa. The goal (later achieved) was to raise $100 million for a five-year R&D programme.

75. Early in the twenty-first century Holling worked with some young scientists associated with the Beijer Institute, a charitable foundation and offshoot of the Swedish Royal Academy of Sciences that supports work on the global environment. Some Swedes at the Institute were interested in extending Holling's ideas on resilience (see chapter 6) to include also human ecosystems. They wanted the work to be interdisciplinary and non-hierarchical. To describe their methodological outlook, and to stress its non-hierarchical nature, they invented the term 'panarchy'. Holling went to New York to make a case to the board of the Rockefeller Foundation for what was called the Resilience Alliance. Some funding came its way. For more on panarchy, (Gunderson and Holling, 2002).

76. The idea for such ground rules had been put forward earlier by the economist Amartya Sen.

77. Some of the research funded by the Rockefeller Foundation in the early 1990s showed that gene transfer from transgenic rice fields to neighbouring weed-like relatives was a possibility. This work makes an interesting comparison with the work of Crawley discussed in chapter 7. Both Conway and Crawley called for case-by-case studies as new plants are bioengineered.

78. Prompted by Conway's speech, Krebs wrote an article on GM foods in Britain and how, in this area, scientists had lost the trust of the public. He wanted scientists to build trust by showing more humility in the face of the unknown, and by showing more clearly the positive, as well as the possible negative aspects, of new biotechnology; (Krebs, 2000b). The paper is a commentary on remarks made by Conway during an address to members of the House of Lords, titled 'Biotechnology and Hunger', given on 8 May, 2000. Conway outlined the situation for poor farmers in Africa and how research in biotechnology could help.

79. Monsanto has allowed free use of its 'mark-rate' technology which allows one to tell whether, and what, genetic material has moved from one plant to another.

80. Conway was proposed in 2002 and elected in 2004. The Royal Society changed its election certificates in the early twenty-first century such that now only two signatures, a proposer and seconder are required. Conway was proposed by Ghillean Prance and seconded by Southwood. I asked Conway about this. He told me that Bob May was very supportive but, as President of the Royal Society, could not sign. Sir Ghillean Prance is a botanist, ecologist and, like Conway, a world traveller. He had recently retired as Director of the Royal Botanic Gardens, Kew.

81. An amusing story Conway told of his time at the Rockefeller was of an invitation to an event at the Alex Haley Farm in Tennessee (Haley was the celebrated author of *Roots*). Conway was invited to sit on a children's literature panel together with Hillary Clinton and the US Poet Laureate, Rita Dove. They were each asked to name their childhood heroes. Conway said that his was Charles Darwin, something that did not go down altogether well with the large Tennessee audience that included several religious leaders. The panellists were also asked which fictional characters they identified with as children. Conway said he identified with Eeyore in the Winnie the Pooh books by A. A. Milne; Clinton (rightly, in my view) stated that Conway was more Tigger than Eeyore.

82. Under his successor, Chris Whitty, an epidemiologist and professor of international health at the London School of Hygiene and Tropical Medicine, the budget has risen further to about £350 million. Whitty is now director of the research and evidence division in the department. Authority in these areas was something Conway fought for.

83. Conway's knighthood is in the order of St. Michael and St. George, an order usually used for rewarding important ambassadors. In Conway's case it recognizes his work with the Rockefeller Foundation and his other work overseas. As a knight in this order Conway ranks (according to diplomatic convention) above the other knights in the Silwood Circle.

84. A copy of the report can be found on the Imperial College website. See also (Conway, 2012).

85. In today's world, scientific authority is, by and large, respected. Religious authority, where tolerant of others, should be too. Further, if religious authorities feel the need to lay claim to truth there is no longer any need to challenge them, provided they acknowledge that most religious truths rest on faith/belief, not on observation. (Routine religious practice, however, is yet something else. As opposed to beliefs and 'truths', practice can have utility not unlike that of some sciences, in that it allows people to function better in the world.) As to belief, Bertrand Russell once said God could have made the world five minutes ago, with the appearance of great age, with all our memories intact, along 'with the holes in our socks'; (Russell, 1997), 70 and *passim*. Russell was making a joke but also a point; namely that logical possibilities cannot be totally discounted, even when highly unlikely and scientifically implausible.

chapter ten

Interlude: My Philosophical Lens

In his *Physiology of Taste* (1826) J. A. Brillat-Savarin wrote, 'tell me what you eat and I will tell you who you are.' What is true of food is true also of ideas. We are exposed to many, and not all are good for us. Leopold von Ranke stated that the historian's task is to represent the past 'as it really was', but even that fine historian was unable to transcend his ideological diet. When writing history there is a need to reflect on one's own experiences and attitudes, not simply on those of one's subjects. This penultimate chapter briefly outlines some theories on the nature of science to which I have been exposed over the past forty years, and sketches my views of them.[1]

I studied chemistry at Imperial College London in the early 1960s. In that period students were expected to take non-credit classes in the arts and social sciences. The classes were introduced partly in response to the negative image of science that had developed as a result of World War II, and to bridge the gap between what C. P. Snow termed the 'two cultures'.[2] In my case the gap was bridged, but probably not in the way intended. On leaving Imperial College I spent a year carrying out post-doctoral research at Harvard University but, for family and medical reasons, was soon to give up work as a scientist. In the mid-1970s I returned to postgraduate studies at the University of Sussex, this time in the history and philosophy of science. It was an exciting period in the development of those fields. Many ideas were in the air, and it was hard to know which of them to take seriously.[3] My doctoral supervisor, Jerzy Giedymin (1925–1993), had been a student of

Kazimierz Ajdukiewicz and, before moving to England, was a professor at the University of Poznan. He was offered a chair in Warsaw but before taking it up was given leave to spend a few months with Karl Popper at the London School of Economics. While in London he asked for political asylum. The ideas that he subsequently brought to the University of Sussex were a mix of those he imbibed while in London, and those of the Polish school of philosophy founded by Kazimierz Twardowski.[4] Ajdukiewicz and Giedymin were specialists in late nineteenth and early twentieth-century French philosophy of science. Both believed that the influence of Henri Poincaré, Pierre Duhem, Edouard LeRoy and Emile Meyerson, among others, on the subsequent development of philosophy, sociology, and history of science, had been underestimated. At the London School of Economics a small nod was made in the direction of those philosophers, but the main focus of discussion was on the more recent anti-positivism of Popper and his contemporaries. Notable among them were Thomas Kuhn, Michael Polanyi, Stephen Toulmin, Imre Lakatos and Paul Feyerabend. Along with the French thinkers, all were on my reading list.

There were several people working in the broad area of science studies at Sussex and I learned much also from others, including from the historian of science Roy McLeod, and the physicist turned science critic, Brian Easlea. Easlea's 1973 book, *Liberation and the Aims of Science: An essay on the obstacles to the building of a beautiful world*, was a major talking point at the time.[5] As mentioned elsewhere, I learned a few things also from the theoretical biologist, John Maynard Smith. I was not much aware of the work being carried out at the science studies unit at the University of Edinburgh. Later I was to learn a little about the 'strong programme' in the sociology of scientific knowledge, but never engaged with it. I should add that Giedymin asked his students to read Nietzsche, a major augur of twentieth-century ideas. He also suggested that we read Ludwick Fleck's 1935 book, *Genesis and development of a scientific fact*. As can be seen from earlier chapters, it influenced how I look at science.

One of the oldest problems in philosophy is when does something become knowledge — at the point of discovery or at the point of acknowledgement. Neither the positivists nor their critics were interested in the discovery or genesis of ideas — that could be left to historians and neuroscientists. For the logical positivists, as for Descartes earlier, the problem was how ideas come to be justified, and whether or not they are true. In trying to formally represent the connection between observation and theory, they argued both that scientific observations reflect reality, and that they can verify or falsify theoretical ideas. Popper did not accept the idea of confirmation or verification, but he did believe that theories could be falsified and that by thinking of how this happens we better understand a theory's limits. Falsification, he argued, leads indirectly to an increase in scientific knowledge.

Under Giedymin's guidance I read much late nineteenth-century French philosophy of science and came to share Duhem's view that from a strictly logical, though not commonsense, point of view, scientific theories can be neither verified nor falsified. Duhem reached this conclusion after considering the exchange of views between Galileo and Cardinal Bellarmine. Galileo agreed with Bellarmine that just because a theory can save the appearances does not mean that it is true. But Galileo also believed that the evidence against Aristotelian and Ptolemaic cosmological ideas was conclusive. Duhem did not believe in conclusive evidence, whether confirming or falsifying. As he put it, 'so many illusions have passed for certainties'.[6] He claimed that there was no such thing as a crucial experiment, one that could decide the truth between two competing theories, because there was no way of isolating what was being tested from other beliefs being held at the same time. He also claimed that for any set of observations there were many consistent hypotheses. To put it another way, all theories were underdetermined by the evidence. Duhem's was a type of methodological holism picked up later by the philosopher Willard van Orman Quine. Quine, like Duhem, believed that knowledge is possible, but that there are no universal criteria for what counts as knowledge. As he put it, 'epistemology, or something like it, simply falls into place as a chapter of psychology and hence of natural science'.[7] Perhaps he was correct in thinking that epistemology should be approached in a scientific way. For Duhem, scientific claims were established on the basis of both consistency with data and utility. But he understood that knowledge does not count as such until communally acknowledged.

Henri Poincaré held similar views but based them on more recent developments in mathematical physics.[8] He held that the axioms of geometry and the laws of motion were conventions and that newer theories, in many respects observationally indistinguishable from older ones, were grounded in different conventions.[9] He claimed that in the seventeenth-century people conceived of physical space in a manner that made new ways of thinking in spatial terms possible, and that the way this was done was contingent on a combination of human genetic inheritance, historical circumstance, and the actual physical environment. Euclidean space was not, as Kant had claimed, an *a priori* form of sensibility.[10] Popper was critical of both Duhem and Poincaré, unfairly in my view. He wrongly claimed that they both saw theories merely as instruments for prediction, and that they were nominalists. Popper was especially dismissive of instrumentalism. Like Polanyi, he was anxious to maintain the distinction between pure and applied science, something he saw instrumentalism as collapsing.[11] He also claimed that instrumentalism had gained ground because it allowed people not to have to face inconvenient theories, and that it allowed those who were religious, people like Bellarmine and Duhem, to argue that essential truth comes only through revelation.[12] But Duhem, like Poincaré, rejected nominalism; neither of them claimed that scientific theories have no descriptive content.[13] They did, however, believe that there was no one-on-one relationship between theory and observation.

Poincaré's position can be seen more clearly in some exchanges he had with his student Edouard LeRoy. LeRoy, also a student of Henri Bergson, combined elements of Bergsonian irrationalism, Spencerian (rather than Darwinian) ideas on evolution, with extreme conventionalism.[14] Since I identify with Poincaré's views, but not with those of LeRoy, it is worth outlining some aspects of their exchange. LeRoy claimed that all observations are theoretical constructs, and that there is no such thing as bare observation. Like others later, he believed that not only ideas in the mind, but also material culture, play a role in how things are observed. Poincaré agreed with LeRoy (and Duhem) that ideas already in the mind can play a role in how things are observed, but held that the contingencies of the external world play the more important role.[15] According to LeRoy, when conventions change nothing remains invariant. Poincaré disagreed. LeRoy was an instrumentalist, claiming that while science can be viewed as a guide to action, it is not a route to truth about the natural world. Earlier, positivists had placed much emphasis on action in their theories but, for them, science was also a route to truth. Poincaré, too, believed that science can lead to genuine knowledge, and that were it solely conventional it could never serve as a basis for action. For him, however, the main argument against the earlier positivism was that it restricted theorising and refused recognition to ideas that he thought within the realm of science. He also refused to brand as nonsense plausible metaphysical speculation. For example, with respect to the kind of mathematical modelling that was to be carried out later by Robert May, he argued that if theoretical models have predictive value then they, too, are facts to be explained. He also had much to say on the pragmatic front, on when it was rational to assert a theory, when to suspend judgement, and when merely to entertain an idea. Much of what he had to say still makes sense to me.

LeRoy, on the other hand, wanted a completely new theory of scientific knowledge. He argued that any visit to a laboratory shows that scientists, themselves, have the impression of constituting facts from the amorphous materials that make up the objects they are observing. Reality is never directly accessible to them. Like Bruno Latour later, LeRoy argued that reality can be accessed only through the mediation of material objects, existing concepts, schemes contingent on the earlier experience of the observer as an individual and, as he put it, 'as a member of a race'. Further, he argued, there were no isolated objects in nature; 'everything is diffused in everything'. The process of separating, of objectifying, is simply a convenient way of thinking, and is an expression of human perceptual weakness.[16] According to LeRoy, the observations and classifications of science are relative to some viewpoint chosen in advance, as well as to the material conditions under which the observations are made. The scientist, he stated, needs laws and finds them 'with the help of ingenious violence to which he subjects nature'. In doing so the scientist 'creates the fact'.[17] He believed that the spiritual life was more powerful than the rational one, and that it can lead to profound (as opposed to commonplace) action. Poincaré did not see facts as created, nor did he share LeRoy's views on material dependence. Further, he was not prepared to cede superiority to 'religious experience'. When it came to understanding the world, he was faithful to older ideas of scientific rationality and objectivity. However he did acknowledge that it was not always possible to determine where the empirical content of a scientific theory ends and its conventional content begins.

As it happened the early French antipositivists were soon displaced from centre stage by logical positivism which came to the fore in the 1920s. As a result many of the earlier ideas were forgotten. Historical memory is poor but the French ideas continued to be discussed in a few places — as they were in Poland. Imre Lakatos, who left Hungary during the uprising of 1956, was another much influenced by Duhem's ideas. Like Giedymin he, too, brought the French ideas to London. Popper, as mentioned, was a critic of both Poincaré and Duhem. But he read Emile Meyerson with interest and suggested to Kuhn that he, too, read Meyerson's work.[18]

Poincaré's conventionalism did not mean denying the common-sense view that scientific observation and reasoning is the best way of understanding the natural world. For him science was a cumulative enterprise leading to genuine knowledge — useful, and with some truth content. He believed that science has direction, and that over time theories cover increasingly more data in increasingly satisfying ways. I, too, see more recent science as epistemically privileged, while acknowledging that much of it could be false. The view that science is getting better was unpopular with many in the science studies world of the 1970s. However, allowing that science builds on its past does not imply that we were bound to have arrived at the ideas we hold today. In this, Duhem was correct. Nor does it mean that our lives are necessarily better as a result of progress in science. Living with science means living with intended and unintended consequences, some good, some bad.

Scientific theories need to be consistent with observation but socio-cultural-political criteria also play a role in their acceptance. Further, we do not need to share the ideas of others in order to understand why they believe what they do. We can appreciate for example why Kepler, having arrived at the idea of elliptical planetary orbits, imagined that something like a magnetic force emanating from the sun was needed to keep the planets moving in those orbits. And, despite their having very different views on the nature of the heavens, we can appreciate why Kepler, following the twelfth-century philosopher Moses Maimonides, believed that it was God's love that makes the world go round. Similarly we can understand why phlogiston theory once made sense to people trying to understand combustion and other chemical reactions. Today we view God as extraneous to science, and ideas such as phlogiston as similarly extraneous, but not necessarily irrational.[19] Presumably future generations will make similar judgements about what is extraneous in the science of today.

The debates in France at the end of the nineteenth and beginning of the twentieth century were in part stimulated by the collapse of Newtonianism and the rise of relativistic physics. If even the best of scientific theories can be shown to be false, what should we believe? However, neither the scientific nor the historical enterprise would get very far if logical and methodological problems, were to be taken overly seriously by working scientists and historians. Poincaré was a brilliant exception, able to turn his mind profitably from mathematics and mathematical physics to the philosophy of science. Duhem, a gifted theoretical physicist, was a major contributor to the history of science. More generally, working scientists do not spend much time thinking about historical or philosophical matters. Kuhn recognized this division of labour as, in his way, did Isaac Newton earlier. In his fourth rule of reasoning, Newton stated,

> In experimental philosophy we are to look upon propositions inferred by general induction from phenomena as accurately or very nearly true, not withstanding any contrary hypotheses that may be imagined, till such time as other phenomena occur, by which they may either be made more accurate, or liable to exception.
>
> This rule we must follow, that the arguments of induction may not be evaded by hypotheses.[20]

This conservative position makes practical sense. Although Duhem was right to argue that we can always come up with alternative hypotheses, it is a waste of time trying to do so unless forced by some glaring inconsistency, or inspired by a brilliant new idea. Kuhn, in stating that scientists cannot think about all contingencies in deciding what to believe, was making a similar point. But it led him to introduce what, in my view, were the misguided ideas of normal and revolutionary science.[21] Why these ideas caught on in the way they did is a bit of a mystery. There are few, if any, major revolutions in science, and scientists cannot afford to remain in testing mode while waiting for anomalies to build up, and for the next big idea to come along. Science is continually in flux and it is also cumulative — increasingly so. That is why it is so hard to keep up with it. The ecologists who are the subjects of this book did not waste their time imagining alternatives to the Darwinian theory of evolution. Rather, they looked to the consequences of taking the theory seriously. Population theory was already well established but sufficiently nebulous to allow lively debates on interpretation. It continues to be suggestive.

All kinds of theoretical offshoots have been, and still are, entertained.[22] To call this activity 'normal' in the Kuhnian sense is to undervalue the innovation, the debates, and the human exchanges that go on.[23] It ignores also the changing cultural and political contexts in which scientists operate. Robert May gave ecology a shake-up and his ideas were much debated. I would not choose 'normal' or 'revolutionary' to describe his innovative contributions. The first undervalues and the second misrepresents his achievements.

The general agreement that something becomes knowledge only at the point when it is accepted by others opened the door to sociologists of knowledge. Here, too, late-nineteenth century French thinkers, incorporating a few ideas from John Dewey and William James, had something to say, though none ventured far along the sociological path.[24] Dewey was famously opposed to the idea of 'justified true belief', and to any philosophy incorporating that positivist idea. Poincaré, Duhem and LeRoy agreed with him. As to James, he believed knowledge converged toward truth over time. He referred to interim knowledge as being 'half-true', something determined in part by reality and in part by existing ideas about what was being observed. Like Poincaré, he argued that, over time, knowledge converges toward truth by a process dependent on historical, psychological and experiential factors. LeRoy, although never moving far toward a sociology of scientific knowledge, anticipated some of the constructionist views of the later twentieth century. His message did not come through directly. But the distinction between the internal and external conditions for scientific activity, made early in the twentieth century, was taken up later by both Robert Merton and Thomas Kuhn.

Since I was trained first as a scientist it is perhaps not surprising that I chose to adopt an empirical approach in writing my DPhil thesis. With an intellectual inheritance derived heavily from the Enlightenment, and primed by exposure to some more recent views, I looked at a few ideas in the philosophy of science. Along with early twentieth-century French conventionalism and instrumentalism, were more recent positivist ideas, Polish ideas in the philosophy of science, mid twentieth-century methodological views of some antipositivists, and the ideas of some contemporary sociologists of science.[25] My purpose was to see whether any of the methodological ideas I knew of could describe, even roughly, the work of my scientist subjects, or could be used in the rational reconstruction of their ideas.[26] This kind of work, bringing together the history and philosophy of science, had been suggested by Lakatos as worthwhile. Roughly speaking, however, I came to the conclusion that we cannot easily objectify the aims of science. None of the methodological ideas examined were sufficient in either a descriptive or prescriptive sense. They had little to teach scientists about how to proceed. Something that surprised me was how small a role the predictive power of theories played among the scientists I studied. This made me question not only the Kuhnian model, but also some of Popper's views. However, things have changed since 1978 when I submitted my thesis. For example, quantum chemistry was, for many years, used more for making sense of what was already known than for predicting chemical behaviour. With new computing power much has changed in this regard, though the predictions are still primitive. The same is true of the population and epidemiological models of concern to the Silwood Circle. New computing technology has allowed the outcomes of complex theories to be spelled out, and for prediction to play a bigger role in informing research paths than it once did. Lakatos also stated that to understand how scientific problems are addressed it was necessary to know more about the ways in which research programmes functioned. A stream of studies of research schools followed this suggestion. It was not something that I took up. I was, however, influenced by his emphasis on heuristic factors as drivers of science, an influence that can be seen in the earlier chapters of this book.[27]

Overall, however, working on my thesis taught me that I was following a methodological dead end, and that not much could be learned from looking at history in such a programmatic way. Although I learned something about late nineteenth and early twentieth chemistry, little had been gained from putting Popper, Lakatos, Feyerabend and Kuhn to the empirical test. But how should one study science in order to understand it better? Was a better way forward to conduct ethnographic studies of contemporary science, in the manner of some people working in the social studies of science? Such studies have taught us something about the nature of field and laboratory work, and about the roles of different workers, but not much about science as a humanistic enterprise — something I wanted to get at.

Another interest of Giedymin's was hermeneutics. I did not pay much attention to this when working with him and later wished I had done so. The Polish school took seriously the hermeneutic ideas of Wilhelm Dilthey, ideas associated with nineteenth-century German Romanticism. Ajdukiewicz, for example, saw the knowing subject as historically embedded, not as some timeless Cartesian being. The situating of people, ideas and texts within defined traditions of workers, writers and readers became a major field of study later in the century. In telling the story of the Silwood Circle, I have situated my subjects, albeit with a light touch. We have seen something of the socio-political context in which they worked, and of how members of the circle discussed and read much in common. We have seen also how their various papers and books display common themes, themes of interest to many of their scientific contemporaries.

On leaving Sussex, I joined the history department at Simon Fraser University.[28] There I was exposed to yet more ideas, many of them also from France. Having been introduced to an early French school, I continued reading work by French intellectuals. It was impossible to be a historian in the second half of the twentieth century without hearing about deconstruction, constructionism, structuralism, post-structuralism, and a number of other 'isms. The older left-wing structuralist realism was by then under attack, and the ideas of Michel Foucault were becoming fashionable.[29] In the history of science, Bruno Latour's work was causing some ripples that would soon become waves. Latour used Foucault's concept of the episteme in his analyses of scientific work, something I, too, found useful.[30] Indeed, I am indebted to both for having enlarged the conceptual repertoire of historians, even though not always agreeing with their ideas. It was no accident that Latour's provocative book *Nous n'avons jamais été modernes* appeared in 1991.[31] Although the late 1980s did not mark the end of history, for Latour they marked the end of European communism (and, after Tiananmen, possibly the beginning of the end for Chinese communism). He also claimed it was the end of capitalism as we know it. Like others, some of whom labelled themselves postmodern rather than nonmodern, Latour connected this outlook to the abandonment/disregard of the Enlightenment view that science gives us objective knowledge of the world.[32] More interesting, however, was his way of looking at the scientific enterprise. To my mind his model is far-fetched, but one of its underlying messages is sound. The same message was delivered earlier by Darwin, and voiced also by members of the Silwood Circle. *We are all part of nature and we had better not forget it.* Latour was not alone in recognizing that there is resentment toward the West for having used more than its share of the earth's resources, and for having led the world in a scientific and technological direction. There is now some resentment toward China for similar reasons. Latour's originality, however, lay in how he visualized the technological world and our relationship to it. He pictured the world in cyborg-like terms, a world teeming with hybrid subjects and objects.[33] In such a world scientific practice was said to entail people with specialist knowledge working alongside new kinds of computational tools, complex apparatus and materials; and their carrying out experiments and making observations under artificially contrived laboratory conditions. This was not unlike LeRoy's earlier view of science, one that supposedly makes it difficult to determine where any one subject or object ends and the next begins. It is not clear whether, like LeRoy, Latour saw objectification as a sign of human perceptual weakness. He clearly saw too much jumping to conclusions, too much premature and enthusiastic reification of ideas among scientists.

It is true that much modern science is driven by complicated pieces of apparatus that in part determine what can and cannot be known. But such complexity need not undermine the knowledge that is produced. Latour opposes his hybrid subject-object mediators to the modernist idea of a knowing subject. His mediators are involved in the construction of knowledge — including self-knowledge. Since he believes that this has always been the case, he sees knowledge as having a status that he terms a-modern. Latour is making an ethical-political point. But his weakness is in conceptualizing how, practically speaking, we will act differently simply because we have become more anthropological and less metaphysical in our thinking about 'subjects', 'objects' and 'hybrids'. For Latour, scientists exist on the same level as things, the tools and apparatus being used, and the material objects being studied in their laboratories.[34] This is surely a mistake, and a dehumanising one. After all it is the humans, not the objects, who do the representing, and knowledge is expressed in a transcendental human language.[35] Further, as Poincaré argued, it is because scientific representation captures some of reality that it is also useful. It appeals to our common sense, to our tacit knowledge, and thus ensures that our beliefs are not entirely arbitrary. We share reality with each other, although the many ways we talk about it are not always clear. For that reason we need leaders, people like Richard Southwood and Robert May. They feature in my history as identifiable icons of representation — reliable sources of a certain type of knowledge.[36] They, like other leading figures in science, are not knowledge hegemons, but rather discoverers of our common reality. Nonetheless, historians should recognize that each area of science has its own coherence, that scientists work under contingent circumstances, that everything is socio-anthropological, and that the problem in describing scientific activity is to find the correct anthropology — something I have attempted in this book.

Confusion between subject and object is something Latour ascribes also to pre-modern times which is why he states we have never been modern. Just like our ancestors, he claims, we live in a world of interpenetrating networks. Like them we struggle to make sense of our lives from within a continuum of ideas and objects. There is something to this, and to the idea that today's techno-industrial world consists of interpenetrating networks. But we should not deny that the human mind is able to make sense of the world — at least to a degree. Nor should we deny the revolutionary impact of Enlightenment philosophy. Science gives us genuine and cumulative knowledge. If we make the effort we can even look objectively at ourselves. However, there is a catch. Poincaré was correct in stating that theories have both empirical and conventional content, and that it is not always possible to know, except possibly in retrospect, where the empirical ends and the conventional begins.[37] Sociologists of science have shown what is behind some of the idealizations used by scientists in presenting their results. But it does not follow that social construction — unavoidable in what is necessarily a social activity — disallows the acquisition of genuine scientific knowledge.

Two of Foucault's ideas (basically, modern formulations of Enlightenment ideas) seem useful: that we live in a world where old hierarchies no longer make sense, and that we are all agents and mediators in the making of history.[38] Nonetheless we still need leadership of the kind I have attributed to Southwood and May. When such people are well connected their ideas can travel fast. Science has captured much of the world and, in that sense, is an imperialist activity. And, since the Silwood Circle has had considerable influence, it can be seen as a minor imperial power. Some of our intellectual ancestors were similarly imperial, important agents in the making of the modern world. That they happened to have been of largely European descent was a historical accident that has led to some modern problems.[39] But we cannot turn the clock back. By now much of humanity appears to have adopted their ideas, albeit changing and adding to them along the way. How the future will turn out cannot be predicted. Today's agents and mediators need to make sensible choices so that collectively, even if not consciously acting together, we maintain a livable world.

Finally, I should note that during the 1960s and 70s I was caught up in the feminist movement and thought much about women and science, about why there were so few women scientists, and about how more could be encouraged to enter the field.[40] A hot topic of the period was whether fundamental differences exist in the way men and women view the world, and whether such differences are biological or cultural in origin. Would the nature of science change with more women practitioners? Would it have made a difference to primate science, for example, had Louis Leakey selected three men rather than Jane Goodall, Dian Fossey and Biruté Galdikas for research on wild chimpanzees, gorillas and orangutans? We still have much to learn about science and gender, and whether there is any basic incommensurability in how men and women see the world. Looking around at science today there seems little evidence to support the idea. Differences, if they exist, are subtle. This book is about an all-male fraternity, the formation of which conformed with the gender conventions of its day. Were such a group to form today it would probably include women. Thinking about women and science, however, helped me to recognize some political aspects of science. What gets funded, who gets appointed, who gets heard, and who gets rewarded are all, though not exclusively, socio-political matters flowing from the conventional power structures of their day. The distribution of responsibility for the consequences, good and bad, of scientific achievement is similarly political.

One idea that had much traction within the academy during the 1980s and 90s was that of situated knowledge.[41] Among feminists a major contributor to the discussion was Donna Haraway, a former student of G. E. Hutchinson.[42] I cannot engage with the idea here, but I accept that knowledge is situated. As mentioned above, local knowledge can travel, sometimes slowly and sometimes very quickly. Unless it does so, it is lost to the larger world. There is little doubt that much of value has been lost. Pushing one's ideas beyond the local is difficult. In a working democracy it should, in principle, be equally easy for all to be heard. But, just as with other biological populations, the human one has its struggles. Hierarchies abound and, in the human case, evolve historically — something that has brought countless gifts as well as resentment. Not everyone will be heard. Being heard, however, is not enough for one's ideas to count as scientific. For that they must also meet some widely accepted consistency criteria.

So, what role did the ideas described above play in writing this book? Their role has been largely suggestive, helping me make sense of the voluminous data that had to be considered. Historians need to be aware of theoretical possibilities in order to frame their ideas. It is not enough simply to record the events of the past, one must attempt to explain and give them meaning. However, when writing I use theory only lightly, not as anything approaching dogma. While suggestive, theories are never totally explanatory and, when wielded as weapons, are destructive of our common humanity.

Endnotes

1. For an account of methodology in the history of science, including some major twentieth-century developments, see Rob Iliffe, 'History of Science', http://www.history.ac.uk/makinghistory/resources/articles/history_of_science.html. This is on the website of the School of Advanced Study, University of London.

2. The instructors were excellent and we had a wide choice of courses. Among others, I attended classes in the history of art given by Ernst Gombrich and on music by Anthony Hopkins. C. P. Snow, *The Two Cultures and the Scientific Revolution* (Rede Lecture, Cambridge, 1959).

3. This was the heyday of critical theory or *critique*. It was applied to all kinds of strongly held beliefs in an attempt to 'deconstruct' them, to show how they function in society and can serve special interests.

4. (Lapointe, Woleński, Marion and Miskiewicz, 2009).

5. And, in a similar vein, so was (Ravetz, 1971). In 1970 MacLeod joined with David Edge, of the science studies unit in Edinburgh, to found the journal *Science Studies* later renamed *Social Studies of Science*.

6. (Duhem, 1969), 113.

7. (Quine, 1969).

8. Poincaré left three collections of essays in which he outlined his philosophy: (Poincaré, 1902, 1905 and 1908). All three books were translated and appeared in English editions published by Dover in 1953, 1958, and 1956 respectively.

9. The term 'conventionalism' refers to two different approaches, both anti-inductivist. For Duhem, conventions were associated with research procedures and

their products. For Poincaré it was scientific concepts that were conventional. For example, Newtonian ideas of space and time were, he believed, conventional and metaphorical. He also believed that Euclidean geometry was based on conventions deeply anchored in human biology and that it would be difficult, if not impossible, to replace with the new geometries then being discussed. Duhem was an instrumentalist, but not a radical one since he believed that over time science gets closer to the truth. However, absolute truth was for him largely transcendental and the domain of theology not science. I am not religious but think that since science is now so well established it could perhaps afford to forego fruitless battles with those who believe in other routes to truth — in the literal truth of holy texts, intelligent design, and so on. The war over authority in matters having to do with the natural world has surely been won.

10. Poincaré did accept Kant's view that the axioms of arithmetic were synthetic *a priori* ideas.

11. As noted in chapter 5, in 1940 Polanyi joined with John Baker and Arthur Tansley to found the Society for Freedom in Science. Opposed to those, notably J. D. Bernal, who stressed the social function of science, they wanted to defend the idea of 'pure' science, namely science as a strictly intellectual pursuit.

12. (Popper, 1959), 769–779 and (Popper, 1963), 74, 99, 104.

13. (Giedymin, 1982). In my view this is the best account in English of the conventionalist ideas of Poincaré and of some of his French contemporaries.

14. LeRoy published often in *Revues de Métaphysique et des Morales*. See especially (LeRoy, 1899), part I, 375–425 and part II, 503–562 and (LeRoy, 1901), 138–153. See also (Poincaré, 1958), 122–124. LeRoy, a relativist, claimed we were free to see the world from different perspectives, all sound in their respective ways.

15. At roughly the same time the Swiss linguist, Ferdinand de Saussure, suggested that it was language that structured our sense and understanding of reality. His influence has been immense in many areas of scholarship.

16. (LeRoy, 1899), part II, 515 and (LeRoy, 1901), 145. Similar ideas are evident in much recent French theory.

17. (LeRoy, 1899), part II, 523 and (LeRoy, 1901), 143. Later Latour, in a similar vein, stated that much work was 'required in order to establish the persistent, stubborn, data'. He writes about ideology, laboratories, instruments, polemics etc. — all involved in the hard work of creating facts; (Latour, 2004), 95. In this respect both LeRoy and Latour are writing in a Baconian style, though not necessarily sharing Bacon's views on experiment as a route to truth. Francis Bacon, in advocating experiment, wrote of 'compelling' nature to give up her secrets.

18. For Kuhn and Popper, (Fuller, 2000), esp. 392. Meyerson was born in Poland before moving to Paris in his twenties. A former student of Robert Bunsen, Meyerson was a chemist as well as philosopher of science. He corresponded with some of the major scientists of his day, including Einstein and Louis de Broglie, on the nature of science. *Avant l'heure*, he was a proponent of the study of scientific practice as a route to understanding science, and he anticipated later interest in tacit knowledge. His ideas were acknowledged by Popper and Polanyi, and picked up later by philosophers, historians and sociologists of science, including Kuhn. For some later work in a similar vein see, for example, (Hacking, 1983). Hacking made a distinction between how we see and represent the world, further encouraging work on scientific practice. Peter Galison has looked at some modern physics with its distinct forms of practice, theoretical and practical, and how different practitioners trade ideas and practices with each other; (Galison, 1987 and 1997).

19. This point was made well in (Chang, 2009). In a further paper (Chang, 2010), Chang claimed that Alan Musgrave, inspired by Lakatos' theory of scientific method, has given what is perhaps the best account for what happened in the switch from the phlogiston to the oxygen theory — that the transition was slow, that phlogiston theory generated much good chemistry, and that it failed only when it ceased being progressive. The oxygen theory took over because it was highly suggestive of further work (Musgrave, 1976). This is not unlike what I have taken from Lakatos, namely that it is heuristic power that makes theories attractive to scientists. However, heuristic power is not enough for a theory to become well established over the longer term. For that it has to show consistency with much observation in areas deemed important, and have predictive power.

20. Newton added the fourth rule of reasoning to his list of hypotheses in the 3rd edition of *Principia*.

21. (Kuhn, 1970).

22. Feyerabend, an epistemological realist, argued for the proliferation of theories since he believed that it was easier to find fault with a theory if others were around to compare it with. Despite giving his book the title, *Against Method*, his was only a minor difference with Popper. It was not necessary, he argued, to falsify all theories put forward in order to progress. Indeed, the more theories put forward the merrier, and the more likely we are to know what not to believe. (Feyerabend, 1975).

23. As to the paradigms that Kuhn saw as guiding normal science, I have little to say. I refer back to chapter one and my discussion of Fleck's 'superindividual idea', less definitive than 'paradigm' and, in my view, more useful since he applied it to relatively small groups of people who knew each other. Kuhn's

importance was in giving license, and some guidance, to those wishing to take a sociological approach to scientific practice and to the acquisition of scientific knowledge.

24. One of Giedymin's suggestions was that I read issues of the *Revue de Metaphysique et des Morales* from the late 19th and early 20th centuries. At the time there was much discussion in the journal of the ideas of James and Dewey, alongside those of Poincaré, Duhem and LeRoy. LeRoy published many articles in the journal (see, for example, citations above). Others who joined the debates, continuing into the twentieth century, include Henri Bergson, Emile Meyerson, Abel Rey, Emile Boutroux and Pierre Boutroux. For James see (Putnam, 1997).

25. In addition to Merton (see chapter 1), I read some other sociological works of the 1960s and 70s. For example, (Barber, 1962; Crane, 1972; Price, 1963; Ziman, 1968).

26. See, for example, (Gay, 1976). A later version of this essay formed a chapter of my doctoral thesis.

27. For Lakatos on scientific method, (Lakatos 1970a and 1970b). The latter paper appeared in a book that was the outcome of a 1965 colloquium organized by the British Society for the Philosophy of Science in which the views of Kuhn and, to a lesser degree those of Popper, were discussed. Lakatos, dismissing the Kuhnian idea of revolution and wanting something more nuanced than Popper's falsification to describe the way science advances, argued that theories change incrementally in light of new evidence (whether confirming or falsifying) and that they should be viewed as 'progressive' provided they are suggestive of further observations to make and of new ideas to test, etc. However, reasonably, he did not think theories were simply heuristic and without any truth content.

28. I ended my career at the Centre for the History of Science, Technology and Medicine, Imperial College London.

29. This is understandable. For many historians structuralism was too simplistic an approach. For example, in the case of the Silwood Circle there is no simple answer to the question of how or why it succeeded. Using structural ideas such as class, gender, economic situation, religion or anti-religion does not get us very far. The only structures that makes sense in this context are biological ones such as innate ability, but even they are insufficient to explain success. In addition to ability, the success of the circle's members depended on a set of complex social interactions. In that sense it can be seen as socially constructed. Foucault appealed to many because he wanted a break from the past, was seeking ways to escape the ideological mistakes that led to WWII and its atrocities, ways to understand the modern world, and was seeking also a new vision of humankind. But he could just as readily be seen as continuing a tradition of seeing history

as a conspiracy of knowledge and power-concepts — a more abstract form of 'structuralism'. For Foucault and some other postwar French philosophers, (Gutting 2011).

30. Episteme (Gr. knowledge) was used by Foucault to describe discursive systems of knowledge and thought. He associated different epistemes with different historical periods and claimed that they defined the conceptual possibilities of the period — what was possible and not possible to think. Foucault claimed that the rules governing epistemes operate below the level of consciousness; people are largely unaware of how their views and attitudes are culturally determined. The episteme was an idea I found useful, but only when used lightly.

31. It was soon translated into English: (Latour, 1993).

32. This relates to (Shapin and Schaffer, 1985). This book features Boyle's experiments with the air pump, and Hobbes' account of political authority. The authors consider the rise of experimental science, and its role in challenging existing authority. Who or what decides what counts as knowledge in the modern world?

33. Bruno Latour, Michel Callon and their colleagues at the Centre for the Sociology of Innovation at L'École Nationale Supérieure des Mines in Paris introduced something they called actor-network theory that incorporates this type of view. The actor-network is something out of which both individual identity and social organization is supposedly constructed — an overweening idea.

34. See also chapter one, note 26, for the similar views of Hans-Jörg Rheinberger.

35. The question of other animals as knowing subjects is set aside here.

36. By reliable, I am not implying that such knowledge is true in any absolute sense.

37. Latour put forward some methodological rules; the third begins 'the settlement of a controversy is the *cause* of Nature's representation, not its consequence ... '. In other words, scientists do not study nature, they *decide* what nature is; (Latour, 1987), appendix. My view is that they do study nature and that nature in combination with other, conventionally held, beliefs gives us what counts as science. Although scientists cannot detach themselves totally from what goes on in the rest of their lives, they can learn something about the natural world. Sometimes, alas, they do not fully detach themselves from the interests of their funders. Scientific views on the nature of women and their aptitudes have changed considerably over the past 100 years, a history that exemplifies well the socially driven change in conventional content.

38. This relates to the idea of reflexivity, an idea with a complicated twentieth-century history. Roughly speaking, being aware of a situation and adjusting to

it can mean that the original situation becomes altered — hence the idea of human agency in history. Both Robert Merton and Karl Popper recognized this problem as it applies to scientific behaviour. Since then reflexivity has been heavily theorized by sociologists, notable among them Pierre Bourdieu.

39. These historical agents held both imperialist and anti-imperialist ideas, a paradox constitutive of Western civilization more generally. Based on this premise, an argument could be made that the success of the West was not a simple historical accident.

40. I should add that I remain a feminist. I had many discussions on women and science with my friend Margaret Lowe (Maggie) Benston (1937–1991). She was a leading theorist in the field; (Franklin, Gay and Miles, 1993).

41. Much emphasis was on the knowledge of those seen as subjugated: the poor, the marginalized, slaves, women, aboriginal peoples etc. As Aeschylus put it, 'from suffering, knowledge'. It is an idea later picked up by Marx, and then by feminists. But there is a geographical as well as social dimension to situation. In politics, as in poetry, all should be heard since anyone can have an anthropologically significant insight. In science, however, a specialized education and observational experience count for something.

42. See, for example, (Haraway, 1988).

chapter eleven

Conclusion

We have seen how the Silwood Circle formed and something of its history. In following its activities this book has also traced some major developments in the field of ecology in the second half of the twentieth century. Except in passing, that history, and the many scientific contributions of members of the circle, will not be recapitulated here. Rather, the chapter will suggest some answers to questions raised in the introduction. The first five sections of the chapter summarize conclusions reached in previous chapters, and do so from different perspectives. The sixth section contains some brief concluding comments.

11.1 Intellectual history: tradition and novelty

The story told by John Lawton, that his students of the early 1990s had never heard of Robert MacArthur's work on warblers, is instructive.[1] Each generation has to learn things anew and can do so only incompletely. Although huge bodies of observational data, as well as theoretical ideas, accumulate around the world, in any one place and at any one time little is actively remembered. Ideas have to be very compelling to spread widely, and to survive for more than a generation; and they need effective champions to have a life other than as records in libraries and data bases. Among the few secure ideas in science today are the theory of evolution, genetic inheritance, the laws of thermodynamics and electrodynamics, and the existence of atoms and molecules. But even seemingly secure ideas can vanish almost overnight.[2]

MacArthur's warblers may not be as well remembered as Darwin's finches, but his more general ideas on island biogeography have survived in attenuated form. MacArthur had his champions and his ideas were widely seen as plausible and suggestive.[3] But how did they come about, and in what way did they affect the work of the Silwood Circle?

The discovery of the structure of DNA in the 1950s had many implications. For ecology, then a marginal scientific activity, there was the danger of being further sidelined in the rush toward molecular biology. Some biologists, including G. E. Hutchinson, D. L. Lack and E. O. Wilson, believed that establishing a theoretical foundation for ecology was key to its survival in the academic world. They actively encouraged people with mathematical skills to help in this endeavour. As a result new mathematical expression was given to old ideas such as a dynamical nature, biological community, food chains, and competition and cooperation among and within species. Much of this work was intended to gain a better understanding of population patterns. It helped also in raising ecology's status within the sciences.

Important to the new mathematical modelling was the Popperian mood of the 1950s and 60s. Hutchinson was affected by it as, a little later, were members of the Silwood Circle.[4] Already in the 1950s Hutchinson encouraged his students to adopt a hypothetico-deductive methodology, not only to arrive at testable predictions but to show its efficacy in accounting for what was already known.[5] It was he who suggested to MacArthur that he find some mathematical models for the distribution of species in bird populations for which data was already available. The approach proved fecund and suggested that ecology could become a predictive, not simply an inductive, science. MacArthur was very good at promoting his ideas and, shortly before he died, found a remarkable ally in Robert May. Both men were important to the development of theoretical ecology, and their ideas gave impetus to a wide range of new work. By generalizing the old equations of Lotka and Volterra, and by using mathematics to look anew at populations in dynamical ecological systems, May found a suggestive way forward. In principle, his pioneering monograph of 1973 could have been written earlier. But, as we have seen, it was the contingencies of the period that prompted Hutchinson and others to set the new approach in motion. Setting things in motion relates to another of Popper's interests, namely situational logic. Popper, however, did not focus on the wider cultural context in his analysis; he was interested in the internal logic of scientific disciplines and how, once problems are identified, people act similarly in trying to solve them. As mentioned in chapter one, this was something studied empirically by Robert Merton.

The vision of Hutchinson, MacArthur and May, and May's ability in manipulating equations were necessary but not sufficient to change existing ways of ecological thinking. For that to happen May and other theorists had to interact with scientists who had a good understanding of the natural world, and to bring people trained in traditional ecological methods on side. It was fortunate that May visited Britain before going to the Institute for Advanced Study at Princeton; and that Charles Birch wrote to Southwood suggesting that May be invited to Silwood. As a result, May was able to meet other ecologist and mathematical biologists at Silwood and Oxford, as well as John Maynard Smith at the University of Sussex. The various meetings primed him to take full advantage of what MacArthur had to offer. For someone who likes looking for the simple in the complex, and who sees opportunity in opening up new avenues for others to work in, May was in the right place at the right time. Ecological theory was in its infancy and he was able to see a way forward. Other members of the Silwood Circle were, in turn, fortunate in having met May early in their careers.

In an interesting essay Lorraine Daston notes that Robert Hooke recommended mathematics to the natural philosopher because

> it accustoms the Mind to a more strict way of Reasoning ... and to a much more accurate way of inquiring into the Nature of things ... [but] we find that Nature it self does not so exactly determine its operations, but allows a Latitude almost to all its workings though ... it seems to be restrain'd within certain Limits.[6]

The use of mathematics coupled with a relaxed attitude to exactitude is characteristic of much in science. For example, in a 1935 paper John Van Vleck and Albert Sherman discussed two new, but less-than-exact, competing theories of chemical bonding — the molecular-orbital and valence-bond theories. Their paper began with Paul Dirac's classic 1929 statement that quantum mechanics, while not yet a complete theory, would soon embody 'the underlying physical laws of a large part of physics and *the whole of chemistry*'. In its ambition, this was not unlike the claims we saw being made for theoretical ecology by Lawrence Slobodkin and Simon Levin. Overreach is sometimes necessary in order to gain attention or make a point. With respect to quantum mechanics, Van Vleck and Sherman stated that one should be satisfied with 'approximate solutions of the wave equation' and should appeal freely to experimental observation to bolster theoretical models. The new chemical theories, they argued, 'give one an excellent "steer"', a good idea of 'how things go'; and they permit the systematization and understanding of what would otherwise be 'a maze of experimental data codified by purely empirical valence rules'.[7]

MacArthur and May can be seen thinking along similar lines. They, too, saw mathematically informed theory as important, but realized that in a field such as ecology a relaxed attitude to exactitude was necessary. Similarly faced with 'a maze of experimental [and field] data', they looked for theories that would give 'an excellent steer' to their discipline. May stated that, his own models while intended to arrive at general principles, gave qualitative, not quantitative, predictions. Hooke's remark on constraint within limits is especially pertinent to ecology.

For the Silwood Circle it was important that Southwood, himself no mathematician, was sensitive to the direction in which ecology was moving. He saw the need to bring people with mathematical ability to Silwood. The result was the kind of cross fertilization of ideas and practices described in earlier chapters. But Silwood was no island. MacArthur's ideas were fashionable also elsewhere and there were many exchanges among the converted. No site contains all of current knowledge or, to put it more generally, all of the culture. The success of Silwood depended on there being other important sites, and that many people were taking part in the same conversation. But how to have one's voice heard above the crowd? For that one has to be active at conferences and in learned societies, and travel around giving papers at different institutions. One also has to publish often in good journals, and to review publications that bolster or threaten one's point of view. One should also write books directed at serious students and, where possible, have those books reviewed by people holding views similar to one's own. We have seen that members of the Silwood Circle engaged in all these activities.

Among the important texts mentioned in earlier chapters was Lawrence Slobodkin's *Growth and Regulation of Animal Populations* (1961). In reviewing the book, MacArthur divided ecologists into different camps. He threw down the gauntlet by declaring that it was those in his camp, those following in his and Slobodkin's footsteps, who would be the ones to make ecology a true science. The review helped Slobodkin's book become part of established pedagogy in Britain as well as in North America. Joining it was MacArthur and Wilson's book on island biogeography, another major source of inspiration for those in the Silwood Circle. We saw, for example, how it informed Lawton's work on insects and bracken. May's 1973 monograph, and the various collections of papers edited over the next few years by May, John Krebs and Michael Crawley were of similar pedagogical importance. Especially notable were the papers in May's *Theoretical Ecology: Principles and Applications* (1976). The volume connected those at Silwood to the other contributors, many of whom were leading ecologists in the United States. The new literature attracted many young people to the MacArthur camp, and members of the Silwood Circle were among its leaders. A later literary milestone was Anderson and May's major synthesis, *Infectious Diseases of Humans: Dynamics and Control* (1991).

Another way of gaining attention is to throw doubt on a widely held view. May did this by challenging the idea that as ecosystems become more complex they also become more stable. Among the promoters of that idea were Hutchinson, MacArthur, Kenneth Watt, author of another widely read text, and Eugene Odum, leader of the most influential of the competing camps. For many traditional ecologists May's idea that one should look to the properties of non-linear differential and difference equations in order to predict patterns in the real world was distasteful, even crazy. It was, however, these equations that suggested to May that the older idea was likely false. Some ecologists resented mathematicans and physicists telling them how to move their discipline forward. But those at Silwood saw value in May's approach and, by identifying with it, found themselves on the right side of history — at least in the shorter term. They championed ideas that became fashionable during the course of their careers. May's 1973 book influenced thinking far beyond Silwood. The relevance of his models to environmentalism and the preservation of ecosystems came to be recognized worldwide.

Mention should also be made of new computing power, widely available by the 1970s. It allowed large bodies of data to be handled more easily than ever before and made possible the epidemiological analyses carried out by Roy Anderson, the agro-ecosystem modelling of Gordon Conway, the statistical work of Michael Crawley, the remote sensing work of David Rogers, and the fisheries management work of John Beddington. But computing power also allowed the predictions of interesting mathematical equations to be examined in ways not previously possible, something that changed the ecological agenda. The evolution of dynamical systems could be displayed on printer plots (later on computer monitors) and it was soon realized that among the many possible futures are some that are chaotic. Deterministic chaos had been suggested earlier, but with computers its time had finally arrived. It was important that ecologists were able to join in the chaos discoveries of the 1970s. May, as we have seen, was a leader in this area. For the Silwood Circle the association had all the advantages that come with being closely identified with what is new and exciting. Chance and contingency are central themes in modern thought, which perhaps accounts for the enormous public interest in the mathematics of chaos. Having its importance demonstrated in ecology only added to the widespread fear that we are living on the edge. Apocalyptic fears are nothing new. Today, however, they are often expressed in terms of environmental and ecological collapse. The idea that our future existence depends on how we manage the natural environment — in so far as we can manage it — is a central belief of our time. In laying claim to expertise in this area, members of the Silwood Circle were able to catch the moment.

11.2 Institutional history and tradition

The 1907 Imperial College charter mandated that work at the college include the 'most advanced research ... especially in its application to industry'. In line with the college name, industry (including agriculture) was understood as being that of the empire, not simply of Britain. Today the empire is long gone, but an emphasis on applied science still distinguishes Imperial College from universities with more traditional educational mandates. The founding professor of zoology was T. H. Huxley. Because he believed that without improved teaching in schools there was little future for British science, he placed much emphasis on the training of science teachers. However, after 1907, when the Royal College of Science became part of Imperial College, the department underwent a major transformation and its faculty was expected to carry out serious research. When Adam Sedgwick became head of the department in 1910, he focussed, though not exclusively, on training entomologists for work in the empire. Good appointments were made and new staff members carried out research on insect pests, especially those associated with tropical agriculture and food storage. Their legacy was an accumulation of data, and a growing departmental expertise in entomology. Without it the Silwood Circle would not have formed. It helped that the botany department specialized in tropical agriculture and plant pathology.

Other institutional factors played a role in the Silwood Circle's success. Until recently Imperial College had little hierarchical administrative structure. People were given freedom to follow their own interests and were expected to be entrepreneurial. The college had a tradition of early promotion for young people with ability, and of appointing youngish people to head departments. It also pioneered the multi-professorial department in Britain. These policies aided Southwood's career, and the careers of some others in the Silwood Circle who became professors at relatively young ages.

Southwood joined the department as an undergraduate in 1949 because of its reputation in entomology. The head of department was then J. W. Munro who, two years earlier, had organized the purchase of the Silwood estate. Munro laid down some long-lasting patterns for both work and social life at Silwood, patterns that survived well into the 1970s. As to research, Munro furthered the work of H. M. Lefroy in developing new, targeted, insecticides and the technologies for their delivery — something Southwood recognized as important and continued to support when he became head of department. But, along with others of his generation, Southwood also recognized that serious ecological problems were associated with the use of chemicals in agriculture. Some of the early work of the Silwood Circle, especially that carried out by Southwood, May and Conway, should be seen in that context — as well as in that of a growing environmental movement. Munro also initiated research on mosquitoes as human disease vectors, despite such work being seen as the domain of the London School of Hygiene and Tropical Medicine. As discussed in chapter four, Southwood carried out some work in this area, as later did some younger associates of the Silwood Circle such as Charles Godfray. Munro's successor, O. W. Richards, together with Nadia Waloff, introduced ecological field and laboratory work, enthusiastically embraced by Southwood when he returned to Silwood in 1955. Richards was a contemporary of Charles Elton as a student at Oxford

University, and the two remained friends and exchanged ecological ideas despite their methodological differences. Even before Richards moved to Silwood there were links between the two zoology departments. Southwood was to strengthen them and would later have a major career also at Oxford.

During the 1950s new work in parasitology came to Silwood. One result was an increased ecological focus on parasites, especially on nematode (round worm) and insect parasites. The coming together of entomologists and parasitologists meant that scientists at Silwood were well prepared in parasite ecology just as the subject was taking off worldwide. As we have seen, Neil Croll, who worked on helminths (parasitic nematodes) causing human disease, was an important mentor to Roy Anderson. Indeed Anderson followed Croll into medical parasitology and, together with May, was to generalize that field along ecological lines.

When Southwood became head of the zoology and applied entomology department in 1967 he brought in the new talent that formed the core of the Silwood Circle. First to be appointed was Gordon Conway; who had experience with tropical insect pests, and had just completed his PhD with Kenneth Watt at the University of California (Davis). Second was Michael Hassell with strength in the area of insect parasitology. He brought new ideas from Oxford and the University of California (Berkeley). The simplicity of some of the parasite/parasitoid systems being studied at Silwood lent themselves to May's type of modelling. The integration of experimental and theoretical work proved very fecund. As head of the department at Silwood, and later at Oxford, Southwood favoured those working on ecological problems, especially younger people such as Conway and Krebs whom he believed showed both scientific achievement and administrative potential.

By the 1970s Silwood was a site of expertise in entomology, ecology, tropical agriculture, ecological and agricultural sustainability, agricultural pests, pest control, and parasitology (including medical parasites). Not only were people working there expert in these various fields, it was a repository for data and specimens from around the world. The department had an international reputation and many people came to study and work at Silwood. The biological departments had a major presence also in South Kensington, at the main campus of Imperial College. When in London students and members of staff had easy access not only to the international scientific literature, but to the expertise and national collections at the

Natural History Museum. The Silwood Circle benefited from the Silwood Park legacy and was geographically well placed for the work it carried out. The London campus is close to Westminster, and within easy reach of politicians and civil servants seeking technical advice. It was also easy for members of the circle to take an active role in the British Ecological Society. Southwood became president in 1977 and together with his younger associates steered the society's activities in ways they saw fit. Following Southwood others in the circle were also elected president. As a postscript it is worth noting that the circle's legacy is apparent in work carried out at Silwood today. Former students and associates can be found working there in entomology, parasitology, population ecology, extinction risk, landscape ecology, and on how to build ecologically protected areas.

11.3 Biography and psychology

How important are individual histories to career outcomes? On this question one can only speculate. As a child, Southwood was encouraged by his parents to pursue his naturalist activities, to write and publish papers, and to give talks to natural history clubs. The positive feedback he received must surely have boosted his self confidence. Similar support was received by others in the Silwood Circle, though perhaps to a lesser degree. Parental encouragement helps to ease people's way in life, but has no simple connection to what is later achieved. Impediments abound. Being a child naturalist, as were also Conway, Hassell, Anderson, Lawton, Crawley and Krebs, suggests an inborn proclivity. That aside, it is probably correct to assume that knowledge early acquired is of advantage later. It enables students to attract the attention of their instructors, to attract mentors, and later it helps in their careers.

Chance plays a role in whom we meet in life. Parents with ambition try to lessen its role and, where circumstance makes possible, seek good schooling for their children and encourage them to attend good universities. There is little doubt that a good education helps in the making of careers. It is also useful to be able to indentify and befriend those with whom one might profitably collaborate. Southwood was primed for that too. He was early encouraged to make contact with people willing to help in his various endeavours. As we have seen, he and others in the Silwood Circle worked collaboratively with a range of capable people. It is one reason for their success.

One aspect of social skill is a heightened, yet selective, awareness of what others are doing; also an awareness of which activities are seen as good, and which are the best rewarded. When I spoke with members of the Silwood Circle, most made comparative remarks of one sort or another. Views were expressed on what counted as good and bad science. Mention was made of the achievements and failures of others, and of who did or did not deserve recognition, or the receipt of certain awards. Such talk is common in academic life, much exacerbated by today's ranking systems and the many awards that have come into existence. Nonetheless, I sensed a greater than normal awareness among members of the Silwood Circle as to where they stood relative to their peers, including those outside the immediate circle. Such sensitivity can be both a blessing and a curse. It results in greater than normal self-policing and can lead to considerable disappointment. But it can also drive people to work hard and to achieve notable success. I cannot take this causal line further but, when thinking about it, was reminded of an old joke. A man boasts of being at the top of the ladder only to be told that the top of the ladder is half way up the tree. In all walks of life people look up to those whom they admire. To climb the ladder in the manner of one's role models, however, requires not only awareness, but an ability to perform. The kind of awareness just discussed goes part way to explaining the mimetic behaviour described in earlier chapters. It helps us understand how members of the Silwood Circle, as well as those loosely associated with it, successively received many of the same professional awards. Learning how to meet expectations is only possible in a social context, and is more easily accomplished when one belongs to a close-knit group.

Luck, good and bad, plays an ongoing role in people's lives, and lies not solely in parentage and inborn ability. I will refer here to just a few of the examples discussed in previous chapters. Lawton was lucky that his teacher, John Coulson, encouraged him to turn from birds to the study of insects. In all likelihood he would have made a good ornithological career, but he might then have missed the chance of meeting certain people at Oxford, and becoming part of a supportive fraternity working at the leading edge of ecology. It was luck that brought David Rogers together with John Ford, presaging his later work on disease vectors in Africa. John Beddington was lucky that scientists from the fisheries laboratory at Lowestoft were sent to take ecology courses at York University. Their interests suggested how he could forward his own career. Beddington and Conway both strike me as very independent minded, a quality not always helpful in a worldly sense. Nonetheless both men achieved worldly as well as scientific success. Beddington was fortunate in attending one of Conway's talks while still a research student. It alerted him to the work being carried out at Silwood. He was also fortunate in meeting Robert May early in his career and in making a favourable impression on him. Conway was fortunate in having met C. S. Holling who suggested he take time off from his graduate studies to work for the Ford Foundation. John Krebs was lucky in his timing, in starting his career as a behavioural scientist just as new and exciting theories were coming to the fore. Robert May had planned to follow his doctoral supervisor, Robert Schafroth, to Geneva. Sadly Schafroth was killed in a plane crash. Had May gone to CERN he might well have continued life as a physicist, and the Silwood Circle would not have had its major driving force. And, late in his career, Southwood was lucky that clear evidence of v-CJD did not appear until after his appointment and tenure as vice-chancellor of the University of Oxford. Southwood and May helped other members of the group to achieve their goals. As we have seen, attracting and keeping good mentors requires the skill of a courtier. Such skill entails being relatively free of social inhibition. Perhaps luck plays a role there too.

Family situation is another factor in people's careers. We have seen how this played out in the lives of Conway and, more especially, Rogers. Theirs were modern marriages, perhaps a little ahead of their time. Both had wives who wished to further their own academic careers and to share some of the work entailed in bringing up children with their husbands. Today more allowance is made for family situation in the academic world; maternity and paternity leave is common and, for those engaged in childcare, more time is allowed for publication before performance is assessed. However, there will always be trade offs when it comes to family and career ambitions. Predicting the outcome of choices made early in a career is difficult, but for those with ambition early career success is important.

It should finally be mentioned that members of the Silwood Circle worked very hard. This is often the case when people enjoy what they are doing. Being accepted as a peer in a successful group helps. It also stimulates competition as members try to keep up with each other. In the case of the Silwood Circle, internal competition was ameliorated by cooperation, a factor in the circle collectively outperforming some other ecological groups. Success, when it comes, brings its rewards and is addictive. It stimulates people to work even harder.

11.4 The Silwood Circle and sociality

Munro institutionalized much social activity at Silwood, patterns that continued under Richards. Southwood built upon this legacy and thought seriously about how to make life at Silwood as enjoyable and inclusive as possible. He was largely successful, though in the 1960s and early 1970s science at Silwood was still a largely male world. It was less inclusive than it should have been of the few female scientists who worked there. The working environment for men with young families was relatively good, but Silwood could be isolating for young wives and mothers. Prompted by a growing feminist movement, some reflected on their situations with consequences for the social and working lives of their husbands.

People have to learn what is intellectually exciting, and do so only partly through formal education. Much learning comes from social exchange outside the classroom. Among other things it allows people to see what is possible. Roy Anderson, for example, enjoyed the field trips organized for undergraduate students, as well as the kind of sociality he experienced at Silwood. The people he befriended showed him the possibility of a career in ecological parasitology. Later it was the persuasive mix of science and social life that drew him back to Silwood after work experience elsewhere. Similarly, it was social bonds formed earlier that drew Michael Crawley back, and that influenced the type of research he was to take up. Most importantly for the future life of the circle, Robert May was drawn back to Silwood each summer by its lively social and intellectual life.

As has been discussed, Southwood was determined to do something for ecology but was not sure which direction to take. He was highly ambitious and wanted to be associated with research that would capture the attention space. In this his skill in recognizing the public mood, in identifying talent, in seeing where ecology was headed, and in finding connections to environmental science proved useful. His strategic ability showed in how he brought people with related interests together, and suggested the type of problems they could work on. The joint publications of the 1970s set the stage for much that was to follow. Southwood was father to the Silwood Circle, though only some of its members were dependent on him early in their careers. Children, however, seek their independence and his were no exception. All soon had research programmes of their own. Early ties remained strong, however, and were renewed at regular social gatherings. Southwood stayed aloof from much of the fraternizing in pubs and restaurants, and did not join the annual walks. Nonetheless his 'children' recognized his authority, and that he possessed much discretionary power. There was some truth to the joke that FRS stood for Friend of Richard Southwood.

One might imagine that conferences and face-to-face meetings would be redundant in light of the way in which scientists rush their work into print. But the written word alone cannot provide the emotional charge that people need to carry out good work. The group experience of listening to lectures at conferences, talking about new ideas and seeing how they are received, is stimulating. It also directs people's attention to what others see as important. In the early days of the Silwood Circle, the mathematical ecology workshops held at York and Silwood served this purpose. But my protagonists did not rely on conferences alone for their emotional charge. As we have seen, ties made at Silwood, Oxford and York led to the organization of gatherings in a variety of social settings. Members of the circle spoke to me with some emotion of the summer hikes — Bob's walks, as they came to be known. They were seen as challenging and enjoyable, and the earlier ones, especially, are remembered as important life events. During the walks, and in the associated leisure time, much lively exchange took place. Those present discussed which experiments might be worth doing next, papers or books that needed to be written, where papers should be published, and how to further their own and each other's careers. There was some circulation of work prior to publication which gave those in the loop many advantages. These types of behaviour resulted in a tendency to use a common approach to many problems.

In the introduction I used the round table as a metaphor, to suggest a group with a common purpose. Like the knights of Arthurian legend, members of the Silwood Circle each brought different skills to the table. There was an ease of communication among them because of the knowledge they held in common. Much of that knowledge was tacit, the result of similar educational and experiential histories. The one person who had to catch up on ecological ways of knowing was May whose earlier career had been in physics. That he is charismatic helped both in persuading the others of the utility of his ideas, and in energising the group. The fusion of ideas was productive and, together, members of the circle planned ways of keeping their kind of ecology afloat. This entailed gaining and keeping the attention of other ecologists. None of this precluded a genuine intellectual interest in the science — without it, none in the circle would have gone far.

Intense and positive social exchange builds trust. As mentioned in chapter one, the importance of trust was remarked on by Ludwick Fleck and, more recently, by Steven Shapin. But does Fleck's more general analysis fit the Silwood Circle? He was surely correct in claiming that a scientist's apprenticeship leads to specific ways of seeing and doing. He was also correct in claiming that cognition is a social activity. No one can make sense of things in isolation. It is also the case that being part of a group implies a degree of intellectual interdependence. Especially interesting is Fleck's view that when scientists get together to exchange ideas they are able to utter thoughts that would not have occurred to them on their own. I think this is true, and was an important factor in fuelling the work of every member of the Silwood Circle. New ideas, as we have seen, are often a remix of the ideas of others. Fleck was also correct in claiming that friendship and rivalry have epistemological significance. This is not to say that might, or friendship, makes for right. Rather that solidarity makes it easier for new ideas to gain a hearing, for the views of rivals to be effectively criticized, and for some knowledge to become more easily established than others. But, as has already been stressed, for the views of a specific scientific circle to spread, the circle must intersect with many others. Only by spreading through the larger network, can local knowledge become established as part of science.

Finally, having close colleagues with whom one can discuss one's career progress allows for ambitious scenic envisioning, thinking collectively about how to change the scene to one's own advantage. Several within the Silwood Circle mentioned partaking in group discussions on how they could become Fellows of the Royal Society, win new positions and awards, what type of government work would bring them knighthoods, and how to win large grants that would allow the launching of major research projects. Southwood and May were the first to climb the career ladder. Subsequently they were in positions to help the others gain a number of grants and awards, university appointments and, later, also government appointments. They also recognized that having their colleagues join them in the Royal Society required a serial approach, and that nominating just one person every one to two years was the way to go. They thought carefully about the order in which the others should be put forward, and shared their views on this with them. Having trusted confidants who possessed strategic ability undoubtedly helped members of the Silwood Circle to achieve their goals.

11.5 The socio-political and cultural context

Ecological science was boosted by new concerns over the environment that came to the fore in the postwar period. Rachel Carson published *Silent Spring* in 1962; during the next decade organizations such as Greenpeace and Friends of the Earth were founded. In earlier chapters I placed some emphasis on the UN Human Environment Conference held in Stockholm in 1972, and on polemical documents such as *Blueprint for Survival* that appeared at roughly the same time. The cultural mood of the 1960s and 70s nudged the political class to take action on environmental and ecological matters. Many of the recommendations that came from the UN conference were taken up. In Britain one result was new funding for university courses in environmental science and ecology, as well as for research in those areas. This was the case even though the country was in economic recession. From this distance the research monies that Gordon Conway received, $500,000 from the Ford Foundation in the late 1960s, and £450,000 from the Ministry of the Environment in the mid 1970s, seem remarkable. Some of the Ford Foundation money was used to bring people together at Silwood; it helped the Circle consolidate in its early years. A small amount of contract research was carried out by members of the Circle. Overall, however, research decisions were made locally. In other words the larger context provided the opportunity, but did not dictate the direction of research. However, like everyone else scientists are part of the larger culture. What they choose to work on cannot be totally isolated from what happens in the rest of their lives. Besides, as Southwood surely understood, it pays to take heed of society's concerns. Nonetheless, as was stated in the previous chapter, I do not see the socio-political context as determining what counts as knowledge *per se*. It does, however, determine what counts as important at any one time or place. To be successful in the worldly sense, one has to have a voice in the issues of the day. For scientists that means making discoveries that have relevance to problems of current interest.

It was palpable links to matters of public concern that helped to promote the careers of members of the Silwood Circle. The public was worried about climate change and sustainable agriculture. There was much discussion of environmental pollution, and fears were expressed on ecosystem collapse, extinction and loss of biodiversity. There was also fear of new disease epidemics. Some worrying outbreaks had already occurred and, given the huge increase in international travel, and in the international sale of farm animals, more were predicted. There was fear also of bioterrorism, and concern over food safety. None of these fears and concerns have gone away, but already by the 1970s and 80s they resulted in much consulting work for ecologists. We have seen that members of the Silwood Circle were well situated to become favoured consultants on many pressing issues. Their voices were heard on the loss of biodiversity, on international agricultural development, international whaling, global warming, and on epidemiological concerns relating to bovine TB, BSE and v-CJD, SARS, avian flu, sleeping sickness, malaria, and foot and mouth disease. They advised also on GM foods and food safety. As chief scientific advisor to the minister of defence, Anderson gave advice on bioterrorism. Others, too, became senior science bureaucrats, notably Krebs as chairman of the Food Standards Agency, Conway as president of the Rockefeller Foundation and then as chief scientific advisor to the minister of international development, and May and Beddington, as chief scientific advisors to the UK government. As was shown in previous chapters, the advice given on many issues was collective. This is as it should be. Indeed we have seen how May and Beddington advised gathering scientific opinion on a wide front. Nonetheless we have also seen how scientists in positions of authority seek advice from those they have grown to trust. That this is especially true in a crisis was illustrated by what happened during the Foot and Mouth epidemic. More generally, because of their mutual trust, those in the Silwood Circle had much indirect as well as direct influence on political decision making in the late twentieth century. By the 1990s they occupied the political as well as the scientific high ground.

11.6 Concluding comments

All social activity, including scientific activity, is path dependent which implies that certain types of success depend in part on being in the right place at the right time. This book has illustrated what this meant for ecology and for a small group of ecological scientists. Although much of what was discussed in previous chapters was *sui generis*, the historian's task is to draw some lessons and construct a narrative that gives meaning beyond the particular. As we have seen, to understand success in science, whether of a group or of an individual, one has to look in many directions — intellectual, institutional, biographical, political and socio-cultural. Understanding the development of an area of science involves a similarly multidirectional approach. There is little doubt that the socio-cultural context in the second half of the twentieth century was important in stimulating work in ecology. Nonetheless it was the internal history of the discipline that played the greater role in directing people's research. Roughly speaking, what seems to have happened is that new cultural circumstances forced people to address new types of problems and seek new types of data. But the problems were approached with inherited knowledge, many old and a few new ways of thinking, and with some new technology — notably computing technology. Among the new ways of thinking in ecology were those of Robert May. Because his ideas were very suggestive, and because May found common cause with others in the circle, all were to profit. Yet even my tentative generalizations of human behaviour, scientific performance, and the growth of scientific knowledge, are problematic. It is worth noting that the scientists discussed in this book spent their working lives attempting to make sense of ecosystems far less complex than the one to which they themselves belong.

Because activity is path dependent, I devoted much thought to both time and place. Since the past is always present in the present, attention was drawn not only to events of the later twentieth century, but to inherited traditions both within the discipline of ecology, and institutionally. The focus was principally on ever-developing traditions at Silwood and at Oxford University's zoology department. To begin one's career in such places, with exposure to respected ways of thinking and doing, has many advantages. However, Silwood and Oxford were no islands. As mentioned they were nodes in an international network that included also some important North American centres. To progress in a scientific career one should exchange ideas with those who already occupy much of the attention space. The success of the Silwood Circle was in considerable measure due to having its ideas discussed locally, nationally and internationally. Indeed, the many citations to the work of its members is a measure of that success. At the peak of their publishing careers, four were ranked among the top ten cited ecologists in the world, and one of the four was ranked second. Science is highly competitive and it is rare for people to succeed without being well integrated in a social network. For the Silwood Circle such integration enabled them to have informal discussions not only within their own small group, but with other leaders in their discipline.

Worldly success comes from occupying the attention space, and in doing so for as long as possible. This is an abstract way of thinking about that type of success, but the previous chapters have shown what it means in practice. Those in the Silwood Circle, well primed to join the international conversation, worked hard to be taken seriously by their peers. Their ideas came to the attention of other ecologists, to people in government, to those on awards committees, and to journalists who conveyed their ideas to a wider public. It has been claimed that the reception of new ideas depends on the authority of who is defending them. This may have been true in earlier times but, today, authority is attached more to place than to person. Scientists attached to major institutions are more likely to be heard than others. Over the longer term, however, scientific evidence counts for more than either person or place. For example, the acceptance of Robert Boyle's ideas during his lifetime may well have had more to do with his standing in society than with his experimental observations.[8] But today it is the other way round, he is admired for his scientific activities, not for some accident of birth. That this is so is something we owe to our Enlightenment inheritance, an inheritance alluded to in earlier chapters. Although much authority still rests with power elites, by and large they are no longer seen as being entitled to it. Rightly, and perhaps naturally, the struggle against vested interest continues. No one in the Silwood Circle had Boyle's type of social standing, nor did they need it in order to be heard. They did, however, have some educational advantages, and some advantages of place. But, as with all scientists, they had to compete for attention. Because their local knowledge received international acceptance it became part of established science.

That ecological science became more firmly established at the centre of the scientific universe was historically contingent. Among other things it depended on a growing environmental movement. In the late 1960s when members of the circle began their careers, ecological science was relatively impoverished. Southwood, very aware of his historical time and place, aware of a growing environmentalism, played a pivotal role in promoting ecology in Britain. Further, the Silwood Circle was able to ride a wave of theoretically informed ecology that peaked in the late twentieth century. More generally, the personal histories of members of the Silwood Circle support the widely held view that individual ability and effort are important to later success, but that they are not enough. An old and familiar point about careerism has been confirmed yet again. It pays to be well connected, to work and associate socially with like minded and similarly talented people, to have ambitions in common, to be politically aware, and to be outward looking.[9] Major success, however, requires that there be some leaders among one's associates. In this the younger members of the Silwood Circle were lucky. Richard Southwood excelled in bringing people together and having them work collectively. His awareness helped them also politically, and his astuteness helped not only those in his immediate circle, but also the discipline of ecology. Robert May's leadership was outstanding in both the heuristic and strategic sense. His personal qualities also drew others to him, not only those within the circle. That many of those drawn were themselves accomplished theoretical ecologists gave others in the Silwood Circle useful contacts in the larger world of science.

Ecological knowledge has been culturally transformative over the past fifty years, something only hinted at in this book. Awareness of ecological problems among the public has increased enormously, but understanding of ecological systems is still poor. Over the lifetime of the Silwood Circle it became clear that ecological patterns are more complex than earlier imagined. I was interested in the early work that Hassell and Anderson carried out on relatively simple parasite/parasitoid systems, and wondered whether the study of simple systems led them to assumptions about interconnectedness in the larger biological world. I came to the conclusion, possibly incorrect, that ecologists more generally are hesitant to jump from the micro- to the macro-ecological level. But a willingness to do so was shown by members of the Silwood Circle, something that brought them much attention. In this, too, there appears to have been some mimetic behaviour. Members of the circle were serially drawn to consider larger issues — no bad thing since it is important that there are scientists who look beyond their own small areas of expertise, think about larger problems, and consider their wider implications.

I also gave some thought to the role of theory in this story and appreciate its importance. But, in complex sciences, such as ecology, reliable theoretical predictions are hard to come by. Sometimes the best way forward is top down, but at other times bottom up is better. Which approach is chosen is in part determined by aesthetic considerations. Just as among artists, there are scientists who prefer 'lessness' and abstraction, and others who prefer 'moreness' and careful description. We saw an interesting exchange between C. S. Holling and Robert MacArthur. Both were theorists, but MacArthur found Holling's 1967 prizewinning work too encumbered with empirical data.[10] By the late 1960s, as MacArthur understood, it was clearly time for the injection of some new theoretical ideas. And, indeed, his own, Holling's, and May's were among those that were to stimulate the entire field of ecology. Although the main focus of this book has been on population theory, behavioural ecology, too, was briefly discussed. The two came together in MacArthur's work; and, as we saw, the ideas of W. D. Hamilton on inclusive fitness were extraordinarily suggestive. Inclusive fitness remains heuristically important although, as mentioned in chapter nine, Martin Nowak and his colleagues are now crowding the attention space with ideas that challenge it. Even E. O. Wilson has reconsidered some of his earlier ideas.[11] But, as surely he must understand, there are times when observation and measurement are the best ways forward. Although new theoretical ideas attract attention and are essential in keeping a discipline moving, the accumulation of natural historical knowledge remains very important and should not be downplayed.

Special to the Silwood Circle was a highly fecund interchange between theory and designed experiment. The resulting observations allowed new ecological theories to be improved. Epidemiological modelling, especially, was helped by the integration of Anderson and May's theoretical ideas with large banks of existing data on many human infectious diseases. Ecological theory may still be in its infancy but members of the Circle found a particular moment in the mid-twentieth century to take a major step forward.

Whether any of the theories discussed in this book will survive the twenty-first century remains an open question. Nonetheless, the contributions of those in the Silwood Circle were considerable. Some may see its members as hard-nosed careerists, and in certain respects they were — but they were surely much more. They were serious and cooperative scientists. The Silwood Circle produced much useful knowledge, passed its ways of doing things to new generations of ecologists, epidemiologists, agronomists and resource management specialists, gave the United Kingdom improved modes of scientific governance, and left all of us with a richer understanding of the natural world. Collectively they understood their time and place and helped each other, as well as others outside the circle, to move forward. Their legacy goes well beyond worldly success.

Endnotes

1. See chapter 6, note 44. There may be a trade off. It has been suggested that because the young know little of the past they are more inventive and that, as people age, they become more acculturated, and more invested in the status quo; (Lehrer, 2012).
2. This was the case of the so-called aether. After Einstein's theory was accepted it was seen as no longer necessary for there to be a material medium for the propagation of light and electricity.
3. See chapter 6, note 44, for some recent championing of MacArthur by some of his, now older, contemporaries, E. R. Pianka and H. S. Horn. They lament the waning of MacArthur's reputation.
4. Karl Popper translated his 1934 work, *Logik der Forschung*. He also revised it before its publication as *The Logic of Scientific Discovery*, (Popper, 1959). Despite the title, Popper saw no logic in discovery itself. On how ideas arise he had little to say. Rather he saw logic in how science advances.
5. Not everyone approved of this approach. Earlier, Einstein had been ambivalent toward hypothetico-deductivism. He saw some of the hypotheses being put forward by his contemporaries as too speculative, based neither on observation nor on well established principles. But such criticism of hypothetico-deductivism fails to take account of its heuristic value; clever speculation can suggest research that is worth doing. Sometimes supporting evidence follows. Even if it does not, something is learned in the process. For a discussion of Einstein's views, (Hesse, 1974), especially 250–54.

6. (Daston, 1995), Hooke quotation, p.11. As Daston tells us, Leibniz held a view similar to that of Hooke, stating that a 'lack of clarity was at the root of all controversy and could therefore be cured by a goodly dose of numbers' (p.9). I have not pursued Daston's ideas far in this book; but in arguing that, in the European tradition at least, there is a need to quantify in order to have people acquiesce to one's ideas, she makes a point that is relevant to developments in modern ecology. Paul Dirac famously stated that beauty was more important to a scientific equation than having it fit empirical observation. One wonders whether Hooke or Leibniz would have agreed with him. I do not; but it is interesting that simple (beautiful?) mathematical patterns appear to be useful. I can understand the utility of having beauty on one's side — all part of the moral economy that Daston was writing about.

7. I think the empirical rules they had in mind were of the type given by the periodic table. (Van Vleck and Sherman, 1935), quotations, 168–69. (The italics in the Dirac quotation are JVV's and AS's.)

8. (Shapin, 1994).

9. This point was well made by Lee Smolin who discussed the isolation of the very clever physicist founders of string theory. Only when their ideas were taken up by a larger group within the theoretical physics community did their careers flourish; (Smolin, 2006).

10. See chapter 6, note 23.

11. See Nowak, Tarnita and Wilson (2010) and Wilson (2012) for Wilson's more recent views on the limitations of inclusive fitness theory, and his ideas on multi-level selection — views that have been challenged.

appendix one

Verhulst, Volterra and Lotka

This appendix is a further articulation of the narrative in chapter two relating to Lotka and Volterra. The basic mathematical ideas they used are given below.[1]

The early pioneers in mathematical ecology were largely people who approached the field from outside the discipline. With backgrounds in mathematics, demography, physics and chemistry, and beginning roughly in the 1920s, they turned their attention to biological problems. In doing so they looked to an equation that had been introduced in 1838 by the Belgian mathematician and demographer, Pierre-François Verhulst. Verhulst recast Thomas Malthus's idea on population growth in mathematical form. The Malthusian idea is expressed by equation (1) below where N is the size of the population and ε is the difference between birth and death rates (or the coefficient of increase).

$$\frac{dN}{dt} = \varepsilon N \tag{1}$$

On integration this gives the well-known law of geometrical increase. But, since environmental resources can support only a limited number of individuals, in the real world the coefficient ε will not be a constant but rather a decreasing function of N. As Verhulst understood, in the simplest case (though again, not in the real world) the decrease will be linear. Hence in the equation below, which is named for him, ε and λ are constants and $-\lambda N$ represents the effect of competition for resources within a single population.

$$\frac{dN}{dt} = (\varepsilon - \lambda N)N \qquad (2)$$

(Verhulst equation)

Integration of this equation leads to equation (3) with its characteristic sigmoid curve which shows population growth over time.[2]

$$N = \varepsilon / (\lambda + ke^{-\varepsilon t}) \qquad (3)$$

These and equivalent models of population growth became known collectively as the logistic equation.

In a more modern formulation as used in ecology the logistic equation can be written as:

$$\frac{dN}{dN} = rN\frac{(K - N)}{K}$$

where r equals the rate of population growth, and K equals maximum sustainable population (carrying capacity). As is discussed in chapter five, the symbols r and K came into use in evolutionary ecology to describe different reproductive strategies.

The Italian mathematician, Vito Volterra, was among those who saw that this equation could be the starting point for thinking about animal populations, and for more than one such population co-habiting.[3] In this he and Alfred Lotka were the founders of one thread in theoretical ecology. Members of the Silwood Circle acknowledge their debt to both these men.[4] Even before Volterra, biologists recognized that the logistic equation could be useful. It was understood, however, that there were fluctuations in the death rate, and that reproduction is never spontaneous or regular. As a consequence animal populations will fluctuate over time. External causes, such as weather patterns affecting food supply, were thought to be among the causes of such fluctuations. Indeed, some sort of periodic term can be added to the right-hand side of the logistic equation to produce an oscillation in N.

Volterra's approach was conceptually different. After being alerted to the problem of the Adriatic fishery (see chapter two) he came up with two equations envisaged as representations of two fish populations, one the prey and the other the predator. In equation (4), representing the prey population, the coefficient of increase ε is positive since, in the absence of predators, the number of fish would increase. In equation (5), representing the predator population, the coefficient is -ε since the predators would die out in the absence of prey. The number of encounters of prey with predator is proportional to the product of the total numbers of prey and predator individuals, N_1 and N_2 respectively; the coefficients γ_1 and γ_2 relate to the degrees of voracity and susceptibility of the predator and prey. Assuming that the encounters are positive for the predator and negative for the prey, then:

$$\frac{dN_1}{dt} = (\varepsilon_1 - \gamma_1 N_2)N_1 \tag{4}$$

represents the population for the prey species

$$\frac{dN_2}{dt} = (-\varepsilon_1 + \gamma_2 N_1)N_2 \tag{5}$$

represents the population for the predator species

Something equivalent to these equations was arrived at independently by Lotka a few year earlier, using a different approach. Both men are remembered in the eponymous Lotka-Volterra equations (4) and (5) above.

By integrating the equations, Volterra showed that the numbers of prey and predator oscillate with periods depending on the coefficients ε, and on the initial numbers of the two species. Further, he claimed that the equations can apply to interactions more generally, not solely to those of prey and predator.[5] Volterra predicted correctly that normal fishing practice would aid the prey species and that cessation of fishing would aid the predators. He went much further than Lotka in developing his mathematical ideas. For example, he understood that the right hand side of equations (4) and (5) could be manipulated in ways that could be of interest to biologists — the coefficients ε could have seasonal variations, for example. He produced a general theory for the interaction of n species and altered the basic equations to allow for the damping of oscillations and for the introduction of time delays for some interactive effects. The idea that simply by interacting, even favourably, two species could have an effect on each other's populations was an idea now cast in mathematical form. Manipulation of these equations in the ways Volterra suggested has since occurred. These manipulated equations underpin much modern theory.

Lotka's approach was different.[6] As discussed in chapter two, he began with a generalized idea of evolution envisaged in terms of the redistribution of matter. He used the law of mass action expressed in equation (6) as his foundation:[7]

$$\frac{dm_1}{dt} = F(m_1, m_2, m_3, v, T) \tag{6}$$

where m_1, the mass of substance 1, is a function of the other masses in the system, say m_2, m_3 etc. and of volume v, and temperature T. Lotka believed that the habit of thought implied by this equation should be used in contemplating the nature of evolution.[8] In some ways this seems rather trivial since in broad terms evolution is, indeed, simply change over time. But the law of mass action led Lotka forward in his physical biology programme. He began to consider the kinds of systems one might actually find in biology or in demography, systems where time lags are important. In this regard, the work of his mentor, Raymond Pearl, the work of the malaria experts Emilio Martini and Ronald Ross, and that of the parasitologist W. R. Thompson were important to his thinking.[9]

Thinking more generally about the kinetics of evolution Lotka introduced the equation:

$$\frac{dX_i}{dt} = F_i(X_1, X_2, \ldots, X_n, P, Q) \tag{7}$$

where there are n variables $X_1, \ldots, X_n$ and constants P, Q, etc. The equation represents the increase over time of a given component X_i.

But the 'tendency to equilibrium', as Herbert Spencer put it, can be reached in a number of different ways. Recasting equation (7) in terms of the excess x_i of each mass X_i over its equilibrium value, the simplified equation takes the form:

$$\frac{dx_1}{dt} = f_i(x_1, x_2, \ldots, x_n) \tag{8}$$

Lotka then manipulated equation (8), including by Taylor expansion, and arrived at a set of equations which, he claimed, provided analytical 'confirmation and extension' of the types of equilibria envisaged by Spencer.[10] Algebraic manipulation of the simplest of his models (with only one variable X, x) deriving from (7) and (8), resulted in something close to the logistic equation (3) above. It must have been pleasing to be working alongside Pearl and his colleagues who were providing so much biological evidence in its support.[11] Data on the population of the United States, on laboratory colonies of *drosophila*, and on bacterial populations were cited as evidence by Lotka. Lotka took a further step by arguing that if age-specific death and reproduction rates (division rates in the case of bacteria) remain constant then a population would have a stable age distribution — in the absence of any serious environmental disturbance.

But Lotka also considered what would happen with two dependent variables, X_1 and X_2, relating to the interaction of two independent species. His equations for this were basically the same as Volterra's. Namely they were models of non-linear dynamical systems. Up to a point the Lotka-Volterra equations have observational support, though this was easier to claim for Volterra and the fishery than for Lotka's more complicated examples. However, both analyses allowed that there were different ways of approaching population equilibria. This proved interesting. Lotka viewed predation in a more inclusive way than did Volterra. He considered not only the outright killing of the prey, but also parasitism, and interactions of the kind discussed by Ross and others interested in the epidemiology of infectious diseases. He also included a discussion of saprophytic and symbiotic behaviours and, as a subset of the latter, the relationships between humankind and domesticated plants and animals. Thus, while the simplest prey-predator interactions were described by Lotka in a form equivalent to that expressed in equations (4) and (5) above, he was looking in many other directions — too many for clarity. He was probably also trying to please Pearl by emphasizing the demographic and epidemiological applications of the equations. Indeed, Ross's malaria patterns with their hills and valleys of infection related to mosquito populations, and the examples of population oscillations between parasite and host collected by Thompson, showed just the kind of oscillatory, cyclical, behaviours that Lotka found in his equations. But equally important for Lotka was that Spencer's intuitions had been given both empirical and theoretical support.

Endnotes

1. I have chosen to summarize Volterra's mathematical modelling from the information given in (Whittaker, 1941) and have used the original notation.

2. The equation is intuitively persuasive and can be used to describe many variables increasing over time to an upper limit. But there are reasons why the equation is not truly descriptive of population behaviour and has to be adjusted; for example, to remove impossibilities such as negative birth rates. Demographers have proposed a range of other adjustments to the equation.

3. (Whittaker, 1941); Whittaker notes that Volterra was interested in some biological applications of mathematics already by 1901, but that his first major paper on the subject was published by the *Accademia dei Lincei* in 1926. It was translated, with editing and small additions, and published in *Nature*; (Volterra, 1926). See also (Volterra, 1931; Vito Volterra with Umberto d'Ancona, 1935) for a more generalized version of his theory. It is Whittaker's summary of the ideas in this second book that I am using here.

4. See, for example, (May, 1974b; Hassell, 1978).

5. As Whittaker noted, and as is clear from (Volterra, 1926), Volterra had been working on the general problem of population interaction even before being alerted to the wartime fisheries problem. The data provided by his son-in-law confirmed one possible result implied by the integration of the equations. Volterra's daughter was also a fisheries biologist and probably deserves some credit in this story.

6. Lotka's 1924 book, *Elements of Physical Biology* was later renamed; (Lotka, 1956). References in the text refer to this later edition.

7. This law was first formulated in 1864 by C. M. Guldberg and P. Waage, working at the University of Oslo. Wilhelm Ostwald provided supporting evidence in 1877 and the law was arrived at independently in 1879 by the Dutch chemist J. H. Van't Hoff who later acknowledged Guldberg and Waage's priority. The law has had enormous impact in chemical kinetics and in epidemiology. Much historical information on this can be found in (Bastiansen, 1964).

8. (Lotka, 1956), 48.

9. (Lotka, 1956), 79–88. See also (Lotka, 1923). W. R. Thompson was a Canadian who worked in France in a laboratory supported by the United States Bureau of Entomology before becoming director of the Farnham Royal Laboratory of the Imperial Institute of Entomology in 1927. For reference to Lotka, (Thompson, 1923).

10. Using equation (8) rather than equation (7) in the expansion is to simplify the mathematics. The result of expansion is the same in both cases.

11. Pearl and Lotka overestimated their empirical support for the logistic curve — though clearly it is roughly descriptive and predictive.

appendix two

This appendix relates to material in chapter 6.

1. C. S. Holling: functional response and resilience

Holling began thinking about numerical and functional response problems already in the late 1950s when he was working at the Canadian Forest Service's Forest Insect Laboratory at Sault Ste. Marie, Ontario. While there, he studied the predation of pine sawfly pupae by small mammals, and carried out a laboratory study of the predatory behaviour of the praying mantis. In both cases he noted that predator consumption rates increased with prey density, but only up to a point. Holling was interested in modelling this phenomenon, namely the functional response. With that in mind he devised some further experiments to observe the ways in which predators behave with respect to their prey.[1] Some of the predator's time is spent hunting down prey and some is spent eating and digesting it. At low prey densities a predator needs to spend much time searching; at high prey densities it spends less time, but may spend more time handling prey — killing, eating, digesting it. Predators do not (as Volterra assumed in his original model) consume every prey item they come across.[2] Indeed, if prey density is high this would be unlikely. Among the behavioural factors to consider are that predators can learn to hunt more efficiently (the prey can also learn to hide), or they could switch to some other, more abundant, prey. Holling considered his experimental observations and thought more generally about how predators behave. He then produced models, variations on the Lotka-Volterra equations, to show how the prey and predator populations responded in light of their different behaviours. Experimentalists and field biologists were able to tinker with the models in their attempts to understand a wide range of predatory behaviour in both field and laboratory studies. In the case of sawflies, mammals destroy most of the pupae when the sawfly population is sparse but make only a small impression on them when the density is high. Holling also noted that how a prey population distributes itself in the environment determines its longer-term survival. The models also suggested that under some conditions populations could become extinct.

Holling's 1973 paper was a synthesis of his work on predation processes over the previous fifteen years. By the 1970s Holling, like May, had become interested in the question of stability. His 1973 article expressed appreciation for May's early papers (from 1971–2) and for the analytical models put forward. But Holling was a simulator and wanted models that more closely resembled the 'real world' situations of interest to him.[3] In particular he wanted more emphasis on realistic fecundity and mortality, especially as these related to predation. He also became interested in what he termed resilience. In that connection he thought about species that, over time, underwent major change or fluctuation in their populations, but nonetheless survived over the long term.

His best illustration of resilience was, perhaps, that of the Spruce Budworm, an insect that he had studied earlier with Kenneth Watt. The Budworm is a pest that has periodic major outbreaks but is otherwise relatively rare. Despite its name, it attacks mainly mature stands of Balsam Fir, a tree species common in eastern and central Canada. Less mature stands are left largely untouched by this pest and Spruce is not seriously affected. Since the outbreaks occur in the wake of several consecutive dry years, Holling assumed there was a random (environmental) element to their occurrence. However, without the periodic infestation and destruction of the firs the co-habiting spruce and birch would be crowded out and not survive. Further, since it takes time for young stands of fir to mature, major fluctuations in budworm population, with fairly long periods between peaks, allow for the budworm's own longer-term survival. One of the many factors that needed to be taken into account for the models to be realistic (realism and generality cannot easily both be achieved) was the presence of insectivorous birds that preyed on the budworm. Holling noted progressively stronger predation (up to a point) by the birds as budworm density increased. In aging forests with increasing foliage, finding prey becomes increasingly difficult and bird predation is reduced.

Budworm populations are highly unstable but at the same time the species is highly resilient. It is also the case that the budworm is an r-strategist in the sense introduced by MacArthur and discussed in chapter five. In discussing real-world examples, such as this one, Holling came to some of the same conclusions as did May. Further, the idea of resilience led him to think more about limit cycles and to conceptualize persistence in terms of changed configurations, as in the moving from one domain of attraction to another (see below). As it later turned out, the idea of resilience took Holling into new areas. The lesson of the Budworm is that the collapse into qualitatively different states need not be viewed negatively since it allows for new arrangements and opportunities for co-habiting species. Holling began thinking about this also in connection with larger ecosystems including social and economic systems, and in how we need to adapt to environmental change. His ideas have caught on and can be seen, for example, as a motivating force among people working at the Stockholm Resilience Centre, an international centre co-sponsored by the University of Stockholm and the Beijer Institute of Ecological Economics (associated with the Royal Swedish Academy of Sciences).

2. Robert M. May: stability, complexity, and chaos

It is probably fair to say that May was the principal contributor to the mathematics of cyclic and chaotic population dynamics which became central to so much ecological theory in the later twentieth century. In this section I will summarize what I see as some important points in a way that will be accessible to readers who, like me, have minimal mathematical understanding. As with Holling's work, anyone with a serious interest will need to consult the original material cited below and in chapter six.

Differential and difference equations can be used to describe the interaction of entities over time. In the case of prey-predator dynamics, the entities are species populations. Their interactions can be represented in different ways. One way is in the manner of Lotka and Volterra, namely by plotting the two populations against time. Such representation shows how prey and predator populations oscillate and how they do so out of phase. Another way of displaying the equations is to plot the population of predator against the population of prey over time.

If one rules out chaos, that the prey and predator populations remain constant, or that one or other of the populations dies out, then the populations will end up in so-called limit cycles (see illustration below).[4] A cycle is a closed curve in the graph of one population against another. A limit cycle is one toward which populations move, provided they start fairly close to each other spatially, and interact. Irrespective of the initial populations of, say, prey and predator, their populations will oscillate and eventually reach levels that conform to the limits prescribed by the cycle.

In the illustration below the population of a prey is plotted against the population of its predator. Points A, B and C are arbitrary starting points for the population sizes. At A, the predator population is relatively high compared to the prey population; at B the predator population is low and the prey population somewhat higher; at C the prey population is relatively high compared to the predator population. The figure illustrates that, regardless of starting point, over time the populations converge on the limit cycle as shown and oscillate within certain limits.

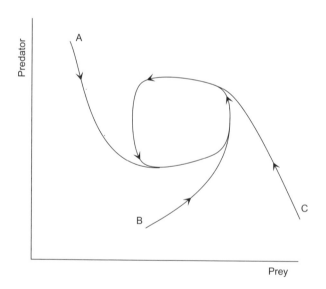

One of the best introductions to the basic idea of limit cycles in simple predator-prey systems can be found in May's 1972 article in *Science*.[5] Much other theory builds from this. As May noted in his 1973 book, the

limit cycle as a description of the terminal interaction between a prey animal and a predator was already envisaged in 1936 by the Russian polymath
Andrei Kolmogorov (1903–87).[6] Kolmogorov was drawn to the earlier
work of Lotka and Volterra. Interestingly, like Lotka, he had studied chemistry before turning to mathematics and mathematical biology. He recognized that the Lotka-Volterra equations were too simple to represent
real-world situations because the oscillations they described did not change
over time. The cycle for Lotka and Volterra is a closed loop from which
there is no departure. Populations simply go round and round (or uniformly up and down) for ever. More generally however — and under the
more plausible conditions Kolmogorov envisaged — two populations in a
prey-predator system may well not start in a limit cycle. Over time, however, they should end up either at a dynamical equilibrium point or in a
limit cycle. He also understood that in the real world such systems will
always be perturbed. Therefore interacting populations may well depart
from regular cycling or closed limit cycles. Nonetheless, the tendency is to
move back, either toward an equilibrium point or toward a limit cycle. The
point and the cycles are said to be attractors in that populations move
toward them.

Kolmogorov's equations are given by May as:

$$\frac{dH}{dt} = HF(H,P) \qquad \text{and} \qquad \frac{dP}{dt} = PG(H,P)$$

where H is prey population, P is predator population, and where F and
G can be any chosen function.[7] The Lotka-Volterra equations are a special
case of these equations. As May noted, natural ecosystems which display
reasonably regular oscillatory behaviour look much like stable limit cycles.
In chapters four and five of his book May went beyond Lotka, Volterra and
Kolmogorov by looking at the possibilities for F and G in the Kolmogorov
equations. The equations as given above are in historical form, namely
intended to account for prey-predator interactions. However, with appropriate F and G functions, the equations could apply to a number of other
types of interaction, such as populations competing for a particular
resource. An example of competing species is given in chapter five of May's
book.

$$\frac{dN_1(t)}{dt} = N_1(t)[k_1 - N_1(t) - \alpha N_2(t)]$$

and

$$\frac{dN_2(t)}{dt} = N_2(t)[k_2 - N_2(t) - \alpha N_1(t)]$$

where the populations are represented by $N_1(t)$ and $N_2(t)$; (by including (*t*), May is making the point that the populations are functions of time); k_1 and k_2 are environmental constants — they become fluctuating environmental variables in a later iteration of the model. α is a competition coefficient, symmetrical in this model but, in later iterations, it is modified to reflect unequal competition between two populations. (The apparent inconsistency in units is due to *t*, on the left hand side of the equations, being a rescaled dimensionless time.)

May showed that movement toward equilibrium points and toward limit cycles becomes more difficult for populations within complex, as opposed to simple, ecosystems. Also, depending on the equation used to describe the situation, there could be more than one limit cycle. This allows for Holling's speculation that resilience could entail populations moving from one attractor to another. How populations in complex ecosystems actually behave in the face of disturbances of one sort or another remains an open question — though the Budworm/forest ecology situation is fairly well understood.

Connected to this is another interesting feature of the type of equations used by May, namely that they are sensitive to changes in initial conditions. In principle, they can end neither in equilibrium points nor in limit cycles, but in chaos — something more likely as complexity increases. In other words, the equation end-points can be unpredictable. However, we tend not to see chaotic situations overall, and so must conclude one of three things: that the equations being used are not fully descriptive of ecological situations, that the kinds of perturbation needed for chaos to ensue do not generally occur, or that such perturbations do occur but are normally damped out for one reason or another. Ecologists appear mainly to assume that the third of these is most likely. Since it is better to be safe than sorry we should not rely on damping in all situations and must be careful to avoid causing major perturbations to our environment. Some things, such as a Krakatau or major meteorite collisions, are fortunately rare and beyond human control. But life did return after both. And, as Holling recognized, some species are resilient and can survive under new limit cycle/ecological conditions. New conditions allow for new arrangements among species and, in this limited sense, can be seen in a positive light. There is no guarantee that the human species will be so lucky.

In his introduction to the 2001 edition of his book, May includes a brief discussion of chaos, something he first discovered in the early 1970s when working with simple difference equations:

> For a population, such as many temperate insects, with discrete non-overlapping generations ... a simple metaphor is $x_{t+1} = rx_t(1 - x_t)$. Here x_t is population in year t, scaled so that *if* x ever gets as large as 1 it extinguishes itself, and r is its intrinsic growth rate at low density (when x is close to 0). ... If r is between 1 and 3 this equation describes a population which settles to a constant equilibrium value .... If r is above 3 but below about 3.57 we see self-sustained cycles. For r bigger than 3.57 but below 4 there is "chaos": apparently random fluctuations generated by this trivially simple deterministic equation.[8]

Endnotes

1. One of Holling's early equations is known as the disc equation because it was a simulation of a student experiment searching for discs. He asked blindfolded students to search as effectively as they could for sand paper discs (the prey) arranged in different ways on a table. The disc equation was the jumping off point for further work; (Holling, 1959, 1965 and 1966). In his 1966 paper, Holling discussed some experiments and observations on the behaviour of the praying mantis. He also carried out some studies on the predatory behaviour of some Hawaiian fish. These studies (with simulations carried out using an early IBM mainframe computer) led Holling to classify predation in terms of four types of functional response and three types of numerical response. Holling's 1965 paper won him the George Mercer Award of the Entomological Society of America. Holling told me that he was surprised that his article, in what was an outlier journal, had been noticed. The prize prompted a wider readership and helped to forward his career. Late in his career Holling was awarded the Eminent Ecologist Award (1999) by the same society and, in 2008, the Volvo Environmental Prize.

2. This assumption was not necessary and was later abandoned. All that is needed is the assumption that the per capita intrinsic growth rate in the predator population contains a term which increases linearly with prey abundance. Conversely the per capita growth rate of the prey decreases linearly with predator abundance. See appendix one.
3. (Holling, 1973).
4. The simple Lotka and Volterra equations, being two dimensional continuous time systems cannot show chaos. But if a third variable is introduced or, alternatively, a time delay is introduced, chaos may well follow.
5. (May, 1972a).
6. (May, 2001), 86.
7. (May, 2001), 86.
8. (May, 2001), xiv

List of Acronyms

ARC	Agricultural Research Council
BBC	British Broadcasting Corporation
BES	British Ecological Society
CABI	Centre for Agricultural Bioscience International, formerly Commonwealth Agricultural Bureau International
CBE	Commander of the Order of the British Empire
CERN	European Centre for Nuclear Research
CMG	Commander of the Order of St. Michael and St. George
BSE	Bovine Spongiform Encephalopathy
v-CJD	Creutzfeldt Jakob Disease
DEFRA	Department of Environment, Food and Rural Affairs
DFID	Department for International Development
DNA	Deoxyribonucleic acid
DSIR	Department of Scientific and Industrial Research
EU	European Union
FRS	Fellow of the Royal Society
FSA	Food Standards Agency
GCRI	Glasshouse Crops Research Institute
GDP	Gross Domestic Product
GM	Genetically Modified
HEFC	Higher Education Funding Council
HIV-AIDS	Human Immunodeficiency Virus-Acquired Immunodeficiency Syndrome
HPA	Health Protection Agency
HSS	Health and Social Security Department

ICA	Imperial College Archives
ICI	Imperial Chemical Industries
ICL	Imperial College London
ICCET	Imperial College Centre for Environmental Technology
IIED	International Institute for Environment and Development
IPARC	International Pesticide Applied Research Centre
IPM	Integrated Pest Management
IPCC	Intergovernmental Panel on Climate Change
IWC	International Whaling Commission
LSE	London School of Economics
MAFF	Ministry of Agriculture, Fisheries and Food
MIT	Massachusetts Institute of Technology
MTCT	Mother to Child Transmission
NERC	Natural Environment Research Council
NFU	National Farmers Union
NGO	Non-Governmental Organization
NSF	National Science Foundation
OECD	Organization for Economic Cooperation and Development
OPEC	Organization of the Petroleum Exporting Countries
PROSAMO	Planned Release of Selected and Modified Organisms
RAE	Research Assessment Exercise
R&D	Research and Development
RCEP	Royal Commission on Environmental Pollution
SSK	Sociology of Scientific Knowledge
SARS	Severe Acute Respiratory Syndrome
TALA	Trypanosomiasis and Land-use in Africa
TB	Tuberculosis bacillus
UBC	University of British Columbia
UNEP	United Nations Environmental Programme
UNFAO	United Nations Food and Agriculture Organization
UNICEF	United Nations Childrens Fund (formerly education fund)
USAID	United States Agency for International Development
WHO	World Health Organization

Bibliography

(Archival sources are referenced in the endnotes)

Allen, David Elliston. 1976. *The Naturalist in Britain.* London: Allen Lane.

Anderson, Christian N. K., Chih-hao Hsieh, Stuart A. Sandin, Roger Hewitt, Anne Hollowed, John Beddington, Robert M. May and George Sugihara. 2008. 'Why fishing magnifies fluctuations in fish abundance'. *Nature*, 452:835–839.

Anderson, Roy M. 1974a. 'Population-dynamics of *Cestode caryphyllaeus-laticeps* (Pallas, 1781) in bream (*Abramis-brama L*)', *Journal of Animal Ecology*, 43: 305–321.

_____1974b. 'Mathematical models of host-helminth parasite interactions', in M. B. Usher and M. H. Williamson (eds.), *Ecological Stability.* London: Chapman and Hall, 43–69.

_____1981. 'Strategies for the control of infectious disease agents' in Gordon Conway (ed.), *Pests and Pathogen Control: Strategy, Tactics and Policy Models.* International Institute for Applied Systems Analysis.

_____(ed.). 1982. *Population Dynamics of Infectious Diseases: Theory and Applications.* London: Chapman Hall.

_____1984. 'Strategies for the control of infectious agents' in Gordon R. Conway (ed.), *Pest and Pathogen Control: Strategic, Tactical and Policy Models.* Chichester: Wiley.

_____1986. 'Rabies control: Vaccination of wildlife reservoirs'. *Nature,* 322: 304–305.

_____1994. The Croonian Lecture. 'Populations, infectious disease and immunity: A very nonlinear world'. *Philosophical Transactions of the Royal Society of London B*, 346: 457–505.

Anderson, R. M., S. P. Blythe. G. F. Malley *et al.* 1987. 'Is it possible to predict the minimum size of the Acquired-Immunodeficiency-Syndrome (AIDS) epidemic in the United Kingdom?'. *Lancet*, 9 May, 1073–1075.

Anderson, R. M., C. A. Donnelly, N. M. Ferguson, M. E. T. Woolhouse, T. R. E. Southwood, *et al.* 1996. 'Transmission dynamics and epidemiology of BSE in British cattle'. *Nature*, 382:779–788.

Anderson, R. M., D. M. Gordon, M. J. Crawley and M. P. Hassell. 1982. 'Variability in the abundance of animal and plant species', *Nature*, 296:245–248.

Anderson, Roy M., Helen C. Jackson, Robert M. May and Anthony M. Smith 1981. 'Population-dynamics of fox rabies in Europe'. *Nature,* 289:765–771.

Anderson, Roy M. and Robert M. May. 1979. 'Population biology of infectious diseases': Part 1 and Part 2 (Order of authors' names reversed in Part 2), *Nature,* 280:361–367 and 455–461.

_____1981. 'The population dynamics of microparasites and their regulation of natural populations of invertebrate hosts', *Philosophical Transactions of the Royal Society of London B*, 291:451–524.

_____(eds.). 1982. *Population Biology of Infectious Diseases.* Berlin: Springer Verlag.

_____1991. *Infectious Diseases of Humans: Dynamics and Control.* Oxford University Press, 1991.

Anderson, R. M., B. D. Turner and L. R. Taylor (eds.). 1979. *Population Dynamics* (20th Symposium of the British Ecological Society). Oxford: Blackwell.

Anderson, R. M., P. T. Whitfield and C. A. Mills. 1977. 'An experimental study of the population dynamics of an ectoparasite *Transversotrema patialense*: The cercarial and adult stages', *Journal of Animal Ecology*, 46:555–580.

Anon. 2009. Lawrence Slobodkin, (1928–2009). Obituary, *The Times*, 13 November.

Anker, Peder. 2001. *Imperial Ecology: Environmental Order in the British Empire, 1895–1945.* Cambridge, MA: Harvard University Press.

Axelrod, Robert M. 1984. *The Evolution of Cooperation.* NY: Basic Books.

Balfour-Browne, W. A. F. 1940, 1950, 1958. *British Water Beetles.* 3 Vols. London: Ray Society.

Barber, Bernard. 1962. *Science and the Social Order.* New York: Collier.

Barlow, Robert B. Jr, John E. Dowling and Gerald Weissman (eds.). 1993. *The Biological Century: Friday Evening Talks at the Marine Biological Laboratory.* Wood's Hole, MA: The Marine Biological Laboratory.

Bartlett, Maurice S. 1960. *Stochastic Processes in Ecology and Epidemiology.* London: Methuen.

Barton, Ruth. 1990. '"An influential set of chaps": The X-Club and Royal Society politics, 1864–85,' *British Journal for the History of Science*, 23:53–81.

_____1998. "'Huxley, Lubbock and half a dozen others': Professionals and gentlemen in the formation of the X-Club, 1851–1864', *Isis* lxxxix:410–444.

_____2003. "Men of science': language, identity and professionalization in the mid-Victorian scientific community', *History of Science*, xli:73–119.

Bastiansen, Otto (ed.). 1964. *The Law of Mass Action: A centenary volume, 1864–1964*. Oslo: Norske Videnskaps-Akademi.

Beddington, J. R. 1974. 'Age distribution and the stability of simple discrete time population models'. *Journal of Theoretical Biology*, 47:65–74.

_____1980. 'How to count whales'. *New Scientist*, 87:194–196.

Beddington, J. R., D. J. Agnew and C. W. Clark. 2007. 'Current problems in the management of marine fisheries'. *Science*, 316:1713–1716.

Beddington, J. R., C. A. Free and J. H. Lawton. 1975. 'Dynamic complexity in predator-prey models framed in difference equations'. *Nature*, 255:58–60.

_____1976. 'Concepts of stability and resilience in predator-prey models'. *Journal of Animal Ecology*, 45:791–816.

Beddington, J. R. and C. A. Free. 1976. 'Age structure effects in predator-prey interactions'. *Theoretical Population Biology*, 9:15–24.

Beddington, J. R., M. P. Hassell and J. H. Lawton. 1976. 'The components of arthropod predation: The predator rate of increase'. *Journal of Animal Ecology*, 45:165–185.

Beddington, J. R. and R. M. May. 1977. 'Harvesting natural populations in a randomly fluctuating environment', *Science*, 197:63–65.

Beddington, J. R., A. A. Rosenberg, J. A. Crombie and G. P. Kirkwood. 1990. 'Stock assessment and the provision of management advice for the short-fin squid fishery in Falkland Islands waters'. *Fisheries Research*, 8:351–365.

Beddington, J. R. and D. B. Taylor. 1973. 'Optimum age specific harvesting of a population'. *Biometrics*, 29:801–09.

Beddington, J. R., C. M. K. Watts and W. D. C. Wright, 1975. 'Optimal cropping of self-reproducible natural resources', *Econometrica*, 43:789–802.

Bertalanffy, Ludwig von. 1968. *General System Theory: Foundation, Development, Application*. New York: George Braziller.

Bezemer, T. M., T. H. Jones and K. J. Knight. 1998. 'Long-term effects of elevated CO_2 and temperature on populations of the peach potato aphid *Myzus persicae* and its parasitoid *Aphidius matricariae*'. *Oecologia*, 116:129–35.

Bijker, Wiebe E., Roland Bal and Ruud Hendriks. 2009. *The Paradox of Scientific Authority: The Role of Scientific Advice in Democracies*. Cambridge MA: MIT Press.

Birmingham, Karen. 2003. 'Roy Anderson', *Nature Medicine*, 9:492.

Bernal, J. D. 1939. *The Social Function of Science*. London: G. Routledge.

————1950. 'Peace or War', *Modern Quarterly*, 5:291–294.

Bloor, David. 1976. *Knowledge and Social Imagery*. London: Routledge and Kegan Paul.

Borman, F. Herbert and Gene E. Likens. 1979. *Pattern and Process in a Forested Ecosystem: Disturbance, Development, and the Steady State Based on the Hubbard Brook Ecosystem Study*. New York: Springer-Verlag.

Brady, John. 1995. 'John Stodart Kennedy, 1912–1993'. *Biographical Memoirs of Fellows of the Royal Society*, 41:244–260.

Bramwell, Anna. 1989. *Ecology in the Twentieth Century: A History*. Yale University Press.

Brock, William. H. 1992. *The Fontana History of Chemistry*. London: Fontana.

Brown, Andrew. 2005. *J. D. Bernal: The Sage of Science*. Oxford University Press.

Brown, J. H. and M. V. Lomolino. 1989. 'On the nature of scientific revolutions: Independent discovery of the equilibrium theory of island biogeography', *Ecology*, 70:1954–1957.

Brown, J. L. and G. H. Orians. 1970. 'Spacing patterns in mobile animals', *Annual Review of Ecology and Systematics*, 1:239–262.

Browne, Janet. 1983. *The Secular Ark: Studies in the History of Biogeography*. Yale University Press.

————2002. *Charles Darwin: The Power of Place*. London: Pimlico.

Brundtland, Gro Harlem (ed.). 1987. *Our Common Future*. Oxford University Press.

Bryden, Harry L., Hannah R. Longworth and Stuart A. Cunningham. 2005. 'Slowing of the Atlantic meridional overturning circulation at 25° N'. *Nature*, 438:655–657.

Butler, Edward Albert. 1923. *A Biology of the British Hemiptera-Heteroptera*. London: H. F. and G. Witherby.

Carr-Saunders, A. M. 1922. *The Population Problem: A Study in Human Evolution*. Oxford: Clarendon Press.

Carson, Rachel. 1962 and 1963. *Silent Spring*. Boston: Houghton Mifflin and London: Hamish Hamilton.

Cassirer, Ernst. 1956. (transl. W. H. Woglom and Charles W. Hendel). *The Problem of Knowledge: Philosophy, Science and History Since Hegel*. Yale University Press.

Chambers, R. and G. Conway. 1991. 'Sustainable Rural Livelihoods: Practical Concepts for the 21st Century'. IDS Discussion Paper 296. Institute of Development Studies, University of Sussex.

Chang, Hasok 'We have never been Whiggish (about phlogiston)', *Centaurus* 51 (2009), 239–264.

_____'The hidden history of phlogiston: How philosophical failure can generate historiographical refinement' (2010) www.hyle.org/journal/issues/16-2/chang.

Charnov, Eric L. 1976. 'Optimal foraging, the marginal value theorem'. *Theoretical Population Biology*, 9:129–136.

_____1989. 'Citation Classic'. *Current Contents*, 44:22.

Charnov, E. L. and J. R. Krebs. 1974. 'On clutch-size and fitness'. *Ibis*, 116: 217–219.

_____1975. 'The evolution of alarm calls: altruism or manipulation?' *American Naturalist*, 109:107–112.

Cherfas, Jeremy. 1987. 'Foreign office plays safe on Falkland fish'. *New Scientist*, 5 February

Clark, C. W. 1976. *Mathematical Bioeconomics*. New York: Wiley.

Clark, C. W. and R. Lamberson. 1982. 'An economic history and analysis of pelagic whaling'. *Marine Policy*, 6:103–120.

Clements, Frederick E. 1916. *Plant Succession: Analysis of the Development of Vegetation*. Publication 242. Carnegie Institution.

Clements, Frederick E. and Victor E. Shelford. 1939. *Bio-Ecology*. New York: Wiley.

Code, Lorraine. 1987. *Epistemic Responsibility*. Hanover N.H. University Press of New England.

Cody, Martin L. and Jared Diamond (eds.). 1975. *Ecology and Evolution of Communities*. Cambridge MA, Belknap.

Cohen, Joel. 1971. 'Mathematics as metaphor', Review of R. Rosen, *Dynamical System Theory in Biology*; *Science*, 172:674–675.

Collins, Harry and Robert Evans. 2007. *Rethinking Expertise*. University of Chicago Press.

Collins, Randall. 1998. *The Sociology of Philosophies: A Global Theory of Intellectual Change*. Cambridge MA: Belknap.

Comins, Hugh N., William D. Hamilton and Robert M. May. 1980. 'Evolutionary stable dispersion strategies', *Journal of Theoretical Biology*, 82:205–230.

Comins, H. N. and M. P. Hassell. 1987. 'The dynamics of predation and competition in patchy environments'. *Theoretical Population Biology*, 31:393–421.

Conant, Jennet. 2002. *Tuxedo Park: A Wall Street Tycoon and the Secret Palace of Science that Changed the Course of World War II*. New York: Simon and Schuster.

Constantino, R. F., J. M. Cushing, B. Dennis, and R. A. Desharnais. 1995. 'Experimentally induced transitions in the dynamic behaviour of insect populations', *Nature*, 375:227–230.

Conway, Gordon. 1976. 'Man versus Pests', Chapter 14 in Robert M. May (ed.), *Theoretical Ecology: Principles and Applications*. Oxford: Blackwell.

_____1977. 'Mathematical models in applied ecology'. *Nature*, 269:291–297.

_____(ed.). 1984. *Pest and Pathogen Control: Strategic, Tactical and Policy Models.* Chichester: Wiley.

_____1985. 'Agroecosystem analysis'. *Agricultural Administration.* 20:31–55.

_____1998. *The Doubly Green Revolution: Food For All in the Twenty-first Century.* Ithaca N.Y: Cornell University Press.

_____2003. 'Sustainable Agriculture', in Nigel Cross (ed.), *Evidence for hope: the Search for Sustainable Development: The Story of the International Institute for Environment and Development.* London: Earthscan.

_____2012. *One Billion Hungry: Can we feed the world?* Ithaca, NY: Cornell University Press.

Conway, Gordon R. and David S. McCauley. 1983. 'Third World development: intensifying tropical agriculture: The Indonesian experience'. *Nature*, 302:288–289.

Conway, Gordon R. and Edward B. Barbier. 1990. *After the Green Revolution.* London: Earthscan.

Conway, Gordon and Jules Pretty. 1991. *Unwelcome Harvest: Agriculture and Pollution.* London: Earthscan.

Conway, Simon. 2010. *A Loyal Spy.* London: Hodder and Staughton.

Crane, Diana. 1972. *Invisible Colleges: Diffusion of Knowledge in Scientific Communities.* University of Chicago Press.

Crawley, Michael J. 1975. 'Numerical responses of insect predators to changes in prey density'. *Journal of Animal Ecology*, 44:877–892.

_____1983. *Herbivory: The Dynamics of Animal-Plant Interactions.* Oxford: Blackwell.

_____1985. 'Reduction of oak fecundity by low-density herbivore populations'. *Nature*, 314:163–164.

_____1989. 'Insect herbivores and plant population dynamics'. *Annual Review of Entomology*, 34:531–564.

_____(ed.). 1986a. *Plant Ecology.* Oxford: Blackwell.

_____1986b. 'The population biology of invaders', in Hans Kornberg and M. H. Williamson (eds.). 'Quantitative aspects of the ecology of biological invasions'. *Phil. Trans. Roy. Soc. B*, 314:711–731.

_____1990. 'Rabbit grazing, plant competition and seedling recruitment in acid grassland'. *Journal of Applied Ecology*, 27:803–820.

_____(ed.). 1992. *Natural Enemies: The Population Biology of Predators, Parasites and Diseases.* Oxford: Blackwell.

_____1993. *GLIM for Ecologists.* Oxford: Blackwell.

_____1999. 'Bollworms, genes and ecologists'. *Nature*, 400:501–502.

_____2002. *Statistical Computing: An Introduction to Data Analysis Using S-Plus*. Chichester: Wiley.

_____2005. *The Flora of Berkshire: Including Those Parts of Modern Oxfordshire That lie South of the River Thames*. Harpenden: Brambleby Books.

_____2007a. *The R Book*. Chichester: Wiley.

_____2007b. 'Plant population dynamics', in Robert M. May and Angela McLean (eds.). *Theoretical Ecology: Principles and Applications*. 3rd edition. Oxford University Press, ch. 6.

Crawley, M. J., S. L. Brown, R. S. Hails, D. D. Kohn and M. Rees. 2001. 'Transgenic crops in natural habitats'. *Nature*, 409:682–3.

Crawley, Michael J. and John R. Krebs. 1992. 'Foraging theory' in Michael. J. Crawley (ed.), *Natural Enemies: The Population Biology of Predators, Parasites and Diseases*. Oxford University Press, 90–114.

Crawley. M. J. and R. M. May. 1987. 'Population-dynamics and plant community structure — competition between annuals and perennials'. *Journal of Theoretical Biology*, 125:475–489.

Cressey, Daniel. 2010. 'Quango bonfire kindles advice fears'. *Nature*, 467: Naturenews online.

Croll, N. A., R. M. Anderson, T. W. Gyorkos and E. Ghadirian 1982. 'The population biology and control of *Ascaris-lumbricoides* in a rural community in Iran'. *The Royal Society of Tropical Medicine*, H76:187–197.

Croll, Neil A. and John H. Cross. 1983. *Human Ecology and Infectious Diseases*. New York: Academic Press.

Cross, Nigel (ed.). 2003. *Evidence for Hope: The Search for Sustainable Development: The Story of the International Institute for Environment and Development*. London: Earthscan.

Crowcroft, Peter. 1991. *Elton's Ecologists: A History of the Bureau of Animal Population*. University of Chicago Press.

Cruwys, Liz and Beau Riffenburgh. 2002. 'Bernard Stonehouse: biologist, writer and educator'. *Polar Record*, 38:157–169.

Daston, Lorraine. 1995. 'The moral economy of science' in Arnold Thackray (ed.), *Constructing Knowledge in the History of Science*; *Osiris*, 10:1–24.

Daston, Lorraine and Elizabeth Lunbeck (eds.). 2011. *Histories of Scientific Observation*. University of Chicago Press.

Davies, R. G. 1995. *A Short History of Entomology at Imperial College*. (typescript) Imperial College archives.

Dawkins, M. S., T. R. Halliday and R. Dawkins (eds.). 1991. *The Tinbergen Legacy*. London: Chapman and Hall.

Dawkins, Richard. 2000. Obituary of W. D. Hamilton, *The Independent*, 3 October.

Dawkins, R. and J. R. Krebs. 1978. 'Animal signals: Information or manipulation?' in J. R. Krebs and N. B. Davies (eds.), *Behavioural Ecology: An Evolutionary Approach*. Oxford: Blackwell, 282–309.

————1979. 'Arms races between and within species'. *Proceedings of the Royal Society B*, 205:489–511.

Desmarais, Ralph J. 2004–11. 'Tots and Quots (Act.1931–1946)', *Oxford Dictionary of National Biography* (on-line edition). Oxford University Press.

Desmond, Adrian. 2001. 'Redefining the X Axis: "Professionals", "amateurs" and the making of mid-Victorian biology — a progress report', *Journal of the History of Biology, xxxiv*:3–50.

Diamond, Jared M. 1975. 'Assembly of species communities' in M. L. Cody and J. M. Diamond (eds.), *Ecology and Evolution of Species Communities*. Cambridge, MA: Harvard University Press, 342–444.

Diamond, Jared M. and Robert M. May. 1981. 'Island Biogeography and the Design of Natural Resources' chapter 9 in Robert M. May (ed.), *Theoretical Ecology: Principles and Applications*. 2nd ed. Oxford: Blackwell.

Diamond, J. M. and E. Mayr. 1976. 'Species area relationship for birds of the Solomon Archipelago', *Proceedings of the National Academy of Sciences*, USA, 73:262–266.

Dobzhansky, T. G. 1973. 'Nothing makes sense in biology except in the light of evolution', *American Biology Teacher*, 35:125–129.

Douglas, Mary. 1986. *How Institutions Think*. Syracuse University Press.

Drickamer, Lee and Donald Dewsbury (eds.). 2010. *Leaders in Animal Behaviour: The Second Generation*. Cambridge University Press.

Dubin, Louis I. 1950. 'Alfred James Lotka, 1880–1949', *Journal of the American Statistical Association*, 45:138–139.

Duhem, Pierre. 1908 and 1969. *Sozein Ta Phainomena: Essai sur la notion de théorie de Platon à Galilée* (Paris: Hermann, 1908); *To Save the Phenomena*. (Transl. of 1908 edition. E. Doland and C. Maschler) University of Chicago Press.

Easlea, Brian. 1973. *Liberation and the Aims of Science: An Essay on the Obstacles to the Building of a Beautiful World*. Totowa, NJ: Rowman and Littlefield.

Edgerton, David, *Science, Technology and the British Industrial 'Decline', ca., 1870–1970*. Cambridge University Press, 1996.

Egerton, F. N. 1977 'A bibliographic guide to the history of general ecology and population ecology', *History of Science*, 15:189–215.

————1983. 'The history of ecology: achievements and opportunities, part one, *Journal of the History of Biology*, 16:259–310.

Elton, Charles S. 1927. *Animal Ecology*. London: Sidgwick and Jackson.

————1930. *Animal Ecology and Evolution*. Oxford: Clarendon Press.

————1942. *Voles, Mice and Lemmings: Problems in Population Dynamics,* Oxford University Press.

————1958. *The Ecology of Invasions by Animals and Plants*. London: Chapman and Hall.

————1966. *The Patterns of Animal Communities*. London: Methuen.

Emden, Helmut F. van. 1998. 'Whatever happened to integrated control'. *Phytoparasitica*, 26:3–7.

Esch, Gerald W. 2007. *Parasites and Infectious Diseases: Discovery by Serendipity and Otherwise*. Cambridge University Press.

Etkowitz, Henry and Carol Kemelgor. 1998. 'The role of research centres in the collectivisation of academic science', *Minerva*, 36:271–288.

Falls, J. B. and J. R. Krebs. 1975. 'Sequences of songs in repertoires of western meadow larks, *Canadian Journal of Zoology*, 53:1165–1178.

Ferguson, N. M., C. A. Donnelly and R. M. Anderson. 2001a. 'Transmission intensity and impact of control policies on the foot and mouth epidemic in Great Britain', *Nature*, 423:542–548.

————2001b. 'The foot-and-mouth epidemic in Great Britain: pattern of spread and impact of interventions', *Science*, 292:1155–1160.

Feyerabend, Paul. 1975. *Against Method: Outline for an Anarchist Theory of Knowledge*. London: Humanities Press.

Fischedick, Kaat Schulte. 2000. 'From survey to ecology: The role of the British vegetation committee, 1904–1913', *Journal of the History of Biology*, 33:291–314.

Fisher, R. A. 1922. 'On the mathematical foundations of theoretical statistics', *Philosophical Transactions of the Royal Society of London*, Series A, 309–368.

————1930. *The Genetical Theory of Natural Selection*. Oxford University Press.

Fisher, R. A., A. S. Corbet and C. B. Williams. 1943. 'The relation between the number of species and the number of individuals in a random sample of an animal population', *Journal of Animal Ecology*, 12:42–58.

Flahault, Charles. 1893. *La répartition géographique des végétaux dans un coin du Languedoc*. Montpellier: Département de l'Hérault.

Fleck, Ludwik. 1979. (translated from the Basel edition of 1935 by Fred Bradley and Thaddeus Trenn. Thaddeus Trenn and Robert K. Merton (eds.) *Genesis and Development of a Scientific Fact*. University of Chicago Press.

Follett, Sir Brian. 2002. *Infectious Diseases in Lifestock*. Royal Society of London.

Ford, E. B. 1965, *Mendelism and Evolution*. 8th. edition. London: Methuen.

————1971. *Ecological Genetics*. 3rd. edition. London: Chapman and Hall.

Ford, John. 1971. *The Role of Trypanosomiases in African Ecology: A Study of the Tse tse fly.* Oxford: Clarendon Press.

Foster, G. N. 2004. 'Balfour-Browne (William, Alexander, Francis) (1874–1967)'. *Oxford Dictionary of National Biography.* Oxford University Press.

Franklin, Ursula, Hannah Gay and Angela Miles (eds.). 1993. *Women in Science and Technology: The Legacy of Margaret L. Benston. Canadian Woman Studies,* Issue 13.

Fretwell, Stephen D. 1975. 'The impact of Robert MacArthur on ecology'. *Annual Review of Ecology and Systematics,* 6:1–13.

Fuller, Steve. 2000. *Thomas Kuhn: A Philosophical History for Our Times.* University of Chicago Press.

Galison, Peter. 1987. *How Experiments End.* University of Chicago Press.

_____1997. *Image and Logic: A Material Culture of Microphysics.* University of Chicago Press.

Gardiner, J. Stanley. 1913. Obituary of Adam Sedgwick. *The Zoologist.* March issue.

Gardner, Mark R. and W. Ross Ashby. 1970. 'Connectance of large dynamic (cybernetic) systems: critical values for stability', *Nature,* 228:784.

Gause, G. F. 1934. *The Struggle for Existence.* Baltimore, MD: Williams & Wilkins.

Gay, Hannah. 1976. 'Radicals and Types: A critical comparison of the methodologies of Popper and Lakatos and their use in the reconstruction of some 19th century chemistry'. *Studies in the History and Philosophy of Science,* 7:1–51.

_____1996. 'Invisible Resource: William Crookes and his circle of support, 1871–81'. *British Journal for the History of Science,* 29:311–336.

_____1999. 'Explaining the universe: Herbert Spencer's attempt to synthesize political and evolutionary ideas'. *Endeavour,* 23:56–59.

_____2000 'Pillars of the college: Assistants at the Royal College of Chemistry, 1846–71'. *Ambix* 47:135–169.

_____2007. *The History of Imperial College London, 1907–2007: Higher Education and Research in Science, Technology and Medicine.* London: Imperial College Press.

_____2008 'Technical assistance in the world of London science, 1850–1900'. *Notes and Records of the Royal Society,* 62:51–75.

_____2012. 'Before and after *Silent Spring*: From chemical pesticides to biological control and integrated pest management — Britain, 1945–80', *Ambix,* 59:88–108.

Gidney, W. T. 1908. *The History of the London Society for Promoting Christianity Among Jews From 1809–1908.* London: internal centenary history of the society.

Gieryn, T. F. 1999. *Cultural Boundaries of Science: Credibility on the line.* University of Chicago Press.

Giedymin, Jerzy. 1982. *Science and Convention: Essays on Henri Poincaré's Philosophy of Science and Conventionalist Theories.* Oxford: Pergamon Press.

Glacken, C. J. 1967. *Traces on the Rhodian Shore.* University of California Press.

Gleason, H. A. 1926. 'The individualistic concept of plant association', *Bulletin of the Torrey Botanical Club*, 53:7–26.

Gleick, James. 1987. *Chaos: Making a New Science.* New York: Viking.

Golinski, Jan. 1998. *Making Natural Knowledge: Constructivism and the History of Science.* Cambridge University Press.

Golley, Frank Benjamin. 1993. *A History of the Ecosystem Concept in Ecology: More than the Sum of its Parts.* New Haven: Yale University Press.

Goodman, Daniel. 1975. 'The theory of diversity-stability relationships in ecology', *Quarterly Review of Biology*, 50:237–266.

Gordin, Michael D. 2011. 'Seeing is believing: Professor Vagner's wonderful world' in Lorraine Daston and Elizabeth Lunbeck (eds.). *Histories of Scientific Observation.* University of Chicago Press, 136–55.

Gould, Stephen J. 1989. *Wonderful Life.* New York: Norton.

Gould, Stephen J. and Lewontin, Richard. 1979. 'The Spandrells of San Marco and the Panglossian Paradigm: A critique of the adaptationist programme'. *Proceedings of the Royal Society B*, 205:475–488.

Granovetter, Mark. 1983. 'The strength of weak ties: A network theory revisited', *Sociological Theory*, 1: 201–33.

Gray, A. J., M. J. Crawley and P. J. Edwards (eds.). 1987. *Colonization, Succession and Stability.* Oxford: Blackwell.

Grenfell B. T., and R. M. Anderson. 1989. 'Pertussis in England and Wales': An investigation of transmission dynamics and control by means of vaccination'. *Proc. Roy. Soc. Lond. B*, 236:213–252.

Grinnell, Joseph. 1917. 'The niche-relationships of the California Thrasher'. *The Auk,* 34:427–433.

Gunderson, Lance H. and C. S. Holling (eds.). 2002. *Panarchy: Understanding Transformations in Human and Ecological Systems.* Washington DC: Island Press.

Gutting, Gary. 2011. *Thinking the Impossible: French Philosophy Since 1960.* Oxford University Press.

Hacking, Ian. 1983. *Representing and Intervening: Introductory Topics in the Philosophy of Natural Science.* Cambridge University Press.

Haeckel, Ernst H. P. A. 1866. *Generelle Morphologie der Organismen.* 2 Vols. Berlin: Georg Reimer.

Hagen, Joel B. 1992. *An Entangled Bank: The Origins of Ecosystem Ecology.* New Brunswick NJ: Rutgers University Press.

Hairston Sr. Nelson G., Frederick E. Smith and Lawrence B. Slobodkin. 1960. 'Community structure, population control and competition'. *The American Naturalist*, 94:421–425.

Haldane, Andrew G. and Robert M. May. 2011a. 'Systemic risk in banking ecosystems'. *Nature,* 469:351–355.

————2011b. 'The birds and the bees and the big banks'. *Financial Times,* 20 February.

Halfon, Efraim (ed.). 1979. *Theoretical Systems Ecology: Advances and Case Studies.* New York: Academic Press.

Halley, J. M., H. N. Comins, J. H. Lawton and M. P. Hassell. 1994. 'Competition, succession and pattern in fungal communities: Towards a cellular automaton model'. *Oikos,* 70:435–442.

Hamilton, W. D. 1996–2005. *Narrow Roads of Gene Land: the collected papers of W. D. Hamilton.* 3 Vols. Oxford: Spektrum.

Hamilton, W. D. and Robert May. 1977. 'Dispersal in Stable Habitats'. *Nature,* 269:578–581.

Haraway, Donna J. 1981–2. 'The high cost of information in post-World War II evolutionary biology: Ergonomics, semiotics and the sociobiology of communication systems', *Philosophical Forum,* 13:244–278.

————1988. 'Situated knowledges: The science question in feminism and the privilege of partial perspective'. *Feminist Studies,* 14:575–599.

Hardin, Garrett. 1968. 'The tragedy of the commons', *Science,* 162:1243–1248.

Hardy, Allister. 1968. 'Charles Elton's influence in ecology', *Journal of Animal Ecology,* 37:3–8.

Harper, J. L. 1977. Review of May (ed.) *Theoretical Ecology* (1976). *Journal of Ecology,* 65:1009–1012.

————1977. 'The contributions of terrestrial plant studies to the development of the theory of ecology' in C. E. Goulden (ed.), *Changing Scenes in the Life Sciences, 1776–1976,* (Special publication 12, Academy of Natural Sciences, Philadelphia), 139–157.

Harrison, Gordon. 1971. *Earthkeeping: The War with Nature and a Proposal for Peace.* London: Hamilton.

————1978. *Malaria, Mosquitoes and Man: A History of Hostilities Since 1880.* London: John Murray.

Harvey, Paul and T. H. Clutton Brock. 1979. 'Comparison and adaptation'. *Proceedings of the Royal Society B,* 205:547–565.

Harvey, P. H. and J. R. Krebs. 1990. 'Comparing brains'. *Science,* 249:140–146.

Harvey, W. F. 1944. 'Anderson Gray McKendrick, 1876–1943', *Edinburgh Medical Journal,* 50:500–506.

Hassell, Michael P. 1966. 'Evaluation of parasite or predator response'. *Journal of Animal Ecology*, 35:65–75.

_____1969. 'A population model for the interaction between *Cyzenis albicans* (Fall) (*Tachinidae*) and *Operophtera brumata* (L) (*Geomentridae*) at Wytham, Berkshire'. *Journal of Animal Ecology*, 38:567–576.

_____1976. 'Arthropod predator-prey systems', Chapter 5 in Robert M. May (ed.) *Theoretical Ecology: Principles and Applications*. Oxford: Blackwell.

_____1978. *The Dynamics of Arthropod Predator-Prey Systems*. Princeton University Press.

_____1984. 'Host-parasitoid models and biological control' in Gordon R. Conway (ed.), *Pest and Pathogen Control: Strategic, Tactical and Policy Models*. Chichester: Wiley.

_____2000. *The Spatial and Temporal Dynamics of Host-Parasitoid Interactions*. Oxford University Press.

Hassell, M. P., H. N. Comins and R. M. May. 1991. 'Spatial structure and chaos on insect population dynamics'. *Nature*, 353:255–258.

Hassell, M. P., M. J. Crawley, H. C. J. Godfray and J. H. Lawton (paper submitted by R. M. May). 1998. 'Top-down versus bottom-up and the Ruritanian bean bug', *Proceedings of the National Academy of Sciences USA*, 95:10661–10664.

Hassell, M. P., J. H. Lawton and J. R. Beddington. 1977. 'Sigmoid functional responses by invertebrate predators and parasitoids'. *Journal of Animal Ecology*, 46:249–262.

Hassell, Michael P., John H. Lawton and Robert M. May. 1976. 'Patterns of dynamical behaviour in single-species populations'. *Journal of Animal Ecology*, 45:471–486.

Hassell, M. P., S. W. Pacala, R. M. May, and P. L. Chesson. 1991. 'The persistence of host-parasitoid associations in patchy environments'; Part 1 'A general criterion', *American Naturalist*, 138:568–583; S. W. Pacala and M. P. Hassell, Part 2. 'Evaluation of field data'. *ibid.* 584–605.

Hassell, M. P. and R. M. May. 1973. 'Stability in host-parasite models', *Journal of Animal Ecology*, 42:693–726.

_____1974. 'Aggregation of predators and insect parasites and its effect on stability'. *Journal of Animal Ecology*, 43:567–594.

Hassell, M. P. and D. J. Rogers. 1972. 'Insect parasite responses in the development of population models'. *Journal of Animal Ecology*, 41:661–676.

Hassell, M. P., T. R. E. Southwood, and P. M. Reader. 1987. 'The dynamics of viburnum whitefly (*Aleutrachelus jelineki*): A case study of population regulation'. *Journal of Animal Ecology*, 56:283–300.

Hawking, Stephen. 1998. *A Brief History of Time.* New York: Bantam.

Heads, Philip and John Lawton. 1986. 'Beat back bracken biologically'. *New Scientist,* 111:40–53.

Hector, A. *et al.* 1999. 'Plant diversity and productivity experiments in European grasslands'. *Science,* 286:1123–1127.

Heims, Steve J. 1991. *The Cybernetics Group: Constructing a Social Science for Postwar America.* Cambridge MA: MIT Press.

Hesse, Mary. 1974. *The Structure of Scientific Inference.* University of California Press.

Hestmark, Geir. 2000. '*Oeconomia Naturae L.*', *Nature,* 405:19.

Higgins, J. 1967. 'Kinetics of Oscillating Reactions', *Industrial and Engineering Chemistry.* 19 May:19–62.

Holland, John H. 1995. *Hidden Order: How Adaptation Builds Complexity.* Reading MA: Addison-Wesley.

Holling, C. S. 1961. 'Principles of insect predation'. *Annual Review of Entomology,* 6:163–182.

_____1965, 'The functional response of predators to prey density and its role in mimicry and population regulations'. *Memoirs of the Entomological Society of Canada,* 45:1–60.

_____1966. 'The functional response of invertebrate predators to prey density'. *Memoirs of the Entomological Society of Canada,* 48:1–86.

_____1973. 'Resilience and stability of ecological systems'. *Annual Review of Ecology and Systematics,* 4:1–23.

Holt, S. J. and R. J. H. Beverton. 1957. *On the Dynamics of Exploited Fish Populations.* London: H. M. Stationery Office (Reprinted in 2004; Caldwell NJ: Blackburn Press).

Hope Simpson, J. F. Rev. David E. Evans. 2004. 'Tansley, Sir Arthur George (1871–1955)', *Oxford Dictionary of National Biography.* Oxford University Press.

Howard, L. O. and W. F. Fiske. 1911. *Entomological Bulletin* 91. U. S. Bureau of Entomology.

Hsieh, Chih-hao, Christian S. Reiss, John R. Hunter, John R. Beddington, Robert M. May and George Sugihara. 2006. 'Fishing elevates variability in the abundance of exploited species'. *Nature,* 443:859–862.

Hubbell, Stephen. 2001. *The Unified Neutral Theory of Biodiversity and Biogeography.* Princeton University Press.

Hull, David L. 1974. *Philosophy of the Biological Sciences.* Englewood Cliffs, NJ: Prentice Hall.

Huntingford, Felicity. 2001. John Michael (Mike) Cullen (1927–2001). Obituary, *The Independent,* 18 April.

Hutchinson, G. Evelyn. 1948. 'Circular causal systems in ecology', *Annals of the New York Academy of Sciences*, 50:221–223, 236–246.

_____1957. 'Concluding remarks', Conference on population studies, animal ecology and demography. *Cold Spring Harbour Symposia on Quantitative Biology*, 22, 415–427.

_____1957–93. *A Treatise in Limnology.* 4 Vols. New York: Wiley.

_____1959. 'Homage to Santa Rosalia: Or why are there so many kinds of animals?' presidential address to the American Society of Naturalists, *American Naturalist*, 93:145–159.

_____1961. 'The Paradox of the Plankton', *American Naturalist*, 95:137–147.

_____1978. *An Introduction to Population Ecology.* Yale University Press.

_____1975. 'Variations on a theme by Robert MacArthur', in Martin L. Cody and Jared Diamond (eds.), *Ecology and Evolution of Communities.* Cambridge MA, Belknap.

_____1979. *The Kindly Fruits of the Earth.* Yale University Press.

Hutchinson, G. E., E. S. Deevey and A. Wollack. 1939. 'The oxidation-reduction potential of lake waters and their ecological significance', *Proceedings of the American Academy of Sciences*, 25:67–90.

Hutchinson, G. E. and A. Wollack. 1940. 'Studies on Connecticut lake sediments II: Chemical analyses of a core from Linsley Pond, North Branford'. *American Journal of Science,* 238:493–517.

Hutchinson, G. Evelyn and Vaughan T. Bowen. 1950. 'Limnological studies in Connecticut' IX: A quantitative radiological study of the phosphorous cycle in Linsley Pond', *Ecology*, 31:194–203.

Huxley, Julian. 1942. *Evolution: The Modern Synthesis.* London: Allen and Unwin.

Iliffe, Rob. History of Science, on the website of the School of Advanced Study, University of London; http://www.history.ac.uk/makinghistory/resources/articles/history_of_science.

Innis, George S. (ed.). 1975. *New Directions in the Analysis of Ecological Systems* (Society for Computer Simulation) La Jolla CA.

Islam, Z. and M. J. Crawley. 1983. 'Compensation and regrowth in ragwort (*Senecio-jacobaea*) attacked by the cinnabar moth'. *Journal of Ecology*, 71:829–843.

Israel, Giorgio. 1988. 'On the contributions of Volterra and Lotka to the development of modern biomathematics' *History and Philosophy of the Life Sciences*, 10:37–49.

_____ 'The scientific heritage of Vito Volterra and Alfred Lotka in mathematical biology' giorgio.israel.googlepages.com/Art82.pdf.

Jasanoff, Sheila. 1986. *Risk Management and Political Culture*. New York: Russell Sage Foundation.

_____1997. 'Civilization and Madness: The great BSE scare of 1996', *Public Understanding of Science*, 6:221–232.

_____2005. *Designs on Nature: Science and Democracy in Europe and the United States*. Princeton NJ: Princeton University Press.

Jensen, J. Vernon. 1970. 'The X-Club Fraternity of Victorian Scientists', *British Journal for the History of Science*, 5:63–72.

Kant, Immanuel. 1952. (Transl. and ed. James Creed Meredith). *The Critique of Judgement*. Oxford University Press.

Keeling, M. J., M. E. J. Woolhouse, R. M. May, G. Davies and B. T. Grenfell. 2003. 'Modelling vaccination strategies against foot and mouth disease'. *Nature,* 421:136–142.

Kenny, Michael G. 2004. 'John R. Baker on eugenics, race, and the public role of the scientist', *Isis*, 95:394–419.

Kermack, W. O. and A. G. McKendrick. 1927. 'A contribution to the mathematical theory of epidemics', *Proceedings of the Royal Society A*, 115:700–721.

King, David A. 2004. 'The scientific impact of nations: What different countries get for their research spending'. *Nature*, 430:311–316.

Kingsland, Sharon E. 1995. *Modeling Nature: Episodes in the History of Population Ecology.* 2nd. edition. University of Chicago Press.

Kohler, Robert E. 1994. *Lords of the Fly: "Drosophila" Genetics and the Experimental Life*. University of Chicago Press.

Kohn, A. J. 1971. 'Phylogeny and Biogeography of *Hutchinsonia*: G. E. Hutchinson's influence through his graduate students', *Limnology and Oceanography*, 16: 173–176.

Kornberg, Hans and M. H. Williamson (eds.). 1986. *Quantitative Aspects of the Ecology of Biological Invasions. Philosophical Transactions of the Royal Society B*, 314.

Krebs, John R. 1971. 'Territory and breeding density in the Great Tit, *Parus Major* L'. *Ecology*, 52:2–22.

_____1973. 'Social learning and the significance of mixed-species flocks of chickadees (*Parus* spp.)'. *Canadian Journal of Zoology*, 51:1275–1288.

_____1976. 'Habituation and song repertoires in the great tit'. *Behavioural Ecology and Sociobiology*, 1:215–227.

_____1977. 'Communities: ecology and evolution'. *Bioscience*, 27:50.

_____1990. 'Food storing birds: adaptive specialization in brain and behaviour?' *Philosophical Transactions of the Royal Society of London B*, 329:153–160.

_____1991. 'Animal Communication: Ideas derived from Tinbergen's activities' in M. S. Dawkins, T. R. Halliday and R. Dawkins (eds.), *The Tinbergen Legacy*. London: Chapman and Hall.

_____1997. *Bovine Tuberculosis in Cattle and Badgers* (Report by the Independent Scientific Review Group to the Rt. Hon. Dr. Jack Cunningham MP). MAFF.

_____1998. 'The work of the NERC'. *Science in Parliament*, 55:4–6.

_____2000a. (chairman's report) in *Genetically Modified Foods: Widening the Debate on Health and Safety*. OECD Edinburgh Conference on the scientific and health aspects of genetically modified foods, 7–12.

_____2000b. 'GM foods in the UK between 1996 and 1999: Comments on "Genetically modified crops: Risks and promise" by Gordon Conway'. *Conservation Ecology*, 4:11.

_____2005. 'Risk: Food, fact and fantasy'. Croonian Lecture, 2004. *Proceedings of the Royal Society B*, 360:1133–1144.

_____2010. 'Luck, Chance and Choice' in Lee Drickamer and Donald Dewsbury (eds.), *Leaders in Animal Behaviour: The Second Generation*. Cambridge University Press, Chapter 14.

Krebs, J. R., R, M. Anderson, T. H. Clutton-Brock, C. A. Donnelly, S. Frost, W. I. Morrison, R. Woodroffe and D. Young. 1998. 'Badgers and Bovine TB: Conflicts between conservation and health'. *Science*, 279:816–818.

Krebs, J. R. and N. B. Davies (eds.). 1978. *Behavioural Ecology: An Evolutionary Approach*. Oxford: Blackwell.

_____1981. *An Introduction to Behavioural Ecology*. Oxford: Blackwell.

Krebs, J. R., J. T. Erichsen, J. T. Webber and E. L. Charnov. 1977. 'Optimal prey selection in the great tit, *Parus Major*'. *Animal Behaviour*, 25:30–38.

Krebs, J. R. and G. Horn (eds.). 1990. 'Behavioural and neural aspects of learning and memory'. *Philosophical Transactions of the Royal Society of London B*, 329:97–227.

Krebs, John R., Alejandro Kacelnik and Peter Taylor. 1978. 'Tests of optimal sampling by foraging great tits'. *Nature*, 275:27–31.

Krebs, J. R., R. M. May and M. P. H. Stumpf. 2002. 'Theoretical models of sheep BSE reveal possibilities: But we must remember that these theories are based on speculation not fact'. *Nature*, 415:115.

Krebs, J. R. and B. Partridge. 1973. 'Significance of head tilting in the great blue heron'. *Nature*, 243:533–35.

Krebs, J. R., J. Ryan and E. L. Charnov. 1974. 'Hunting by expectation or optimal foraging? A study of patch use by chickadees'. *Animal Behaviour*, 22:953–964.

Krebs, J. R., D. F. Sherry, S. D. Healy, V. H. Perry and A. L. Vaccarino. 1989. 'Hippocampal specialization of food storing birds', *Proceedings of the National Academy of Sciences,* 86:1388–1392.

Krebs. J. R. and S. Sjölander. 1992. 'Konrad Zacharius Lorenz', *Biographical Memoirs of Fellows of the Royal Society,* 38:209–228.

Krebs, J. R. and C. L. Whitney. 1975. 'Mate selection in Pacific Tree Frogs', *Nature,* 255:325–326.

Kruuk, Hans. 2003. *Niko's Nature: The Life of Niko Tinbergen and his Science of Animal Behaviour.* Oxford University Press.

Kuhn, Thomas S. 1970. *The Structure of Scientific Revolutions.* (2nd ed.). University of Chicago Press.

Lack, David *et al.* 1973. 'Memoir' *Ibis,* 115:421–441.

Lakatos, Imre. 1970a, 1976. 'History of science and its rational reconstructions' Reprinted in C. Howson (ed.) *Method and Appraisal in the Physical Sciences.* Cambridge University Press, 1–39.

_____1970b. 'Falsification and the Methodology of Scientific Research Programmes' in Imre Lakatos and Alan Musgrave (eds.), *Criticism and the Growth of Knowledge.* Cambridge University Press, 91–195.

Lankester, K. and J. R. Beddington. 1986. 'An age-structured population model applied to the gray whale (*Eschrichtius robustus*)'. *Reports of the International Whaling Commission,* 36:353–338.

Lapointe, Sandra, Jan Woleński, Mathieu Marion and Wioletta Miskiewicz (eds.). 2009. *The Golden Age of Polish Philosophy, Kazimierz Twardowski's Philosophical Legacy.* Dordrecht: Springer.

Latour, Bruno. 1987. *Science in Action: How to Follow Scientists and Engineers Through Society.* Harvard University Press.

_____1993. *We Have Never Been Modern.* (Transl. Catherine Porter). London: Harvester Press.

_____2004. 'Why has critique run out of steam? From matters of fact to matters of concern', *Critical Inquiry,* 30:225–48.

_____2004. *Politics of Nature: How to bring the sciences into democracy.* (Trnsl. Catherine Porter). Harvard University Press.

Latour, Bruno and Steve Woolgar. 1986. *Laboratory Life: The Construction of Scientific Facts.* Princeton University Press.

Lawton, John H. 1969a. 'Studies on the ecological energetics of Damselfy larvae (Odonata: Zygoptera)'. PhD thesis, University of Durham.

_____1969b. 'Studies on the ecological energetics of damselfly larvae'. *Journal of Animal Ecology,* 38:28–29.

_____1970a. 'Feeding and food energy assimilation in larvae of the damselfly (*Pyrrhosoma nymphula*) (Sulzer) (Odonata: Zygoptera). *Journal of Animal Ecology*, 39:669–689.

_____1970b. 'A population study on larvae of the damselfly'. (*Pyrrhosoma nymphula*) (Sulzer) (Odonata: Zygoptera), *Hydrobiologia*, 36:33–52.

_____1983. 'Plant architecture and the diversity of phytophagous insects'. *Annual Review of Entomology*, 28:23–39.

_____1987. 'Are there assembly rules for successional communities?' in A. J. Gray, M. J. Crawley and P. J. Edwards (eds.), *Colonization, Succession and Stability*. Oxford: Blackwell: 225–244.

_____1988. 'Biological control of Bracken in Britain'. *Philosophical Transactions of the Royal Society of London B*, 318:335–355.

_____1991. 'View from the Park: Warbling in different ways' *Oikos*, 60:273–274.

_____1998. 'Ecological experiments with model systems: The Ecotron facility in context', in W. J. Resetarits Jr. and J. Bernardo (eds.), *Experimental Ecology: Issues and Perspectives*. Oxford University Press, 170–182.

_____2000. *Community Ecology in a Changing World*, (O. Kinne (ed.), Ecology Institute, Oldendorf, Ge.). *Excellence in Ecology*, 11.

_____2004. Japan Prize Commemorative Lecture. 'Biodiversity, Conservation and Sustainability'. *Notes and Records of the Royal Society of London*, 58:321–333.

Lawton, J. H. *et al.* 1993. 'The Ecotron: A controlled environmental facility for the investigation of population and ecosystem processes'. *Philosophical Transactions of the Royal Society of London B*, 341:181–194.

Lawton, J. H., M. P. Hassell and J. R. Beddington. 1975. 'Prey death rates and rate of increase of arthropod predator populations, *Nature*, 255:60–62.

Lawton, John H., Andrew J. Letcher and Sean Nee. 1994. 'Animal distributions: patterns and processes' in P. J. Edwards, R. M. May, N. R. Webb, *Large Scale Ecology and Conservation Biology*. Oxford: Blackwell, 41–58.

Lawton, John H. and Robert M. May (eds.). 1995. *Extinction Rates*. Oxford University Press.

Lawton, J. H. and S. McNeill. 1979. 'Between the Devil and the Deep Blue Sea: On the problem of being a herbivore', in R. M. Anderson, B. D. Turner and L. R. Taylor, *Population Dynamics* (20th. Symposium of the British Ecological Society) Oxford: Blackwell, Chapter 11.

Lawton, J. H. and S. L. Pimm. 1978. 'Population dynamics and the length of food chains'. *Nature*, 272:190.

Lawton, J. H. and D. Schröder. 1977. 'Effects of plant type, size of geographical range and taxonomic isolation on the number of insect species associated with British plants'. *Nature*, 265:329–331.

Lehrer, Jonah, 2012. *How Creativity Works.* Toronto: Penguin.

LeRoy, Edouard. 1899. 'Science et Philosophie'. *Revues de Métaphysique et des Morales,* 7:375–425 and 503–562.

_____1901. 'Un Positivisme Nouveau'. *Revues de Métaphysique et des Morales,* 9:138–153.

Levin (ed.), S. A. 1974. *Ecosystems Analysis and Prediction* (Proceedings of SIAMS-SIMS Conference on Ecosystems) Alta, Utah.

Levin, Simon A. 2002. 'Complex adaptive systems: Exploring the known, the unknown and the unknowable' *Bulletin of the American Mathematical Society,* 40:3–19.

Levins, Richard. 1966. 'The Strategy of model building in population biology', *American Scientist,* 54:421–431.

_____1968. *Evolution in a Changing Environment.* Princeton University Press.

_____1969. 'Some demographic and genetic consequences of environmental heterogeneity for biological control'. *Bulletin of the Entomological Society of America,* 15:237–240.

Levins, R. and R. Lewontin 1980. 'Dialectics and Reductionism in Ecology, *Synthese,* 43:47–78.

Lewontin, Richard C. 1961. 'Evolution and the Theory of Games'. *Journal of Theoretical Biology,* 1:382–403.

_____(ed.). 1968. *Population Biology and Evolution.* Syracuse University Press.

_____1969. 'The meaning of stability' in G. A. Woodwell and H. H. Smith (eds.), *Diversity and Stability in Biological Systems. Brookhaven Symposia in Biology,* 22:13–24.

Li, T. Y. and J. A. Yorke. 1975. 'Period three implies chaos'. *American Mathematical Monthly,* 82:985–992.

Lindeman, R. L. 1942. 'The Trophic-dynamic aspect of Ecology', *Ecology,* 23: 399–418.

Livingstone, David N. 2003. *Putting Science in its Place: Geographies of scientific Knowledge.* University of Chicago Press.

Lomolino, Mark V., James H. Brown and Don F. Sax. 2010. 'Island biogeography theory: Reticulations and reintegration of "A biography of the species"' in Jonathan B. Losos and Robert E. Ricklefs (eds.), *The Theory of Island Biogeography Revisited.* Princeton University Press.

Losos, Jonathan B. and Ricklefs, Robert E. 2010. *The Theory of Island Biogeography Revisited.* Princeton University Press.

Lotka, Alfred J. 1907. 'Studies on the mode of growth of material aggregates', *American Journal of Science,* 24:199–217.

_____1956. *Elements of Mathematical Biology.* New York:Dover.

Lovejoy, Arthur O. 1936, *The Great Chain of Being: A Study of the History of an Idea*. Cambridge MA: Harvard University Press.

Lovelock, James. 1979. *Gaia: A New Look at Life on Earth*. Oxford University Press.

Lowe, Philip and Jane Goyder 1983. *Environmental Groups in Politics*. London: Allen and Unwin.

MacArthur, Robert H. 1955. 'Fluctuations of animal populations, and a measure of community stability'. *Ecology*, 36:533–536.

_____1957. 'On the relative abundance of bird species'. *Proceedings of the National Academy of Sciences*, USA, 43:293–295.

_____1958. 'Population ecology of some warblers of northeastern coniferous forests'. *Ecology*, 39:599–619.

_____1962. 'Growth and regulation of animal populations'. *Ecology*, 43:579.

_____1970. 'Species packing and competitive equilibrium for many species', *Theoretical Population Biology*, 1, 1–11

MacArthur, R. H. and Richard Levins. 1964. 'Competition, habitat selection, and character displacement in a patchy environment'. *Proceedings of the National Academy of Sciences,* 51:1207–1210.

MacArthur, Robert H. and Eric R. Pianka. 1966. 'On optimal use of a patchy environment', *The American Naturalist*, 100:603–609.

MacArthur, Robert H. and Edward O. Wilson. 1967. *The Theory of Island Biogeography*. Princeton University Press.

MacArthur, Robert H. and Robert M. May. 1972. 'Niche overlap as a function of environmental variability', *Proceedings of the National Academy of Sciences*, 69:1109–1113.

Maienschein, Jane (ed.). 1986. *Defining Biology: Lectures from the 1890s*. Cambridge MA: Harvard University Press.

Malthus, Thomas Robert. 1989. *Principles of Population* (1798). Cambridge University Press.

Mannheim, Karl. 1936. *Ideology and Utopia*. London: Routledge and Kegan Paul.

May, R. M. 1971a. 'Stability in model ecosystems'. *Proceedings of the Ecological Society of Australia*, 6:18–56.

_____1971b. 'Stability in multi-species community models'. *Mathematical Biosciences*, 12:59–79.

_____1972a. 'Limit cycles in predator-prey communities'. *Science,* 177:900–902.

_____1972b. 'Will a large complex system be stable?' *Nature*, 238:413–414.

_____1973, 1974, 2001. *Stability and Complexity in Model Ecosystems*. 3 edns. Princeton University Press.

_____1974. 'Biological populations with non-overlapping generations: Stable points, stable cycles and chaos'. *Science*, 186:645–647.

_____1975. 'Patterns of species abundance and diversity', in Martin L. Cody and Jared Diamond (eds.), *Ecology and Evolution of Communities*. Cambridge MA, Belknap, 81–120.

_____1976a. 'Simple mathematical models with very complicated dynamics'. *Nature*, 261:459–467.

_____(ed.). 1976b. *Theoretical Ecology: Principles and applications*. Oxford: Blackwell.

_____1978. 'The dynamics and diversity of insect faunas' in L. A. Mound and N. Waloff (eds.), *Diversity of Insect Faunas*. (*Symposium of the Royal Entomological Society*): Oxford: Blackwell, 188–204.

_____1979. 'The structure and dynamics of ecological communities' in R. M. Anderson, B. D. Turner and L. R. Taylor (eds.). 1979. *Population Dynamics* (20th Symposium of the British Ecological Society). Oxford: Blackwell.

_____1981. 'The role of theory in ecology'. *American Ecologist*, 21:903–910.

_____1987. 'More evolution of cooperation'. *Nature*, 327:15–17.

_____1988. Citation Classic 50, *Current Contents*, 12 December, 1988, 22.

_____1997. 'The scientific wealth of nations'. *Science*, 275:793–796.

_____1998. 'The scientific investments of nations'. *Science*, 281:49–51.

_____2000. 'The role of theory in ecology' in David R. Keller and Frank B. Golley (eds.), *The Philosophy of Ecology: From science to synthesis.* University of Georgia Press, Chapter 11.

May, R. M. and R. M. Anderson. 1983. 'Epidemiology and genetics in the coevolution of parasites and hosts'. *Proceedings of the Royal Society B*, 219:281–313.

_____1987. 'Transmission dynamics of HIV-Infection'. *Nature*, 326:137–42.

_____1988. 'Transmission dynamics of Human-Immunodeficiency Virus (HIV)'. *Philosophical Transactions of the Royal Society of London B*, 321:565–607.

May, Robert M., John R. Beddington, Colin W. Clark, Sidney J. Holt and Richard M. Laws. 1979. 'Management of multispecies fisheries'. *Science*, 205:267–77.

May, R. M., G. R. Conway, M. R. Hassell, and T. R. E. Southwood. 1974. 'Time delays, density dependence, and single species oscillations'. *Journal of Animal Ecology*, 43:747–770.

May, Robert M. and Michael P. Hassell. 2008. 'Thomas Richard Edmund Southwood, 20 June 1931 — 26 October 2005'. *Biographical Memoirs of Fellows of the Royal Society*, 54:333–374.

May, Robert M., Simon A, Levin and George Sugihara. 2008. 'Ecology for bankers'. *Nature*, 45:893.

May, R. M. and G. F. Oster. 1976. 'Bifurcations and dynamical complexity in simple ecological models'. *American Naturalist*, 110:573–599.

May. Robert M. and Angela R. McLean (eds.). 2007. *Theoretical Ecology: Principles and applications*. 3rd edition. Oxford University Press.

Mayfield, Harold F. 1989. 'In Memoriam: Frank W. Preston'. *The Auk,* 106: 714–717.

Maynard Smith, John. 1974. *Models in Ecology.* Cambridge University Press.

———1979. 'Game theory and the evolution of behaviour'. *Proc. Roy. Soc. London B,* 205:475–488.

Maynard Smith, J. and G. R. Price. 1973. 'Logic of Animal Conflict'. *Nature,* 246:15–18.

Mayr, Ernst. 1942. *Systematics and the Origin of Species From the Viewpoint of a Zoologist.* Harvard University Press.

McIntosh, Robert P. 1980. 'The background and some current problems of theoretical ecology'. *Synthese,* 43:195–255.

———1985. *The Background of Ecology: Concept and Theory.* Cambridge University Press.

McNeill, S. and J. H. Lawton. 1970. 'Annual production and respiration in animal populations'. *Nature,* 225:472–474.

Meadows, Donella H., Dennis L. Meadows, Jørgen Randers and William W. Behrens III. 1972. *Limits to Growth.* New York: Universe Books.

Mellanby, K. 1967. *Pesticides and Pollution.* London: Methuen.

Merton, Robert K. 1973. (ed. Norman W. Storer). *The Sociology of Science: Theoretical and Empirical Investigations.* University of Chicago Press.

———1958. *Science, Technology and Society in Seventeenth-century England. Osiris,* 4.

———1988 'The Matthew Effect in science II: Cumulative advantage and the symbolism of intellectual property', *Isis,* 79:606–623.

Merton Robert K. and Harriet Zuckerman. 1968. 'The Matthew effect in science: The reward and communication systems of science are considered', *Science,* 159:56–63.

Mitchell, Peter Chalmers. 1915. *Evolution and the War.* London: John Murray.

Mitman, Greg. 1992. *The State of Nature: Ecology, Community and American Social Thought,* 1900–1950. University of Chicago Press.

Moran, V. C. and T. R. E. Southwood. 1982. 'The guild composition of arthropod communities in trees'. *Journal of Animal Ecology,* 51:289–306.

Moroso, Mario. 2008. 'The Institutionalisation of GMOs: Institutional dynamics in the GM regulatory debate in the UK, 1986–1993'. PhD thesis, University of Exeter.

Morrell, Jack. 1972. 'The Chemist Breeders: The research schools of Liebig and Thomas Thomson'. *Ambix,* 19:1–80.

_____1993. 'W. H. Perkin Jr., at Manchester and Oxford: From Irwell to Isis' in Gerald L. Geison and Frederick L. Holmes (eds.). *Research Schools: Historical Reappraisals. Osiris*, 8:104–136.

Morse, D. R., J. H. Lawton, M. M. Dodson and M. H. Williamson. 1985. 'Fractal dimension of vegetation and distribution of arthropod body lengths'. *Nature*, 314:731–733.

Munroe, E. G. 1963. 'Perspectives in biogeography', *The Canadian Entomologist*, 95:299–308.

Musgrave, Alan. 1976. 'Why did oxygen supplant phlogiston? Research programmes in the chemical revolution' in C. Howson (ed.). *Method and Appraisal in the Physical Sciences*. Cambridge University Press, 181–209.

Naeem, S., L. J. Thompson, S. P. Lawler, J. H. Lawton and R. M. Woodfin. 1994a. 'Declining biodiversity can alter the performance of ecosystems', *Nature*, 368:734–737.

_____1994b. 'Biodiversity in model ecosystems — a reply'. *Nature* 371:565.

_____1995. 'Empirical evidence that declining biodiversity may alter the performance of terrestrial ecosystems'. *Philosophical Transactions of the Royal Society of London B*, 347:249–262.

Nelkin, Dorothy. 1976. 'Ecologists and the public interest', *Hastings Center Report*, 6:38–44.

Nicholson, A. J. 1933. 'The Balance of animal populations', *Journal of Animal Ecology*, 2:131–178.

_____1954. 'An outline of the dynamics of animal populations', *Australian Journal of Zoology*, 2:9–65.

Nicholson, A. J. and V. A. Bailey. 1935. 'The balance of animal populations', *Proceedings of the Zoological Society of London*, 1:551–598.

Nokes, D. J. and R. M. Anderson. 1988. 'Measles, Mumps, and Rubella Vaccine: What coverage to block transmission?' *Lancet*, 10 December, 1374.

Nowak, Martin A. 2008. 'Generosity: A winner's advice'. *Nature*, 456:579.

Nowak, Martin N. with Roger Highfield. 2011. *Super Cooperators: Altruism, Evolution and Human Behaviour or Why We Need Each Other to Succeed*. Edinburgh: Canongate.

Nowak, Martin A., Roy M. Anderson, Angela R. Maclean, Tom F. W. Wolfs, Jaap Goudsmit and Robert M. May. 1991. 'Antigenic diversity thresholds and the development of AIDS'. *Science*, 254:963–969.

Nowak, Martin A. and Robert M. May. 1992, 'Evolutionary games and spatial chaos'. *Nature*, 359:826–829.

_____2008. *Virus Dynamics: Mathematical Principles of Immunology and Virology*. Oxford University Press.

Nowak, Martin A. and Karl Sigmund. 2007. 'How Populations Cohere: Five rules of cooperation', in Robert M. May and Angela R. McLean (eds.). *Theoretical Ecology: Principles and Applications*. 3rd. edition. Oxford University Press.

Nowak, Martin A., Corina E. Tarnita and Edward O. Wilson. 2010. 'The evolution of eusociality'. *Nature*, 466:1057–1062.

Nye, E. R. and M. E. Gibson. 1997. *Ronald Ross: Malariologist and Polymath: A biography*. London: Macmillan.

Nye, Mary-Jo. 2011. *Michael Polanyi and His Generation: Origins of the Social Construction of Science*. University of Chicago Press.

Odum, E. P. 1953. 1971. *Fundamentals of Ecology*. 1st and 3rd editions. Philadelphia: W. B. Saunders.

Okasha, Samir. 2010. 'Altruism researchers must cooperate', *Nature*, 467, 653–655.

Oldroyd, D. R. 1980. *Darwinian Impacts*. Milton Keynes: Open University Press.

Oreskes, Naomi and Erik Conway. 2010. *Merchants of Death: How a Handful of Scientists Obscured the Truth on Issues from Tobacco Smoke to Global Warming*. London: Bloomsbury.

Orians, G. H. 1966. 'The Ecology of Blackbird (*Agelaius*) Social Systems'. *Ecological Monographs* 31:285–312.

Pacala, S. W., M. P. Hassell and R. M. May. 1990. 'Host-parasitoid associations in patchy environments'. *Nature*, 344:150–153.

Packer, Richard. 2006. *The Politics of BSE*. Palgrave Macmillan.

Paddock, William and Paul Paddock. 1967. *Famine 1975: America's Decision Who Will Survive?* Boston: Little, Brown.

Paviour-Smith, Kitty, 2004. 'Elton, Charles Sutherland (1900–1991), *Oxford Dictionary of National Biography*. Oxford University Press.

Payne, Andy, John Cotter and Ted Potter. 2008. *Advances in Fisheries Science: 50 Years From Beverton and Holt*. Oxford: Blackwell.

Peters, R. H. 1991. *Critique for Ecology*. Cambridge University Press.

Pianka, Eric R. and Henry S. Horn. 2005. 'Ecology's Legacy from Robert MacArthur', in Kim Cuddington and Beatrix E. Beisner. *Ecological Paradigms Lost: Routes of Theory Change*. Amsterdam: Elsevier, Chapter 11.

Pimm, S. L. and J. H. Lawton. 1977. 'Number of trophic levels in ecological communities'. *Nature*, 268:329–331.

Poincaré, Henri. 1902. *La Science et l'Hypothèse*. Paris: E. Flammarion.

_____1905. *La Valeur de la Science*. Paris: E. Flammarion.

_____1908. *Science et Méthode*. Paris: E. Flammarion.

_____1958. *The Value of Science*. New York: Dover.

Polanyi, Michael. 1958. *Personal Knowledge: Towards a Post-critical Philosophy*. University of Chicago Press.

_____1966. *The Tacit Dimension.* Garden City NY: Doubleday.

Pope, Alexander. 1733. *Essay on Man.*

Popper, Karl. 1959. *The Logic of Scientific Discovery.* London: Hutchinson.

_____1963. *Conjectures and Refutations.* London: Routledge:1963).

Preston, F. W. 1962. 'The Canonical Distribution of Commonness and Rarity', *Ecology.* Part 1, 43:185–215 and Part 2, 43:410–432.

Pretty, Jules. 1991. *Unwelcome Harvest: Agriculture and Pollution.* London: Earthscan.

Price, Derek J. de Solla. 1963. *Little Science, Big Science.* New York: Columbia University Press.

Pritchard, Peggy A (ed.). 2006. *Success Strategies for Women in Science: A Portable Mentor.* Burlington MA: Elsevier Academic Press.

Putnam, Hilary. 1997. 'James's Theory of Truth' in R. A. Putnam (ed.), *Cambridge Companion to William James* Cambridge University Press, 166–185.

Pycior, Helena M., Nancy G. Slack and Pnina G. Abir Am (eds.). 1995. *Creative Couples in the Sciences.* Rutgers University Press.

Quine, W. Van. Orman 1969. 'Epistemology Naturalized' in *Ontological Relativity and Other Essays.* New York: University of Columbia Press.

Randolph. S. E. and D. J. Rogers. 1978. 'Feeding cycles and flight activity in field populations of tsetse', *Bulletin of Entomological Research*, 68:655–671.

_____1997. 'A Generic Population Model for the African Tick, *Rhipicephalus Appendiculatus*'. *Parasitology,* 115:265–279.

Ravetz, Jerome R. 1971. *Scientific Knowledge and Social Problems.* Oxford: Clarendon Press.

Resetarits, William J. Jr. and Joseph Bernardo (eds.). 1998. *Experimental Ecology: Issues and perspectives.* New York: Oxford University Press.

Rheinberger, Hans-Jörg. 2005. 'A reply to David Bloor: "Toward a sociology of epistemic things"', *Perspectives in Science* 13:406–410.

_____2010. *An Epistemology of the Concrete: Twentieth-Century Histories of Life.* Durham NC: Duke University Press.

Richards, Joan. 2006. 'Focus: Biography in the history of science', *Isis*, 97:302–304.

Richards, O. W. 1948. 'The interaction of environmental and genetic factors in determining the weight of grain weevils, *Calandra granaria* (L) (Col., Curculionidae)', *Proceedings of the Zoological Society of London*, 118:49–81.

_____1978. *The Social Wasps of America Excluding the Vespinae.* British Museum: Natural History.

Richards, O. W. and M. J. Richards. 1951. 'Observations on the social wasps of South America', *Transactions of the Royal Entomological Society of London*, 102:1–170.

Richards, O. W. and G. C. Robson. 1936. *The Variation of Animals in Nature.* London: Longmans Green.

Richards, O. W. and Nadia Waloff. 1946. (a) 'The study of population of *Ephestia elutella* Hübner (Lep. Phycitidae) living on bulk grain', *Transactions of the Royal Entomological Society of London*, 97 (1946) 253–298 and (b) 'Observations on the behaviour of *Ephestia elutella* Hübner (Lep. Phycitidae) breeding on bulk grain', *ibid*, 299–335.

_____1954. 'Studies on the biology and population dynamics of British grasshoppers'. *Anti-Locust Bulletin*, 17:1–182.

_____1961. 'A study of a natural population of *Phytodecta olivacea* (Forster) (Coleoptera: *Chrysomeloidea)*', *Philosophical Transactions of the Royal Society B*, 244:205–257.

_____1977. 'The effect of insect fauna on growth, mortality and natality of broom, *Sarothamnus scoparius*', *Journal of Applied Ecology*, 14:787–798.

Richards, Paul W. 1996. *The Tropical Rain Forest: An Ecological Study.* 2nd ed. Cambridge University Press.

Rogers, D. J. and M. P. Hassell. 1974. 'General models for insect parasite and predator searching behaviour'. *Journal of Animal Ecology*, 43:239–253.

Rogers, D. J., S. I. Hay and M. J. Packer. 1996. 'Predicting the distribution of tsetse flies in West Africa using temporal Fourier processed meteorological satellite data'. *Annals of Tropical Medicine and Parasitology*, 90:1–19.

Rogers, D. J. and S. E. Randolph. 1985. 'Population ecology of Tsetse'. *Annual Review of Entomology*, 30:197–216.

_____1986. 'The distribution and abundance of Tsetse flies (*Glossina* spp)'. *Journal of Animal Ecology*, 55:1007–1025.

Rogers, D. J., S. E. Randolph and S. I. Hay. 2002. 'Satellite imagery in the study and forecast of malaria'. *Nature,* 415:710–715.

Rosenberg, A. A., G. P. Kirkwood, J. A. Crombie and J. R. Beddington. 1990. 'The assessment of stocks of annual squid species'. *Fisheries Research*, 8:351–365.

Rosenzweig, M. L. and R. H. MacArthur. 1963. 'Graphical representation and stability conditions of predator-prey interactions'. *American Naturalist*, 97: 209–223.

Roth, Andrew. 2002. Jeremy Bray MP (1930–2002). Obituary. *The Guardian*, 5 June.

Roy, Mary. 1999. 'Three generations of women'. *Indian Journal of Gender Studies*, 6:203–19.

Ruse, Michael. 1973. *The Philosophy of Biology.* London: Hutchinson.

Russell, Bertrand. 1935 and 1997. *Religion and Science.* Oxford University Press.

Salisbury, E. J. 1961. *Weeds and Aliens.* London: Collins.

Saunders, Claire. 2001. 'Analysis: Science and the family don't mix'. *Times Higher Education Supplement*, 12 October. (*THES* website).

Savill, Peter, Christopher Perrins, Keith Kirby and Nigel Fisher (eds.). 2010. *Wytham Woods: Oxford's Ecological Laboratory.* Oxford University Press.

Schoener, Thomas W. 2010. 'The MacArthur-Wilson equilibrium model: A chronicle of what it said and how it was tested', in Jonathan B. Losos and Robert E. Ricklefs (eds.), *The Theory of Island Biogeography Revisited.* Princeton University Press.

Scudo, Francesco M. and James R. Ziegler (eds.), 1978. *The Golden Age of Theoretical Ecology.* Berlin: Springer-Verlag.

Segestrale, Ullica. 2000. *Defenders of the Truth: The Battle for Science in the Sociology Debate and Beyond.* Oxford University Press.

Seguin, Eve. 2000. 'The UK BSE crisis: strengths and weaknesses of existing conceptual approaches', *Science and Public Policy*, 27:293–301.

Shapin, Steven. 1994. *A Social History of Truth: Civility and Science in Seventeenth-Century England.* University of Chicago Press.

_____2008. *The Scientific Life: A Moral History of a Late Victorian Vocation.* University of Chicago Press.

Shapin, Steven and Arnold Thackray. 1974. 'Prosopography as a research tool in the history of science: The British scientific community 1700–1900', *History of Science* xii:1–28.

Shapin, Steven and Simon Schaffer. 1985. *Leviathan and the Air Pump: Hobbes, Boyle and the Experimental Life.* Princeton University Press.

Sheail, John. 1987. *Seventy-five Years in Ecology: The British Ecological Society.* Oxford: Blackwell.

_____2002. *An Environmental History of Twentieth-Century Britain.* Basingstoke: Palgrave.

Sherry, D. F., J. R. Krebs and R. J. Cowie. 1981. 'Memory for the location of stored food in marsh tits'. *Animal Behaviour* 29:1260–1266.

Shettleworth, S. J. and J. R. Krebs. 1982. 'How Marsh Tits Find Their Hoards: The roles of site preference and spatial memory'. *Journal of Experimental Psychology: Animal Behavioural Processes*, 8:354–375.

Simberloff, D. S. 1969. 'Experimental zoogeography of islands: A model for insular colonization'. *Ecology*, 50:296–314.

_____1984. 'Citation Classics', *Current Contents*, 10:12.

Simberloff, D. S. and E. O. Wilson. 1969a. 'Experimental zoogeography of islands: Defaunation and monitoring techniques'. *Ecology*, 50:267–78;

Simberloff, D. S. and E. O. Wilson. 1969b. 'Experimental zoogeography of islands: The colonization of empty islands'. *Ecology*, 50:278–96.

Slack, Nancy G. 2010. *G. Evelyn Hutchinson and the Invention of Modern Ecology.* Yale University Press.

Slobodkin, Lawrence B. 1980. *Growth and Regulation of Animal Populations.* 2nd. ed. N.Y: Dover.

_____1975. 'Comments from a biologist to a mathematician', in. S. A. Levin (ed.), *Ecosystem Analysis and Prediction.* (Proceedings of the 1974 SIAM-SIMS Conference on Ecosystems) Alta, Utah.

Slobodkin, Lawrence B. and Nancy G. Slack. 1999. 'George Evelyn Hutchinson: 20th-century ecologist', *Endeavour,* 23:24–30

Smith, Charlie Pye and Richard D. North. 2005. Richard Sandbrook (1946–2005), *The Independent,* 12 December.

Smolin, Lee, *The Trouble with Physics: The Rise of String Theory, the Fall of a Science, and What Comes Next* (Boston: Houghton Mifflin, 2006).

Söderqvist, Thomas. 2003. *Science as Autobiography: The Troubled Life of Niels Jerne.* Yale University Press.

_____(ed.). 2007. *The History and Poetics of Scientific Biography.* Aldershot: Ashgate.

Soemarwoto, Otto and G. R. Conway. 1992. 'The Javanese Homegarden'. *Journal for Farming Systems, Research-Extension,* 3:95–118.

Solomon, M. E. 1949. 'The natural control of animal populations'. *Journal of Animal Ecology,* 19:1–35.

Southwood, T. R. E. 1961. 'The number of species of insect associated with various trees', *Journal of Animal Ecology,* 30:1–8.

_____1966, 2000. *Ecological Methods With Particular Reference to the Study of Insect Populations.* London: Chapman and Hall. 3rd edition with Peter Henderson. Oxford University Press.

_____(ed.). 1968. *Insect Abundance.* Oxford: Blackwell.

_____1976. 'Bionomic strategies and parameters', in Robert M. May (ed.), *Theoretical Ecology: Principles and Applications.* Oxford: Blackwell, Chapter 2.

_____1976b. 'Continuing the MacArthur tradition', *Science* 192:670–672; review of Martin L. Cody and Jared M. Diamond (eds.) *Ecology and the Evolution of Communities.* Harvard University Press (1975).

_____1977. 'Habitat the templet for ecological strategies?' *Journal of Animal Ecology,* 46:337–365.

_____1980. 'Ecology — A mixture of pattern and probabilism', *Synthes,* 43: 111–122.

_____1987. 'Owain Westmacott Richards', *Biographical Memoirs of Fellows of the Royal Society,* 33:539–571.

_____1995. The Croonian Lecture. 'Natural communities: structures and dynamics'. *Phil. Trans. Roy. Soc. B*, 351:1113–1129.

_____1997. 'Ecological entomology at Silwood Park'. *Antenna*, 21:115–120.

_____2003. *The Story of Life*. Oxford: Oxford University Press.

Southwood, T. R. E., V. K. Brown and P. M. Reader. 1979. 'The relationship of plant and insect diversities in succession'. *Biological Journal of the Linnean Society*, 12:327–348.

_____1983. 'Continuity of vegetation in space and time: a comparison of insects' habitat templet in different successional stages'. *Research in Population Ecology; Supplement*, 3:61–74.

Southwood, Richard, J. R. Clarke. 1999. 'Charles Sutherland Elton, 29 March 1900–1 May 1991'. *Biographical Memoirs of Fellows of the Royal Society*, 45:129–146.

Southwood, T. R. E. and H. N. Comins. 1976. 'A synoptic population model', *Journal of Animal Ecology*, 45:949–965.

Southwood, T. R. E. and G. R. Conway. 1975. 'Man's effect on the ecology of the Rhine basin'. Chatham House Study Group, *European Series*, 26:11–14.

Southwood, T. R. E. and D. J. Cross. 1969. 'The ecology of the partridge III: Breeding success and the abundance of insects in natural habitats'. *Journal of Animal Ecology*, 38:497–509.

Southwood, T. R. E., M. P. Hassell, P. M. Reader and D. J. Rogers. 1989. 'Population dynamics of the Viburnum whitefly *Aleurotrachelus jelinekii* (Frauenf.)'. *Journal of Animal Ecology*, 58:921–942.

Southwood, T. R. E. and W. F. Jepson. 1962. 'Studies on the Populations of *Oscinella frit* L. (Dipt: Chloropidae) in the Oat Crop', *Journal of Animal Ecology*, 31:481–495.

Southwood, T. R. E. and Dennis Leston. 1959. *Land and Water Bugs of the British Isles*. London: Frederick Warne.

Southwood, T. R. E., R. M. May, M. R. Hassell and G. R. Conway. 1974. 'Ecological strategies and population parameters', *American Naturalist*, 108:747–770;

Southwood, T. R. E., R. M. May and G. Sugihara. 2006. 'Observations on related ecological exponents'. *Proceedings of the National Academy of Sciences USA*, 103:6931–6933.

Southwood, T. R. E. and P. M. Reader. 1976. 'Population census data and key factor analysis for the viburnum whitefly *Aleurotrachelus jelinekii* (Frauenf.) on three bushes'. *Journal of Animal Ecology*, 45:313–325.

Southwood, Sir Richard and Robin Russell Jones (eds.). 1987. *Radiation and Health: The Biological Effects of Low Level Exposure to Ionizing Radiation*. New York: Wiley.

Southwood, T. R. E., G. R. W. Wint, C. E. J. Kennedy and S. R. Greenwood. 2004. 'Seasonality, abundance, species richness and specificity of the phytophagous guild of insects on oak (*Quercus*) canopies'. *European Journal of Entomology*, 101:43–50.

Spencer, Herbert. 1862. *First Principles*. London: Williams and Norgate.

Stephens, D. W. and J. R. Krebs. 1986. *Foraging Theory*. Princeton NJ: Princeton University Press.

Stone, Lawrence. 1971. 'Prosopography', *Daedalus* 100 (1971), 46–79.

Strong, D. R., D. Simberloff, L. G. Abele, and A. B. Thistle. 1984. *Ecological Communities: Conceptual Issues and Evidence*. Princeton NJ: Princeton University Press.

Strong, Donald R., John H. Lawton, and Richard Southwood. 1984. *Insects on Plants*. Oxford: Blackwell.

Sugihara, G. and R. M. May. 1990. 'Non-linear forecasting as a way of distinguishing chaos from measurement error in time series'. *Nature*, 344:734–741.

Sugihara, George, Michael J. Crawley and Robert M. May. 2007. 'Community: Patterns', in Robert M. May and Angela McLean (eds.). *Theoretical Ecology: Principles and Applications*. 3rd edition. Oxford University Press, Chapter 9.

Tansley, A. G. 1923. *Practical Plant Ecology: A Guide for Beginners in the Field Study of Plant Communities*. London: George Allen and Unwin.

_____1939. *The British Islands and their Vegetation*. Cambridge University Press.

Taylor, Peter J. 2005. *Unruly Complexity: Ecology, Interpretation, Engagement*. University of Chicago Press.

Thackray, Arnold (ed.). 1995. *Constructing Knowledge in the History of Science*. *Osiris*, 10.

Thomas, Nicholas, Harriet Guest and Michael Dettelbach (eds.). 1996. J. R. Forster, *Observations Made During a Voyage Round the World*. University of Hawaii Press.

Thompson, W. R. 1924. 'La théorie mathématique de l'action des parasites entomophages et le facteur du hasard', *Annales de la Faculté des Sciences des Marseilles*, 2:69–89.

Thorpe, W. H. 1974. 'David Lambert Lack, 1910–1973'. *Biographical Memoirs of Fellows of the Royal Society*, 20:271–293.

Tickell, Crispin. 1986. *Climatic Change and World Affairs*. revised edition. Harvard University Press.

Tilly, Charles. 2002. *Stories, Identities and Political Change*. London: Rowan and Littlefield.

_____2010 *Why?* Princeton University Press.

Tinbergen, Niko. 1951, *The Study of Instinct*. Oxford: Clarendon Press.

_____1953. *The Herring Gull's World*. London: Collins.

_____1958. *Curious Naturalists*. New York: Basic Books.

Trivers, R. L. 1971. 'The evolution of reciprocal altruism', *Quarterly Review of Biology*, 46:35–57.

———1985. *Social Evolution*. Menlo Park, CA: Benjamin. Cummings.

Trivers, R. L. and H. Hare. 1976. 'Haplodiploidy and the Evolution of Social Insects. *Science*, 191:249–263.

Usher, M. B. and M. H. Williamson (eds.). 1974. *Ecological Stability*. London: Chapman and Hall.

Varley, G. C. 1953. 'Ecological aspects of population regulation', *International Congress of Entomology Proceedings* (Amsterdam) 9:210–214.

Varley, G. C. and G. R. Gradwell. 1968. 'Population models for the winter moth' in Southwood. T. R. E. (ed.), *Insect Abundance*. Oxford: Blackwell.

Varley, G. C., G. R. Gradwell and M. P. Hassell. 1973. *Insect Population Ecology: An Analytical Approach*. Oxford: Blackwell.

Volterra, Vito. 1926. 'Fluctuations in the abundance of a species considered mathematically', *Nature* 118:558–560.

Volterra, Vito, with Umberto d'Ancona. 1935. *Les associations biologiques au point de vue mathématique'*. Paris: Hermann.

Vorzimmer, P. 1965. 'Darwin's ecology and its influence on his theory'. *Isis*, 56: 148–155.

Vleck, John H. Van and Albert Sherman. 1935. 'The Quantum Theory of Valence'. *Reviews of Modern Physics* 7:167–255.

Waloff, Nadia. 1968. 'Studies on the insect fauna on Scotch Broom, *Sarothamnus scoparius* (L) Wimmer', *Advances in ecological research*, 5:87–208.

———1986. 'Owain Westmacott Richards', *Journal of Animal Ecology* 55:393–94.

Warming, Eugenius. 1909. *Oecology of Plants: An Introduction to the Study of Plant Communities* (translation of the 1895 Danish edition by Percy Groom and Isaac Bayley Balfour). Oxford: Clarendon Press.

Watson, Elizabeth L. 1991. *Houses for Science: A pictorial history of Cold Spring Harbour Laboratory*. Cold Spring Harbour Laboratory Press.

Watt, Kenneth. 1968. *Ecology and Resource Management: A Quantitative Approach*. N.Y.: McGraw-Hill.

Weber, Max. 1978. *Economy and Society*. Berkeley: University of California Press.

Weber, Max. 1968. *On Charisma and Institution Building* (ed. S. N. Eisenstadt). University of Chicago Press.

Weindling, Paul. 2004. 'Kenneth Mellanby', *Oxford Dictionary of National Biography*. Oxford University Press.

Wentworth Thompson, D'Arcy. 1917. *On Growth and Form*. Cambridge University Press.

Werskey, Gary. 1988. *The Visible College: A Collective Biography of British Scientists and Socialists of the 1930s.* (London: Free Association Books; first published by Allen Lane, 1978).

White, Gilbert. 1993. *The Natural History of Selborne.* London: Everyman, J. M. Dent.

White, Harrison C. 2008. *Identity and Control: How Social Formations Emerge* (2nd ed.). Princeton University Press.

Whittaker, E. T. 1941. 'Vito Volterra, 1860–1940', *Obituary Notices of Fellows of the Royal Society* 3:691–729.

Wigglesworth, V. B. 1942. *The Principles of Insect Physiology.* London. Methuen.

Williams, George C. 1966. *Adaptation and Natural Selection: A Critique of Some Current Evolutionary Thought* Princeton University Press.

Williams, Shirley. 1971. 'The Responsibility of Science'. *The Times* (Saturday Review), 27 February.

Willis, J. C. 1922. *Age and Area: A Study of Geographical Distribution and Origin in Species.* Cambridge University Press.

Wilson, Edward O. 1992. *The Diversity of Life.* Cambridge MA: Belknap, Harvard University Press.

_____1994. *Naturalist.* Washington DC: Island Press.

_____2012. *The Social Conquest of Earth.* NY: Liveright.

Wilson, E. O. and F. M. Peter (eds.), 1988. *Biodiversity.* Washington DC: National Academy Press.

Wilson Edward O. and Evelyn G. Hutchinson. 1989. 'Robert Helmer MacArthur: April 7, 1930 — November 1, 1972', *Biographical Memoirs, National Academy of Sciences,* 58:319–327.

Wood, R. K. S. and M. J. Way (eds.). 1988. 'Biological control of pests, pathogens and weeds: developments and prospects', *Philosophical Transactions of the Royal Society* B 318.

Woods, Abigail. 2004. *A Manufactured Plague: The History of Foot-and-mouth Disease in Britain.* London: Earthscan, 2004.

Worster, Donald. 1994. *Nature's Economy: A History of Ecological Ideas.* (2nd edition). Cambridge University Press.

Wynne-Edwards, V. C. 1952. 'Zoology of the Baird Expedition: Part 1, The birds observed in central and south-east Baffin Island'. *The Auk,* 69: 353–391.

Yates, F. and R. Mather. 1963. 'Ronald Aylmer Fisher, 1890–1962', *Biographical Memoirs of Fellows of the Royal Society,* 9:109–132.

Ziman, J. M. 1968. *Public Knowledge: The Social Dimensions of Science.* Cambridge University Press.

Index